T5-AGS-970

Asian Rice Bowls: The Returning Crisis?

Prabhu L. Pingali

Director
CIMMYT Economics Program
Mexico
(Formerly, Agricultural Economist at IRRI)

Mahabub Hossain

Head
Social Sciences Division
IRRI
The Philippines

Roberta V. Gerpacio

Research Analyst
Environment and Production Technology Division
IFPRI
Washington DC
USA
and
Senior Research Assistant
Social Sciences Division
IRRI
The Philippines

CAB INTERNATIONAL

in association with the

International Rice Research Institute

CAB INTERNATIONAL
Wallingford
Oxon OX10 8DE
UK

Tel: +44 (0)1491 832111
Fax: +44 (0)1491 833508
E-mail: cabi@cabi.org

CAB INTERNATIONAL
198 Madison Avenue
New York, NY 10016-4341
USA

Tel: +1 212 726 6490
Fax: +1 212 686 7993
E-mail: cabi-nao@cabi.org

A catalogue record for this book is available from the British Library, London, UK.

ISBN 0 85199 162 9

Library of Congress Cataloging-in-Publication Data

Pingali, Prabhu L., 1955–
Asian rice bowls : the returning crisis? / P.L. Pingali, M. Hossain, and R.V. Gerpacio.
p. cm.
Includes bibliographical references and index.
ISBN 0–85199–162–9 (alk. paper)
1. Rice—Asia. 2. Rice—Economic aspects—Asia.
3. Rice trade—Asia. I. Hossain, M. II. Gerpacio R.V.
III. Title.
SB191.R5P56 1997
338.1′7318′095—dc21 96–48886
CIP

Published in association with:
International Rice Research Institute (IRRI)
PO Box 933
1099 Manila
The Philippines

Typeset in Melior by Wyvern 21 Ltd, Bristol
Printed and bound by Biddles Ltd, Guildford and Kings Lynn

Contents

Acknowledgements

This book would not have been completed but for the enthusiastic support of numerous colleagues and friends, both inside and outside IRRI. Our understanding of rice science and our insights on the Asian rice economies came from our interactions with IRRI Scientists and from our participation over the years in hundreds of formal and informal seminars and discussion groups. We also gained enormously from having access to IRRI's farm level and experiment station data, often available in unpublished forms in the personal archives of IRRI scientists. While it is hard for us to mention the names of everyone whose wisdom and advice we gained from, we would like to say that this book would not have been possible if we were not members of the IRRI family.

The conceptualization and initial drafts some of the chapters of this book were done when Prabhu Pingali was on sabbatical at Stanford University Food Research Institute and the Institute for International Studies. Wally Falcon was extremely generous with his time and was always ready to read the initial chapters and discuss the contents at length. Scott Rozelle, also from Stanford University, made very helpful comments on the first set of chapter drafts.

Several staff of the Social Sciences Division of IRRI helped assemble the graphs, charts and other statistical tables for the book. Recognition in particular ought to go to Piedad Moya, Don Pabale, Florencia Palis, Zenaida Huelgas, Josephine Narciso, Aida Papag and Mayette Manza. Lydia Damian and Mirla Domingo helped format several drafts of the manuscript, Juan Lazaro IV provided the illustrations.

Gurdev Khush, Sant Virmani, Swapan Datta and Boru Douthwaite read and commented on draft chapters. Lisa Price and Sushil Pandey encouraged us through the entire process. The writing of this book was wholeheartedly encouraged by IRRI management, the former and current Directors General, Klaus Lampe and George Rothschild, and Ken Fischer the Deputy Director General for Research.

Finally, our heartfelt thanks go to our spouses, Kumari Pingali, Parveen Hossain and Cesar Gerpacio; they released us from our domestic responsibilities so that we could get this book out.

1 Introduction: The State of Rice in Post-Green-Revolution Asia

Less than two decades ago, policymakers, the academic community and the media were focusing on the impending food crisis across Asia. The mood of despondency among intellectuals was widespread, and Asia was given little hope of ever being able to meet its rapidly growing food demand. Today, the words 'Asia' and 'famine' are very rarely used in the same sentence. Despite rapid population growth, widespread poverty and high levels of illiteracy, Asia has made a quantum leap in food production. Technological innovations complemented by public investments and policies that promoted intensive rice production systems helped bring about the change in food supply.

There is a growing sense of complacency about current and future food supplies in Asia. Several factors have contributed to this lethargy: (i) the almost instant success of the Green Revolution makes the problem seem less severe than it was previously thought to be; (ii) the low international rice price creates a perception of abundant supplies; (iii) the highly publicized economic successes of a few (small) Asian countries and their ability to import food are being generalized across the board; and (iv) an overly optimistic assessment of the low income elasticity of demand for rice leads to an over-estimate of the projected rate of decline in aggregate rice demand.

With growing complacency about food supplies, there has been an appreciable reduction in support for sustaining rice productivity growth. This lack of concern is seen in declining national budgetary allocations for maintaining existing rice infrastructure (including irrigation); reduced bilateral and multi-lateral donor support for expanding irrigation infrastructure; a deliberate policy shift towards diversification out of rice; and a substantial cut-back in research funding at both the national and the international level. The intellectual community itself has moved on to other challenges, such as natural resource management and environmental sustainability.

There are growing signs that this complacency is not warranted. While the growth in rice demand is slowing down, it turns out that the growth in supply is slowing even faster. This deceleration is because of increasing withdrawal of agricultural lands for urban uses; degrading irrigation infrastructure due to poor maintenance and upper watershed externalities; declining profitability of rice farming relative to competing agricultural and non-agricultural alternatives; a diminishing gap between the technological potential and farmer productivity in the favorable (irrigated) environments; and the lack of a technological breakthrough in the unfavorable (rainfed) environments.

At the same time, a decline in productivity has been observed in the intensively cultivated irrigated lowlands that is independent of output and factor price effects. The paddy ecosystem that is subject to intensification pressures over the long term exhibits degradation trends in both the soil and the biotic system. Long-term changes in soil fertility and soil health are observed along with an increase in the number and intensity of pest species. Many of these changes are directly related to the way in which intensive agriculture is practiced. The net result is a decline in yields per hectare where inputs are held constant, or, as is more common, a decline in total factor productivity (reflecting the fact that the rate of growth in yields is slower than the rate of growth in the inputs needed to achieve those increased yields). Under these circumstances, profitability would decline even if relative prices were constant. Productivity losses become even larger when the environmental and health externalities associated with modern technology are accounted for fully.

Future productivity growth in rice requires that several concurrent trends be reconciled: withdrawal of land and labor resources from agriculture to non-agricultural uses; increased agricultural commercialization and the consequent competition for resources; and continued degradation of the agricultural resource base. All of these suggest a substantial slow-down in rice productivity growth, while, at the same time, the demand for rice continues to grow because of population growth.

What are the technological and policy options that will lead to a sustained growth of rice supplies that match projected growth in demand, while at the same time enhance farm-level profits and income? For rice to compete with alternative opportunities for investing scarce farm resources would require substantial reduction in the unit cost of production. Land, labor and water are the three factors of production that are becoming increasingly scarce and in need of efficiency gains. This book provides a thorough assessment of the opportunities for increasing land productivity, including crop diversification. Similarly, as the opportunity cost of family labor rises, given alternative employment opportunities, there is a need for labor-saving technologies. Growing water scarcity for irrigated agriculture is a very serious concern, one for which technological solutions exist, but little work has been done on complementary policies. An overriding concern is that the technologies that are promoted must sustain rather than degrade the long-term productivity of the farm resource base. The increasing share of chemical inputs has to be viewed in this context.

Reduction in the unit cost of production can be achieved by shifting the current yield frontier and/or by improving input efficiencies at current yield levels. The search for efficiency gains is not incompatible with the need for sustainable resource use. In many instances, these work together. For example, an increase in the efficiency of water or pesticide use also improves the long-term productivity of the paddy ecosystem. Efficiency gains come at a cost, however, in terms of increased farmer knowledge, skill and judgment. Technology recommendations that have not accounted for these costs have generally failed to be adopted. In addition to the opportunity cost of family labor, ex-ante (forecasting) assessment of rice technologies ought to factor in the increasing trend towards a diversified production system, and hence the increasing opportunity costs of land in a rice monoculture system.

The experience of the Green Revolution has shown that technology alone cannot alleviate the food supply problems of Asia. It must be complemented by appropriate policies, which may include greater farm-level incentives – paradoxically they may come about through an open rather than a closed rice market, and sustained investments in irrigation and research infrastructure – getting away from the cyclical nature of public investments.

While the movement away from the 'rice-specific' policies of the 1960s and 1970s is expected, commodity-specific policy paradigms continue. The policy goal ought to provide the farmer flexibility to choose rather than to make the choice for him/her.

About the Book

This book evaluates the successes and shortcomings of the rice Green Revolution in Asia and presents a vision for the rice sector in 2025. It provides a comprehensive assessment of three decades of technological innovation in the rice production systems of Asia (1965–1995) and considers the prospects for sustaining productivity growth through further technical change over the next two decades. Evidence on the productivity gains made from the rice Green Revolution is presented alongside evidence on associated resource and environmental costs, including human health costs. The strong complementarity between technology and policy in promoting and sustaining productivity growth is emphasized throughout the book. The technology–policy link is also evident in the discussion on the resource-base consequences of the Green Revolution and on the opportunities for sustaining the resource base.

Evidences and data presented in this book came from a variety of sources. The voluminous Green Revolution literature was exploited for evidence on productivity gains and resource-base costs associated with the intensification of rice production systems. Unpublished papers and reports, especially those from the National Agricultural Research Systems (NARS) of Asia, were used for case study evidence on the impact of technologies and policies. Multi-country farm-level time series data sets collected by the Social Sciences Division of the International Rice Research Institute (IRRI) supplemented the evidence gleaned from the literature.

The book is divided into three parts: the first looks back over the last 30 years of rice technology adoption and impact; the second looks at the future of the rice sector; and the last provides an assessment of on-the-shelf and forthcoming technologies that will help meet future rice needs. The phrase 'rice bowls' is used throughout the book to indicate the irrigated and the favorable rainfed lowland rice-production environments. These have been the heartlands of the rice revolution in Asia. Sustaining productivity growth of the rice bowls of Asia is the primary and all-pervasive concern of this book.

Overview of Conclusions

Productivity gains from the exploitation of Green Revolution technologies are close to exhaustion

Modern rice varieties that ushered in the Green Revolution were specifically targeted for the irrigated and the favorable rainfed environments. By 1990, 74 percent of the rice area in Asia was planted to modern varieties, accounting for all of the irrigated areas plus around a third of the rainfed lowlands (Chapter 2). Growth in rice production in the high-potential environments of Asia has been largely responsible for the stability of rice supplies and declining rice prices. The opportunities for further expansion of modern variety use are exhausted in the irrigated lowlands and are severely constrained in the rainfed environments by drought and/or submergence risks.

The potential for further expansion of irrigated area through infrastructural development is limited because of high developmental costs and increasing environmental concerns (Chapter 9). In fact, there has been a net decline in area due to degradation of irrigated lands because of increased salinization and waterlogging (Chapter 4). In the high-potential irrigated lands across Asia, the economically exploitable gap between the technological frontier and farmer yields is very small. Given current technology and relative prices, it is not profitable for farmers to increase yields in the 'rice granary' provinces across Asia. Moreover, productivity growth in the irrigated lands has been slowing down and declining in some instances due to intensification-induced degradation trends (Chapter 4).

In the absence of further technical change, Asian farmers face increasing costs per ton of rice produced

The primary reason for the rapid adoption of modern rice varieties in Asia was that it was enormously profitable, compared to traditional varieties. Once investments in land modification were made (such as irrigation), the move to modern rice technology resulted in a dramatic fall in the unit cost of production. The continual and incremental modernization of rice production systems, after the initial adoption of modern varieties, resulted in the long-term reduction of unit production costs through the 1980s. Thus, the private profitability of rice production has been maintained despite the decline in domestic rice prices throughout Asia (Chapter 3).

Since the beginning of the 1990s, however, unit production costs are beginning to rise and rice farmers are facing declining profits. A stagnant yield frontier and diminishing returns to further intensification are the primary reasons for the current reversal in rice profitability (Chapter 3). Moreover, other contemporaneous changes in factor markets – especially land, labor and water – are driving up input prices. The rapid withdrawal of labor from the agricultural sector, the diversion of land to other agricultural and non-agricultural activities, increased competition for water, and the withdrawal of subsidies for agricultural inputs have all led to increasing costs of rice production, and the prospects are for continued increases (Chapter 9). In the absence of a shift in the yield frontier for rice and/or a substantial improvement in input use efficiencies, the profitability of rice production in Asia may return to its pre-Green-Revolution level.

Emerging trends in declining productivity of ricelands could be the result of government policies aimed at achieving and maintaining food self-sufficiency

Emerging farm-level evidence from the rice bowls of Asia indicates that intensive rice monoculture systems lead, over the long term, to declining productivities of inputs. Over time, farmers have been found to use increasing amounts of inputs to sustain the yield gains made during the Green Revolution years (Chapter 4). Intensive rice monoculture on the lowlands results in the following changes in production systems: (i) rice paddies flooded for most of the year without an adequate drying period; (ii) increased reliance on inorganic fertilizers; (iii) asymmetry of planting schedules; and (iv) greater uniformity of cultivars. Over the long term, the above changes impose significant ecological costs due to negative biophysical impacts (Chapter 4). Adverse biophysical consequences that have reduced productivity have been: the buildup of salinity and waterlogging; declining soil nutrient status; increased incidence of soil toxicities; and pest buildup and reduced resilience of the ecosystem to pest attacks.

Ironically, the very policies that encouraged increased food supply through intensive monoculture systems also contributed to the declining sustainability of these systems (Chapter 4). Massive investments in irrigation infrastructure were essential for the success of the Green Revolution. In retrospect, it is clear that many of the systems were poorly designed, especially in the drainage component which was often left out in order to reduce costs. It was not just the design of irrigation systems that was problematic; system management, water allocation and water pricing also contributed to the environmental costs of irrigated agriculture. The persistence of injudicious and inefficient input use – especially in the case of subsidized water, fertilizers and pesticides – contributed to the ecological consequences discussed above. Where input prices were kept low through government intervention, farmers had little incentive to acquire the knowledge necessary for being more discriminatory in input use decision making.

Adverse agricultural externalities resulted from the lack of a holistic perspective of the farm resource base management

Intensification-induced environmental externalities are soil erosion for the uplands, water and air pollution (methane and nitrous oxide emissions) and pesticide poisoning in the lowlands (Chapter 5). Upland degradation also imposes a negative externality on the lowlands through increased sediment buildup in irrigation infrastructures. Environmental externalities generated within a watershed transcend various sectors of economic activity; for example, pesticide runoff from rice paddies affects fish production in lowland lakes. Similarly, indiscriminate pesticide use can affect drinking-water supplies and human health. Poor forest conservation policies could have detrimental effects on irrigation water supply.

Where multiple uses of the watershed are common, interventions made in one sector ought to be carefully considered regarding the externality costs imposed on other users. Where such costs are not explicitly accounted for, the returns to innovation and development are often overestimated. For example, where upper watershed externalities are not explicitly accounted for, there is an under-investment of research resources in upland conservation. Holistic planning and management of watershed resources would help minimize externality costs of contemporaneous economic activities. Community participation in the decision-making process would help move it away from sector-specific concerns and towards a more holistic watershed-level approach.

Despite an anticipated decline in per capita rice consumption, aggregate Asian demand for rice is expected to increase by 50–60 percent during the 1990–2025 period

As incomes grow, per capita rice consumption is expected to decline as consumers substitute rice with high-cost quality food containing more protein and vitamins, such as vegetables, bread, fish and meat (Chapter 2). East Asia has already made this transition and Southeast Asia is currently making the transition. South Asia, on the other hand, continues to see rising rice consumption with income growth. But the major boost in increasing demand for rice over the next few decades will come from population growth. Asian population is expected to grow by 18 percent in the 1990s, and by an additional 53 percent in the next 30 years. To meet future demands, rice production needs to increase to more than 800 million tons by 2025 from the 1990 level of 480 million tons, a production growth rate of 1.8 percent per year (Chapter 6).

There are early warning signs, however, that a deceleration in rice supply has set in. The annual growth in global rice production was only 1.7 percent per year during the 1985–93 period, compared to 3.2 percent during the 1975–85 period, and 3.0 percent a decade earlier. Rice production has failed to outpace population growth in several Asian countries. Many countries of South and Southeast Asia will find it difficult to maintain self-sufficiency in rice production over the next 10–20 years. The problem will be eased only if there is a further technological breakthrough in shifting the yield frontier for irrigated rice along with continued investments in irrigation infrastructure.

The long-term consequence of rice trade liberalization could be net rice imports into Asia

Given the political importance of adequate and reliable domestic rice supplies, most Asian governments pursued policies for achieving self-sufficiency in rice production. Of the major rice-growing countries in Asia, few are major importers (except Malaysia and Sri Lanka), and it is only during times of natural calamities that they go to the world market (Chapter 6). It is anticipated that over the next two decades Asian countries may substantially increase their rice imports in order to make up for domestic shortfalls. Domestic shortfalls could occur due to: (i) continuing and unabated population growth in the low-income developing countries; (ii) rapid economic growth in the middle- and higher-income countries that makes the maintenance of self-sufficiency unprofitable; and (iii) trade liberalization and trade agreements that could lead to the Asian food sector becoming more open to international competition.

The Uruguay Round of trade negotiations has increased pressures to liberalize agricultural trade and to open up rice markets in the middle- and high-income countries of Asia (Chapter 7). While the short-term impact of the current GATT agreement on rice trade could be modest, the long-term effects could be substantial because of continued trade liberalization (Chapter 7) and the transformation of the agricultural sector (Chapter 8). The movement away from self-sufficiency objectives to market-oriented decision making results in increased competitiveness for production resources, both in the agricultural and non-agricultural sectors. Rice supplies could be expected to be negatively affected by competing demands for land, labor, water and other factors of production. Several countries that are now self-sufficient in rice may find it more profitable to import at least a part of their rice requirements in exchange for diverting production resources to more remunerative activities.

Economic growth and the commercialization of agricultural systems could reduce the competitiveness of rice relative to other crops and other farm enterprises

Economic growth, urbanization and the withdrawal of labor out of the agricultural sector leads to increasing commercialization of agricultural systems (Chapter 8). Commercialization implies the movement out of subsistence food production to a diversified market-oriented production system and the progressive substitution out of non-traded inputs (family labor) in favor of purchased inputs. The process of diversification out of staple food production is triggered by rapid technological change in agricultural production, improved rural infrastructure, and diversification in food demand patterns. Initially, diversification implies the addition of other crops and other enterprises at the farm-household level. As the level of commercial orientation increases, however, mixed farming systems give way to specialized production units that are designed to respond rapidly to market price and quality signals.

Governments have a difficult task to perform: on one hand, continued food security needs to be assured for populations that are growing in absolute terms; on the other hand, research and infrastructural investments need to be made for diversification out of the primary staples. It is important to remember that, even with increased commercialization and diversification trends, rice will continue to be the most important staple food in Asia, in relative and absolute terms. Substantial research efforts have to be made to improve the profitability of rice production in order for it to compete with more lucrative options for farm resources.

An upward shift in the rice yield frontier is necessary to meet future rice requirements and to sustain farm-level profits

The future across Asia is one of increased scarcities for inputs used in rice production, especially land, labor and water (Chapter 9). A substantial shift in the rice yield frontier is needed to meet future rice demands while at the same time reducing the unit costs of rice production. Recent progress in plant breeding research indicates that a significant shift in the rice yield frontier is possible both in the medium term and in the long term (Chapter 10). In the medium term, yield increases of around 20 percent could be possible through the adoption of hybrid rice. The longer-term prognosis is for a new 'super' high-yielding plant type that could yield about 12.5–13 t ha^{-1} and, as a parent of the hybrids, could increase this yield to 15 tons of grain per hectare. While the prospects are good for the generation of new seed technologies, research and adaptation work is by no means completed; continued high levels of research investments are needed in order to make widespread farmer adoption a reality. In order to secure productivity growth of rice, the new seed technologies ought to be complemented by continued investments in irrigation and other rice-related infrastructure.

The above gains in shifting the yield frontier have been achieved using conventional plant-breeding tools. The role of biotechnology in shifting the yield frontier for rice will be limited in the foreseeable future. The greatest contribution of biotechnology tools would be in improving yield stability through generating rice plants with improved resistance to pests and tolerance to abiotic stresses. Even where transgenic rices are available, their farm-level adoption may be constrained by regulatory impediments that may restrict their access.

Intellectual property protection and plant breeders' rights may further prevent the easy access of materials developed through the use of biotechnology tools. Substantial policy reforms ought to be in place before the fruits of biotechnology are available to the rice farmers of Asia.

Significant gains in input efficiencies are possible and contribute to the sustainability of the paddy resource base, but come at the cost of increasingly expensive farmer-time

For post-Green-Revolution Asia, meeting the dual goals of maintaining rice productivity growth while at the same time sustaining the agricultural resource base would require substantial improvements in input use efficiencies (Chapters 3, 9 and 11). Examples of efficiency-enhancing technologies are: improved varietal selection; improved timing and application systems for fertilizers; integrated pest management; and judicious water management. The above 'second generation technologies' are more knowledge intensive and location specific than the modern seed–fertilizer technology that was characteristic of the Green Revolution. Productivity gains accrue to farmers who have the ability to learn about the new technologies and to use them effectively. Profitable adoption of knowledge-intensive input management technologies would depend on the value of input savings relative to the cost of additional time required for learning and decision making (Chapter 11).

The challenge for the research system is in finding cost-effective methods for transferring knowledge-intensive crop management technologies to the farmers. The current practice of providing blanket recommendations on input use to farmers over a large area is no longer a valid mode of operation for the extension service. Unlike input use recommendations, knowledge transfer requires that farmers adapt scientific principles to their own particular circumstances and derive farm-specific practices. Therefore, to be cost effective, any system of knowledge transfer ought to encourage farmer learning through experimentation and adaptation. The complexity and the costs associated with training programs that try to incorporate the above features are very high and often act as impediments to action on the part of the extension service.

Finally, it ought to be recognized that the profitable adoption of knowledge-intensive technologies that enhance input use efficiency will only occur when the price is right. Holding input prices low through the use of subsidies or output prices high through price support programs will reduce the likelihood of adoption of input-efficient technologies.

Is the rice crisis returning?

If the current sense of complacency about the rice situation continues, future rice supplies will fall substantially short of Asian requirements. While the smaller countries of Asia can make up the shortfall through imports, larger Asian countries – such as India, China and Indonesia – cannot effectively utilize this option given the small size of the world rice market. Maintaining rice productivity growth is an enormous challenge facing Asian governments, especially given the increased competition for production resources, such as land, water and labor.

Technological breakthroughs are on the horizon, for shifting the yield frontier for rice and for increasing the profitability of rice production systems. Long-term national and international support for rice research is necessary for the effective use of technological change to sustain and even increase rice productivity growth in Asia. In addition to research funding, governments need to continue high levels of investments in infrastructural development and maintenance, especially for irrigation infrastructure. Finally, improved farmer incentives will play a crucial role in assuring rapid growth in rice productivity and rice supplies. Concerted and consistent efforts made towards increasing and/or sustaining rice productivity growth will help avert any potential crisis in the next three decades.

Rice Productivity Growth: The Case against Complacency

2

In three decades, following the introduction of modern rice varieties to tropical Asia in 1965, rice production almost doubled. Seventy percent of the production increase came from increased yields and increased cropping intensity, but 30 percent resulted from new land brought under cultivation or shifted to rice from other crops (IRRI, 1993a). Food production grew at a rate of 3 percent per annum, faster than the growth in population, mainly through the replacement of traditional with modern rice varieties. Many traditional rice-importing countries have achieved self-sufficiency, and average per capita rice consumption in 1993 was 25 percent higher than it was in 1965. The introduction and rapid spread of modern rice varieties is often referred to in Asia as the Green Revolution.

The success of the Green Revolution has led to a growing sense of complacency about current and future food supplies in Asia. Several analysts have made highly optimistic projections of Asia's capability to produce and meet its future food requirements (see Mitchell and Ingco, 1995, and Avery, 1988 for example). Several factors have contributed to this optimism: (i) the misperception that the viability of modern rice technology transcends biophysical and economic boundaries, and that there is still a large technology gap that can be exploited; (ii) an overly optimistic assessment of the low income elasticity of demand for rice leads to an over-estimate of the projected rate of decline in aggregate rice demand; (iii) the low international rice price creates a perception of abundant supplies; and (iv) the economic successes of a few (small) Asian countries and their ability to import food are being generalized across the board.

In this chapter we argue the following:

1. There are very definite boundaries to the extent of productivity growth in rice, even with modern technology.
2. The economically exploitable technology gap is small, i.e. where productivity gains can be made given current technology, these have already been made.
3. The rate of consumption growth, for over 70 percent of Asian rice consumers, has not leveled off and will not for the next several decades.
4. International rice price trends, over the long term, do not reflect domestic supplies and shortfalls.

Boundaries of the Productivity Impact

The widespread adoption of modern varieties and their appeal in the development literature has created several misperceptions about their potential productivity impact. This section specifies the agroecological and economic boundaries within which productivity impact of modern varieties can be expected. The Green Revolution was not one 'big bang' that had a uniform impact on all rice-growing areas of Asia, as is often thought to be. Worse yet, the regions that did not gain from it have not all failed due to the lack of technical knowhow or political will. The gains from the Green Revolution varied (and they were expected to vary) by rice ecologies, agroclimatic zones, demographic pressures, and policy environments. Countries that witnessed large gains in rice productivity have invariably had a congruence of several of the above factors. The yield difference between countries and between regions within a particular country is a reflection of the differences in the above factors and not necessarily an exploitable productivity gap.

Rice ecologies and technological innovation

Rice is grown in four very different ecologies in Asia, and these are: the irrigated lowlands; the rainfed lowlands; deepwater and flood-prone environments; and the upland systems. These ecologies differ in terms of hydrology and water control (see Box 2.1 for a description of the rice ecologies). The type of rice plants that are grown are different for each ecology. There are fundamental differences in plant characteristics and physiology. Plants bred for the irrigated lowlands, for instance, cannot be grown in the uplands or in floodprone and deepwater environments. The modern semi-dwarf, high-yielding varieties that ushered in the Green Revolution were developed for the irrigated and the favorable rainfed lowlands.

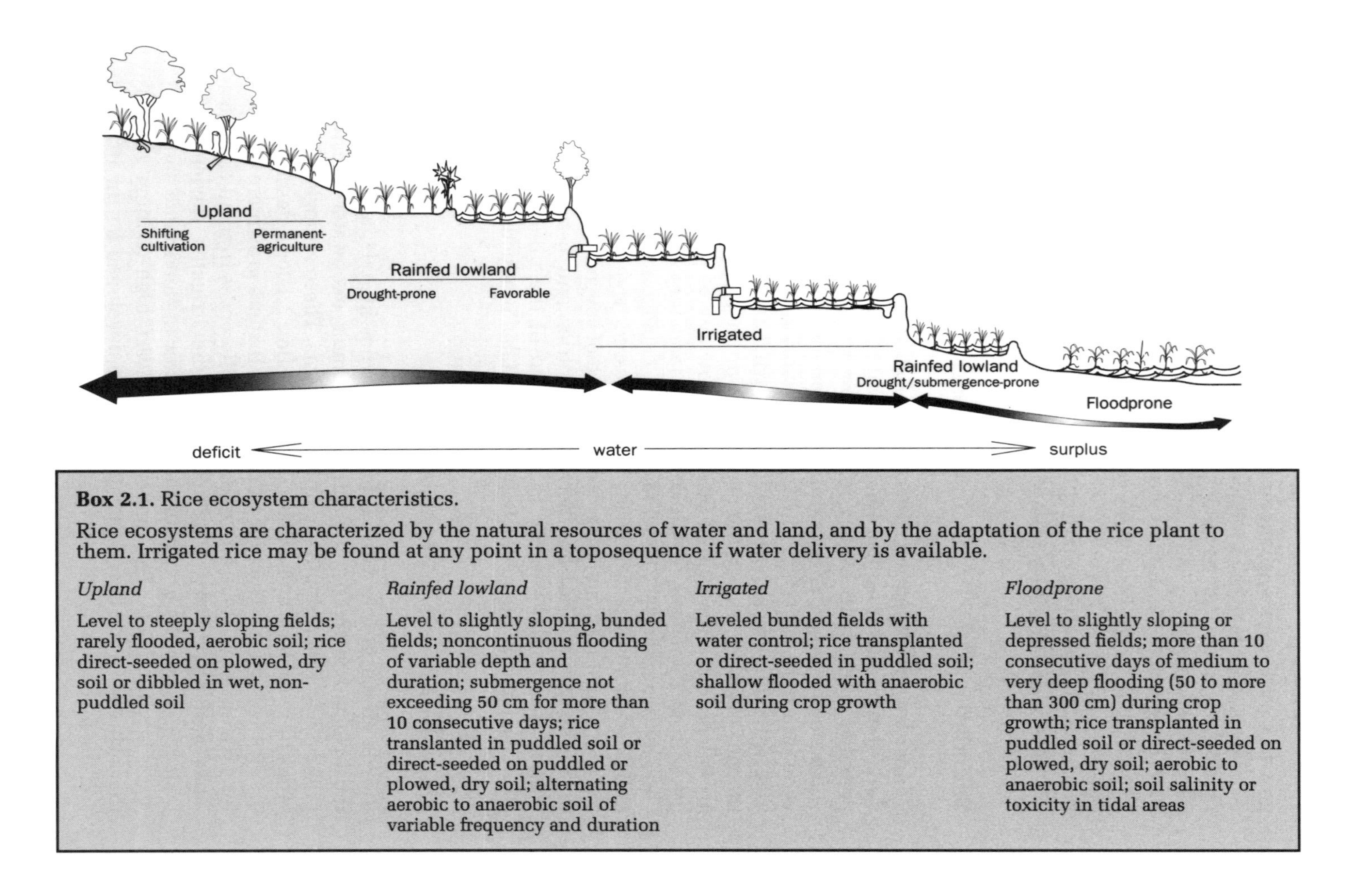

Box 2.1. Rice ecosystem characteristics.

Rice ecosystems are characterized by the natural resources of water and land, and by the adaptation of the rice plant to them. Irrigated rice may be found at any point in a toposequence if water delivery is available.

Upland

Level to steeply sloping fields; rarely flooded, aerobic soil; rice direct-seeded on plowed, dry soil or dibbled in wet, non-puddled soil

Rainfed lowland

Level to slightly sloping, bunded fields; noncontinuous flooding of variable depth and duration; submergence not exceeding 50 cm for more than 10 consecutive days; rice translanted in puddled soil or direct-seeded on puddled or plowed, dry soil; alternating aerobic to anaerobic soil of variable frequency and duration

Irrigated

Leveled bunded fields with water control; rice transplanted or direct-seeded in puddled soil; shallow flooded with anaerobic soil during crop growth

Floodprone

Level to slightly sloping or depressed fields; more than 10 consecutive days of medium to very deep flooding (50 to more than 300 cm) during crop growth; rice transplanted in puddled soil or direct-seeded on plowed, dry soil; aerobic to anaerobic soil; soil salinity or toxicity in tidal areas

Table 2.1. Rice area, production and yield in Asia by rice ecosystem, 1991.

Regions	Irrigated	Rainfed lowland	Floodprone	Upland
Total Asia				
Area (000 ha)	73,943	38,753	10,016	10,539
Production (000 tons)	361,553	89,250	15,284	11,180
Yield (t ha^{-1})	4.89	2.30	1.53	1.06
Southeast Asia				
Area (000 ha)	14,050	13,632	3,202	2,091
Production (000 tons)	64,317	25,045	4,190	2,857
Yield (t ha^{-1})	4.58	1.84	1.31	1.37
South Asia				
Area (000 ha)	24,989	23,001	6,759	7,599
Production (000 tons)	89,306	57,221	10,994	6,189
Yield (t ha^{-1})	3.57	2.49	1.63	0.81
East Asia				
Area (000 ha)	32,339	1,888	0	762
Production (000 tons)	192,133	5,876	0	2,019
Yield (t ha^{-1})	5.94	3.11	0	2.65

Source: IRRI (1993a).

Irrigated rice is planted on 81 million hectares (harvested area, taking into account double and triple cropping) worldwide. Asia alone has 74 million hectares (about 90 percent of the total irrigated area) and produces more than 70 percent of the world's rice. Growth in rice production on the irrigated farms has been largely responsible for the recent stability of urban rice supplies and prices. The potential for further expansion of irrigated area is limited (Chapter 3), and in fact there has been a net decline in area due to degradation of irrigated lands because of increased salinization and waterlogging (Chapter 4). Productivity growth in the irrigated lowlands has been slowing down and declining in some instances due to intensification-induced degradation trends (Chapter 4). (Table 2.1 shows the areas, yields and production from each of the rice ecologies.)

The rainfed lowland rice area worldwide is approximately 37 million hectares and accounts for 17 percent of total rice production. Most rainfed lowland rice is grown in South and Southeast Asia. More than 25 million hectares of rainfed lowland areas are favorable or are affected only moderately by drought or submergence. Productivity gains in the favorable rainfed lowlands have occurred because of spillover of varieties from the irrigated ecosystems.

More than 10 million hectares of riceland in South and Southeast Asia are subject to uncontrollable flooding. The floodprone areas are generally in the delta areas of large rivers such as the Mekong, Irrawaddy and the Ganges. Rice plants adapted to this environment are tolerant to submergence and prolonged flooding. Yields in this environment are low and therefore the floodprone ecosystem accounts for only 3.2 percent of Asian rice production.

The uplands of Asia, Africa and Latin America grow about 19 million hectares of rice. The upland topography ranges from sloping terraces to flatlands. Upland rice is low yielding and generally prone to serious soil erosion and soil degradation problems. The uplands, however, host some of the world's poorest farmers. With improvements in lowland productivity growth there has been a net decline in upland rice area across Asia. As transport infrastructure improves, the Asian uplands have been moving out of subsistence rice production to commercial cultivation of maize, vegetables and fruit.

IR-8, the new plant that ushered in the Green Revolution, was not meant for adoption across all rice ecologies. The irrigated environment was seen as the best bet for raising the yield potential because of assured water supply and the ability to plant in the dry season when solar radiation is at its highest (Chandler, 1982). The probability of research success was high for the irrigated lowlands and the spillover benefits to the rainfed lowlands were expected to be large. By 1990, 74 percent of the rice area in Asia was planted to modern rice varieties. That accounts for all of the irrigated areas plus around a third of the rainfed lowlands.

Agroclimatic zones and productivity impact

The productivity gains for irrigated rice varied substantially by agroclimatic environments. The differences were not due to input levels or technological knowhow but rather differences in physiological growth processes due to climatic differences. The yield potential of the same rice variety is greater in the temperate zones than in the tropics. Varieties with yield potential of 10 tons in the Philippines have been reported to yield over 14 tons in China, Southern Australia and California. The reported differences in yield potential are mainly caused by differences in solar radiation and temperature. The temperate zone has a higher number of sunny daylight hours during the summer months and hence a higher level of photosynthesis relative to the tropics. Lower temperatures in the more temperate latitudes result in lower respiration losses and long duration of all crop growth phases, particularly the grain filling period (Kropff *et al.*, 1993).

Homogeneous agroclimatic areas can be quantified and mapped using the concept of agroecological zones (AEZs) developed by FAO,

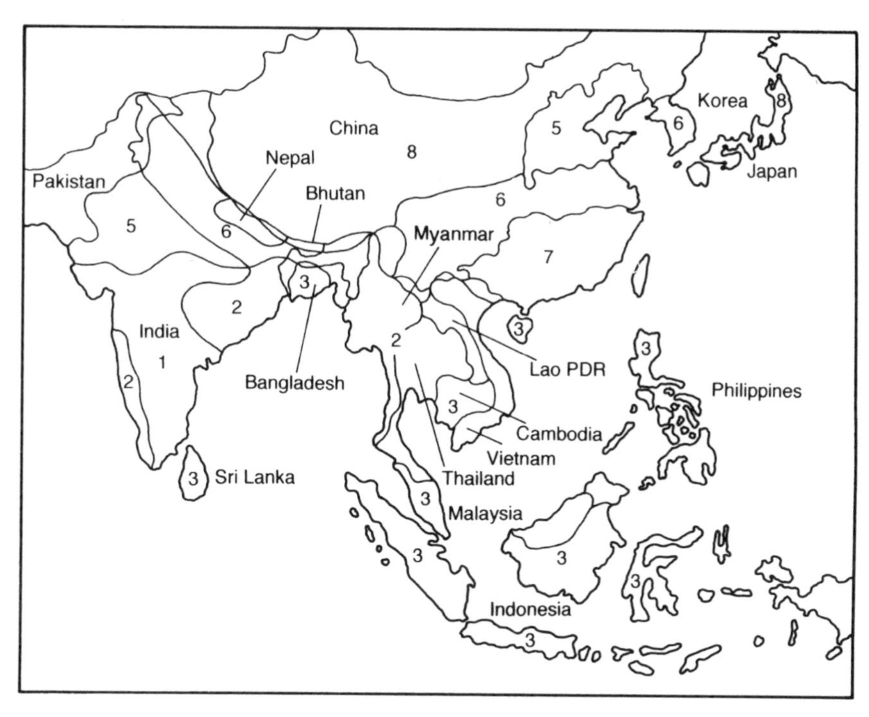

Fig. 2.1. Regional agroecological zones in Asia. Dominant agroecological zones: 1, warm semi-arid tropics; 2, warm subhumid tropics; 3, warm humid tropics; 5, warm semi-arid subtropics with summer rainfall; 6 warm subhumid subtropics with summer rainfall; 7, warm/cool humid subtropics with summer rainfall; 8, cool subtropics with summer rainfall. (Source: IRRI Medium-Term Plan 1994–1998.)

Table 2.2. Sources of growth in rice production in different agroecological zones in Asia, 1975–1991.

Agroecological zone	Share of Asian rice area (%)	% of irrigated rice area	Trend rate of growth (1975–1991) (% per year)		
			Area	Yield	Production
Semi-arid tropics	7.3	68.2	–0.2	2.2	2.0
Subhumid tropics	30.0	26.8	0.2	2.3	2.5
Humid tropics	26.4	48.2	0.7	3.1	3.8
Subhumid subtropics	18.4	76.8	0.1	3.4	3.5
Humid subtropics	14.2	92.1	–1.1	3.6	2.5

Sources of basic data: IRRI (1991) and subnational-level statistics obtained from country sources.
Source: Hossain and Laborte (1993).

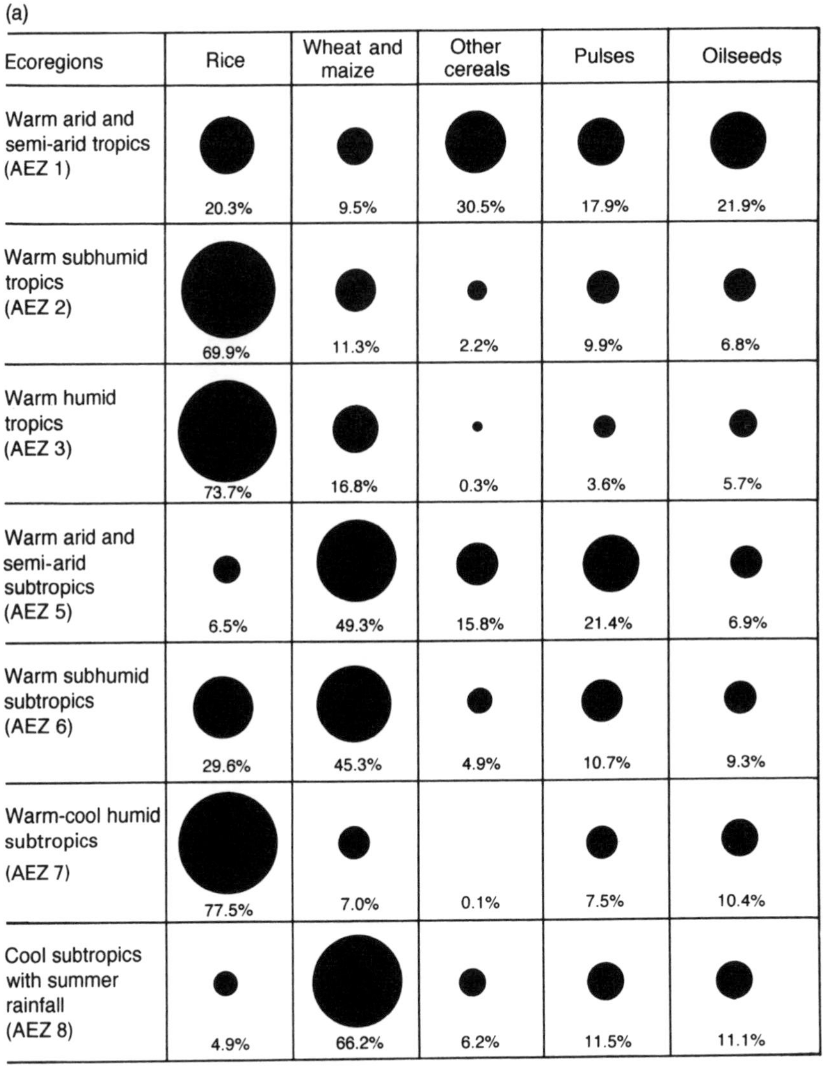

(a)

Ecoregions	Rice	Wheat and maize	Other cereals	Pulses	Oilseeds
Warm arid and semi-arid tropics (AEZ 1)	20.3%	9.5%	30.5%	17.9%	21.9%
Warm subhumid tropics (AEZ 2)	69.9%	11.3%	2.2%	9.9%	6.8%
Warm humid tropics (AEZ 3)	73.7%	16.8%	0.3%	3.6%	5.7%
Warm arid and semi-arid subtropics (AEZ 5)	6.5%	49.3%	15.8%	21.4%	6.9%
Warm subhumid subtropics (AEZ 6)	29.6%	45.3%	4.9%	10.7%	9.3%
Warm-cool humid subtropics (AEZ 7)	77.5%	7.0%	0.1%	7.5%	10.4%
Cool subtropics with summer rainfall (AEZ 8)	4.9%	66.2%	6.2%	11.5%	11.1%

Fig. 2.2. (a) Importance of foodcrops in agroecological zones (Key: Figures in percent of area under foodgrain for each agro-ecoregion). (Source of basic data: IRRI, 1991). (b) (*opposite*) Importance of rice ecosystem in Asian ecoregions. (Key: Figures in percent of total rice area.) (Source of basic data: IRRI, 1991.)

the Food and Agriculture Organization (Higgins, 1982). The agroecological map of Asia is presented in Fig. 2.1. The distribution of arable land, population and the production of foodcrops, by AEZs of Asia, is shown in Fig. 2.2(a) and Table 2.2. The production potential of food-

(b)

Ecoregions	Irrigated	Rainfed lowland	Upland	Floodprone
Warm arid and semi-arid tropics (AEZ 1)	5.8%	3.0%	2.0%	0.4%
Warm subhumid tropics (AEZ 2)	7.2%	11.3%	3.6%	4.0%
Warm humid tropics (AEZ 3)	12.7%	7.1%	3.0%	3.6%
Warm arid and semi-arid subtropics (AEZ 5)	3.1%	0.0%	0.0%	0.0%
Warm subhumid subtropics (AEZ 6)	14.1%	2.5%	0.9%	0.9%
Warm-cool humid subtropics (AEZ 7)	13.0%	0.9%	0.2%	0.0%
Cool subtropics with summer rainfall (AEZ 8)	0.6%	0.0%	0.0%	0.0%

grain per hectare of arable land is about 5.7 times higher in the subhumid tropics (Southern and Southwestern China) than in the semi-arid tropics (Southern and Western India). The semi-arid tropics have the lowest rice production potential of all AEZs (IRRI, 1993).

Rice is the dominant crop in the humid and subhumid tropics (Southeast Asia, and the eastern parts of South Asia), and in the humid subtropics (Southern and Southwestern China). Not surprisingly, there is a close match between the dominance of rice

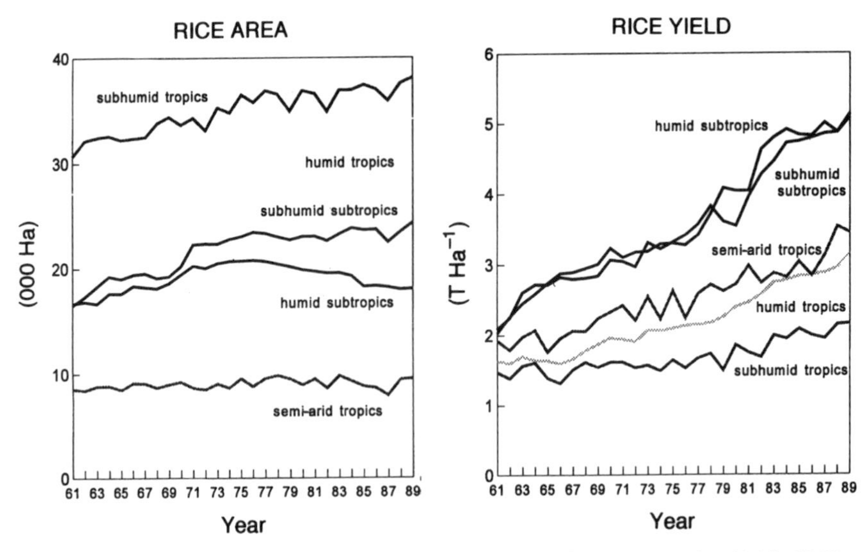

Fig. 2.3. Trends in rice area and yield in agroecological zones in Asia, 1961–1989. (Source: Hossain and Laborte, 1993.)

within an AEZ and the irrigated area in that zone (Fig. 2.2(b)), indicating that rice investments were made in areas of highest production potential.

Yield performance across the AEZs attests to the physiological differences in plant growth discussed above. In the pre-Green-Revolution period, the difference in average yields between the tropical and subtropical zones was less than half a ton; by 1990, this difference was around 2.5 tons. For the same technology level, average yields in the subtropics are twice that of the tropical zones (Fig. 2.3). Given temperature and humidity differences, average yields from East Asia will generally be higher than those from Southeast Asia and South Asia, for the same varieties and input levels. Research efforts aimed at narrowing the gap in yield potential between the temperate and tropical zones could have low payoffs because of the fundamental physiological differences governing plant growth in the two zones.

Demographic pressures and agricultural intensification

The demand for intensification is very closely related to demographic pressures – land/labor ratios. Where opportunities exist for area expansion these would be superior to increasing production through an increase in yield per hectare. Massive investment in irrigation infrastructure, increased cropping intensities and the widespread adoption of modern high-yielding technology have generally occurred only under high land pressure[1].

Kikuchi and Hayami (1978) document this process of land augmentation as the cultivation frontier closes for Japan, Taiwan, Korea and the Philippines.

> As population pressure pushes the cultivation frontier into marginal areas, we expect the marginal cost of agricultural production via expansion of cultivated area to rise relative to the marginal cost of production via intensification. Eventually the economy will reach a stage at which internal land augmentation becomes a less costly means of increasing agricultural output than external land augmentation.
>
> (Hayami and Ruttan, 1985, p. 310)

Table 2.3. Population pressure, use of modern inputs and rice yields in selected Asian countries, 1990.

Region/ country	Rice yield (t ha^{-1}) (1989–1991)	Adoption of modern varieties (%)	% of irrigated area	Fertilizer use[a] (kg NPK ha^{-1})	Arable land per capita (m^2)
East Asia					
China	5.8	100	93	278	819
Japan	6.1	100	99	460	366
Korea, DPR	7.8	100	67	483	1117
Korea, Republic of	6.3	100	99	475	501
Southeast Asia					
Cambodia	1.4	11	8	1	3638
Indonesia	4.3	77	72	152	1220
Laos	2.3	2	10	1	2254
Malaysia	2.6	90	66	104	2757
Myanmar	2.9	50	18	23	2373
Philippines	2.8	89	61	119	1308
Thailand	1.9	18	27	46	3938
Vietnam	3.2	80	53	98	980
South Asia					
Bangladesh	2.7	51	22	98	865
India	2.6	66	44	71	1995
Nepal	2.4	36	23	25	1377
Pakistan	2.3	42	100	91	1853
Sri Lanka	2.9	91	77	102	1109

[a] Fertilizer consumption in all crops per hectare of arable land. Data on fertilizer consumption in rice cultivation are not available.
Sources of basic data: FAO (1992) and IRRI (1991).
Source: Hossain and Laborte (1993).

A cross-sectional comparison across Asian countries shows that the proportion of irrigated area and the percentage of area under HYVs and fertilizer use is positively related to population densities (Table 2.3). It was not a historical accident that research for enhancing the yields of the two major foodcrops of Asia, rice and wheat, began in the 1950s. Research investments for enhancing yields were made in response to declining opportunities for area expansion, which was the main source of output growth in the pre-Green-Revolution Asia[2]. A comparison of sources of rice output growth during the early and late Green Revolution periods shows that the overall rate of growth was similar for the two time periods (Table 2.4). In the early period, one-third of the growth came from an increase in cropped land, while in the latter period almost all of it was due to yield increases. For the

Table 2.4. Rice: annual growth rates of area, production and yield, Asia, 1957/59–1988/90.

Countries/ regions	1957/59– 1988/90	1957/59– 1965/67	1965/67– 1973/75	1973/75– 1981/83	1981/83– 1988/90
Total Asia					
Area	0.73	0.85	1.09	0.24	0.25
Production	3.08	2.60	3.37	3.09	2.16
Yield	2.36	1.74	2.27	2.86	1.91
Southeast Asia					
Area	0.93	1.73	0.35	1.51	0.72
Production	3.24	3.17	3.29	4.28	2.29
Yield	2.32	1.46	2.94	3.22	1.57
South Asia					
Area	0.89	1.26	0.61	0.88	0.25
Production	2.33	3.13	1.63	2.57	2.31
Yield	1.45	1.89	1.02	1.71	2.03
China					
Area	0.52	–0.58	2.25	–1.07	–0.38
Production	3.55	2.62	3.92	2.98	1.25
Yield	3.03	3.21	1.68	4.06	1.63
India					
Area	0.67	1.21	0.74	0.46	0.34
Production	2.49	1.95	2.90	2.22	3.62
Yield	1.81	0.74	2.15	1.57	3.23

South Asia includes Bangladesh, Nepal, Pakistan and Sri Lanka, excluding India.
Southeast Asia includes Myanmar, Indonesia, Kampuchea, Laos, Malaysia, Philippines, Thailand and Vietnam.
Source of basic data: IRRI (1991).
Source: Pingali and Rosegrant (1993).

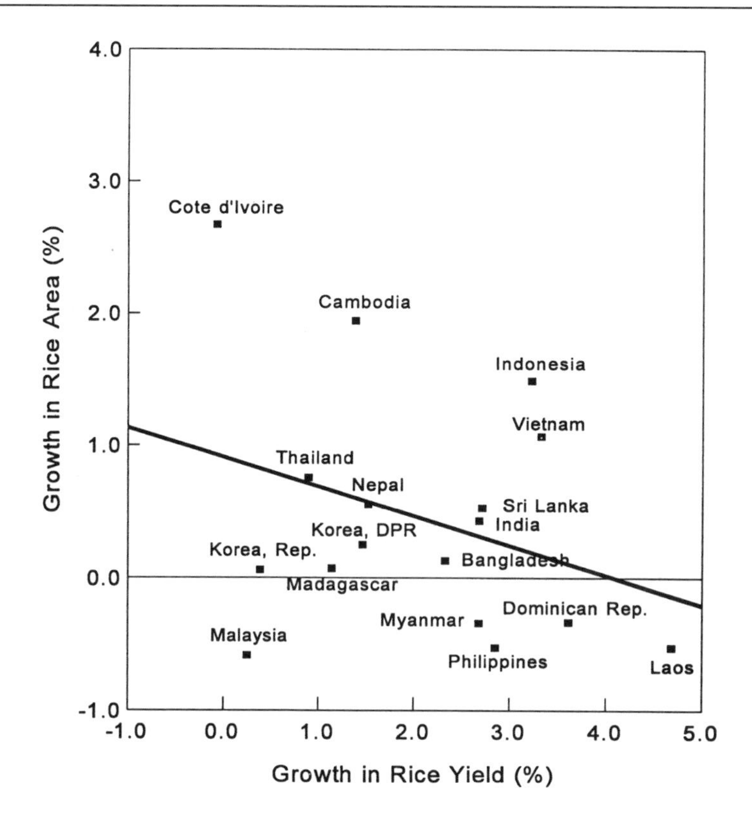

Fig. 2.4. (a) Relationship between growth in rice area and growth in rice yields, major rice-growing countries, 1975–1992. (Source of basic data: FAO *Agrostat Database,* various issues.)

densely populated countries of Asia, continued productivity growth will only come from further yield improvements (Fig. 2.4(a)).

Where population densities are low, the returns to intensification are high only if export markets are available and accessible (the Central Plains of Thailand is an example). If not, the labor and other costs associated with intensive agriculture are substantially higher than the incremental returns to intensification. The last statement holds true even with the adoption of labor-saving technologies.

Exploitable Yield Potential?

A large part of the current complacency comes from the perception that there is a significant gap between the technological potential and farmer yields, and that further gains in productivity will accrue through the exploitation of this gap. The argument is as follows: The proven rice yield potential for India, for example, is 8 t ha^{-1}. At the current national average yield of 2.6 t ha^{-1}, there is a yield gap, therefore, of around 5 tons that can be exploited. In a trivial sense this argument is correct; that is, if there are no biophysical or economic constraints to technology adoption. Once these constraints are explicitly accounted for, the exploitable gap becomes much smaller and, in many instances, non-existent.

To start with the commonly used notion of technology potential; what exactly does this mean? The best yields obtained on experiment stations are often used as the indicator of technology potential. Experiment stations are often located on the best lands within a region and have a very reliable irrigation water supply. Agronomic yield potential determined on experiment stations is the maximum achievable yield with no physical, biological and economic constraints. Maximum experimental yields also reflect the knowledge frontier and best-known management practices at any given point in time[3]. It is interesting to note that when the objective of the experiment is changed from maximizing yields to maximizing profits, the yields obtained by university researchers are significantly lower than the maximum attainable yields (Herdt, 1988). For a given location and set of prices, the latter yield could be referred to as the exploitable yield potential.

Table 2.5. Comparative national average yields, irrigated rice yields and experimental station rice yields, Asian countries, 1991.

Countries	National average rice yield (t ha^{-1})	Irrigated rice yield (t ha^{-1})	Average potential rice yield (t ha^{-1})
Bangladesh	2.6	4.6	5.4
China	5.7	5.9	7.6
India	2.6	3.6	5.9
Indonesia	4.4	5.3	6.4
Nepal	2.5	4.2	5.0
Myanmar	2.7	4.2	5.1
Philippines	2.8	3.4	6.3
Thailand	2.0	4.0	5.3
Vietnam	3.1	4.3	6.1

Sources: IRRI (1993).
Average potential yield data cited from Dey and Hossain (1995).

Is the national average yield an appropriate indicator of farm-level performance and potential for growth? It is not, because the national average yield is an average across agroclimatic zones, across rice ecologies and rice types, across technologies, across soil quality and across socioeconomic differences inherent in the various segments of a population within a country. Inherent in the comparison of technological potential and national average yields is either a presumption that modern technology transcends biophysical and agroclimatic differences, or a limited understanding of the magnitude and nature of these differences. Recall the boundaries within which productivity impact of modern technology can be expected (first section of this chapter). Table 2.5 presents data for several countries on experiment station yields, national average yields, and irrigated farmer yields.

Comparisons of national average yields across countries is commonly done to indicate the level of technology adoption. This is generally not a useful exercise since these averages pool away differences in agroclimatic and biophysical conditions that may explain the yield differences.

To get a more reasonable measure of the performance of a technology in its target environments one ought to use yield data from provinces or districts where these environments dominate. Pooling away heterogeneity is less of a problem when examining district/provincial yields, especially if the geographic areas are small. In the example for India, consider the state of Punjab, the most important agricultural state in India and the center of the Green Revolution for rice and for wheat. The average rice yield in Punjab is around 5 t ha^{-1}, which is more than twice the national average and only around 2 tons lower than the technological potential. Punjab is relatively homogeneous in climatic and biophysical conditions and almost all rice is grown on irrigated lands, therefore the trends in state yields are a good indication of the level of technology exploitation. (Fig. 2.4(b) shows the yield trends for India and for the state of Punjab.)

A comparison of experiment station yields with farmer yields in the same geographic area, at minimized agroclimatic and biophysical differences, shows that the gap between what is achievable and what is actually achieved is very small. In fact, studies in the Philippines have shown that over a third of the farmers in the 'rice bowl provinces' of Laguna and Nueva Ecija have been matching the best yields on neighboring experiment stations since 1980 (Box 2.2) (Pingali *et al.*, 1990).

The bottom line is that, in the high-potential irrigated areas across Asia, the economically exploitable gap between the technology frontier and farmer performance is very small. Given current technology and relative price levels it is not profitable for farmers in these envi-

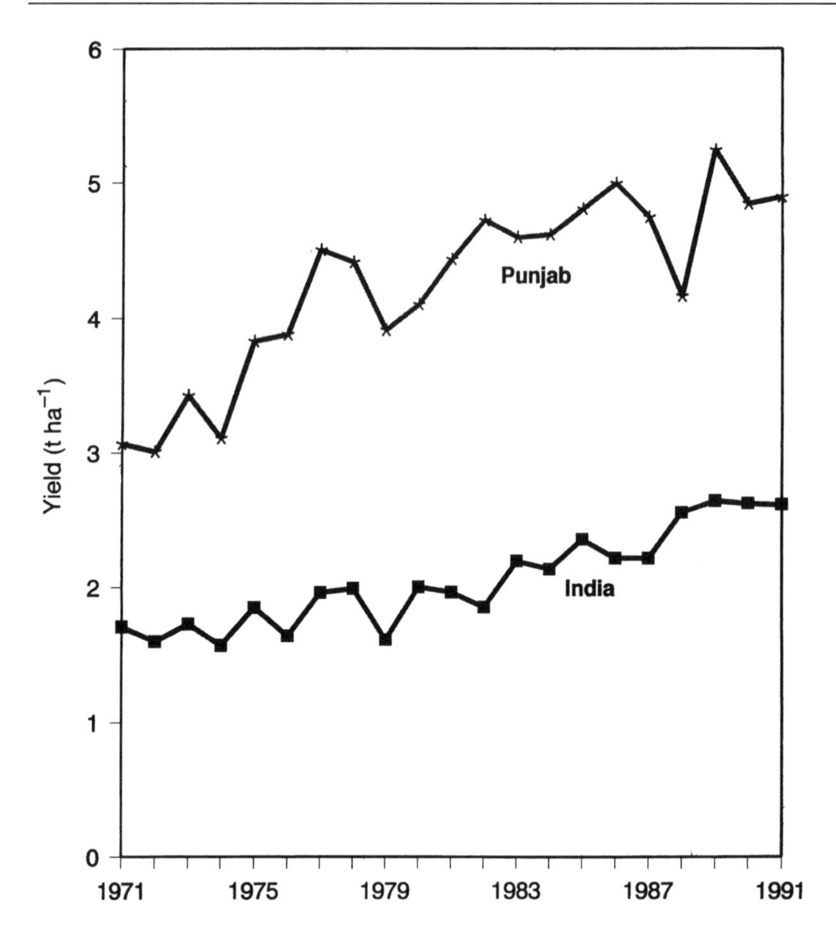

Fig. 2.4. (b) Trends in rice yield, State of Punjab and the whole of India, 1971–1991. (Source of basic data: IRRI, 1995.)

ronments to further bridge this gap. One accordingly sees that the yields in the 'rice granary' provinces across Asia have leveled off. Dramatic shifts in productivity in these high-potential areas, as had happened in the 1960s and 1970s, would require new rice varieties with substantial improvement in yield potential (Chapter 10).

The recent leveling off of yields in the high-potential irrigated environments shows up in national yield trends. Table 2.6 shows that the rate of growth in rice yields has started to decelerate in China, Korea DPR, Indonesia, Myanmar, the Philippines, Thailand, Pakistan and Sri Lanka. In Japan and the Republic of Korea, yields have been stagnant at 6–6.5 t ha^{-1} since the late 1960s. Yields are continuing to rise only in India, Bangladesh and Nepal and this is due to the further expansion of irrigated area using tubewells.

Box 2.2. The Philippine farmer and the irrigated rice yield gap.

The Philippine national average rice yield in 1987 was 2.65 t ha^{-1}. When compared with the technological potential of 6 t ha^{-1}, this suggests a yield gap of 3.5 t ha^{-1}. However, yield comparisons on randomly sampled irrigated rice farms with yields on neighboring experiment stations show a very different picture. Farmers in the neighborhood of IRRI in Laguna and Philippine Rice Research Institute (PhilRice) in Nueva Ecija were monitored over a two-decade period. Experimental yield-potential data from the IRRI farm was compared with the farmer sample in Laguna, while the yield-potential data from PhilRice was compared with the Nueva Ecija farmer sample. These comparisons reveal a yield gap of around 1.2 t ha^{-1}.

Figures 2.5(a) and 2.5(b) provide a comparison of the top one-third and the remaining two-third farm yields with the experiment station yields for Laguna (1966–87) and for Nueva Ecija (1970–86) respectively. Consider first the farmer and experiment station yields in Laguna for the years 1966–1987 (wet season). In 1966 the gap between the IRRI farm and the top third of the Laguna farm yields was approximately 2.2 t ha^{-1}. As Fig. 2.5(a) shows, this difference declined rapidly and by 1978 the top-third yields on the Laguna farms were matching the yields on the IRRI farm. Farmer and experiment station yield comparisons for Nueva Ecija for the years 1970–86 (wet season) show a similar pattern. In 1970 the gap between the top-third farms in Nueva Ecija and the MRRTC (PhilRice) farm was approximately two tons. This gap diminished to less than half a ton within a decade. Since 1986, the top-third farm yields have matched experiment station yields.

How have the yields on the remaining two-thirds of the sample fared, relative to the top-third farm yields during this time period? Wet-season yield comparisons for Laguna and Nueva Ecija show a significant yield difference between the two groups of farms for each of the years studied. In Laguna the top-third and the remaining two-third farms started off with a yield difference of 1.8 t ha^{-1} in 1966 (Fig. 2.5(a)). Through the 1970s and until 1981 the gap between the two groups remained at around 2 t ha^{-1}. Since then the gap seems to have widened, becoming 2.8 t ha^{-1} in 1984 and 3 t ha^{-1} in 1987. The difference between the top-third and the remaining two-third farm yields for wet season Nueva Ecija has remained at around 2 t ha^{-1}, over the years 1970–1986 (Fig. 2.5(b)).

The yield gap in Laguna and Nueva Ecija is not between the farmer and the experimental potential but rather between farmers themselves. This yield gap between farmers is explained by structural differences such as differences in land quality, reliable irrigation water access and differences in farmer ability to use the technology. There were no significant differences in varieties or input levels between the two groups of farmers for both sites. The highest-yielding farmer group has incomes that average 40 percent higher than the remaining farmers.

Source: Pingali *et al.* (1990).

(a)

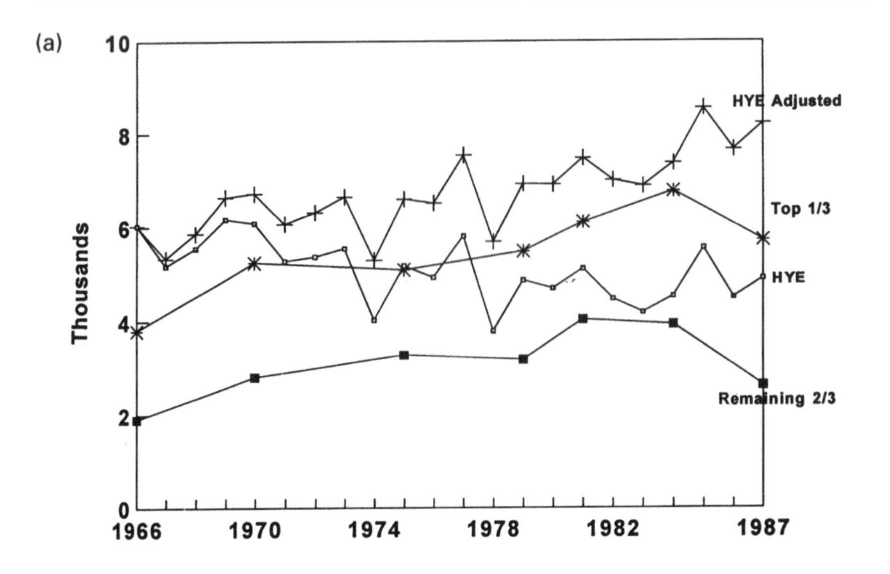

(b)

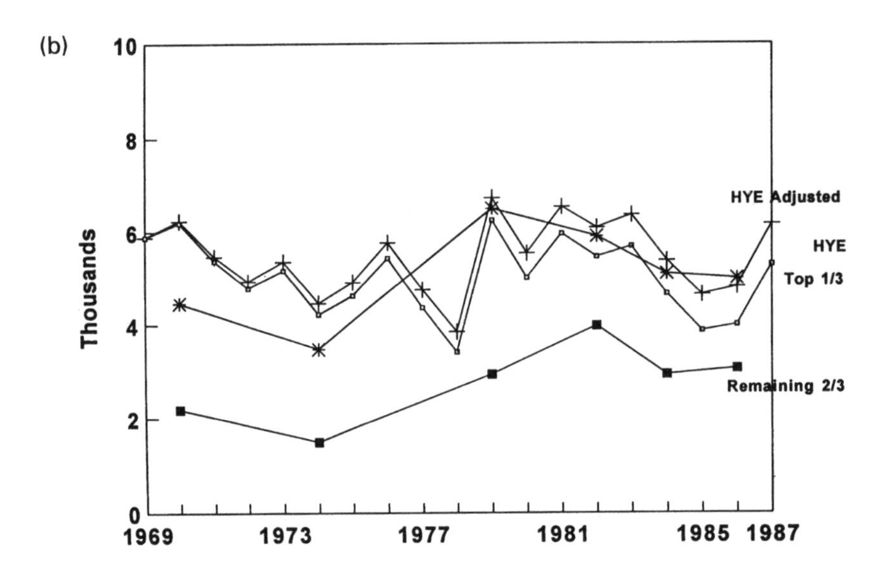

Fig. 2.5. (a) Trends in HYE and farmers' yield, Laguna. (b) Trends in HYE and farmers' yield, Nueva Ecija.

Table 2.6. Deceleration of rice yields in Asia, 1975–1991.

Region/country	Constant	Regression coefficients of Time	Regression coefficients of $Time^2$	R^2
East Asia	3.74	0.0565**	–0.0016**	0.98
		(1.2)	(–5.33)	
China	3.16	0.0659**	–0.0019**	0.98
		(11.2)	(–6.33)	
Japan	5.91	–0.0051	0.0060	0.32
		(–0.61)		
Korea, DPR	5.86	0.507**	–0.0020**	0.96
		(11.3)	(6.67)	
Korea, Rep.	5.90	0.0029	0.0001	0.16
		(0.20)	(0.13)	
Southeast Asia	1.89	0.054**	–0.0014	0.97
		(8.6)	(4.7)	
Indonesia	2.42	6.22**	–0.0017**	0.98
		(10.0)	(–4.3)	
Myanmar	1.52	0.1247**	–0.0054**	0.99
		(24.9)	(20.0)	
Philippines	1.66	0.0623**	–0.0020**	0.99
		(24.9)	(20.0)	
Thailand	1.69	0.0274	–0.0013**	0.81
		(4.6)	(–3.67)	
Vietnam	1.95	0.0162**	0.0010	0.90
		(7.19)	(1.3)	
South Asia	1.78	0.0111	0.0008**	0.98
		(2.60)	(4.01)	
Bangladesh	1.82	0.0052	0.0011**	0.97
		(1.02)	(3.75)	
India	1.73	0.0114	0.0010	0.98
		(2.20)	(3.31)	
Nepal	2.03	–0.0386**	0.0032**	0.99
		(6.50)	(10.67)	
Pakistan	2.20	0.0299**	–0.0017**	0.85
		(8.47)	(8.50)	
Sri Lanka	1.74	0.0852**	–0.003**	0.98
		(13.90)	(10.10)	

Numbers in parentheses are t-statistics.
**, * indicates significance at the 1 and 5 percent levels, respectively.
Source: Hossain and Laborte (1993).

Income, Population Growth and Rice Consumption

Complacency regarding future rice supplies can also be attributed to the widely held perception that rice consumption is declining as incomes grow in Asia. The argument commonly made is that rice is an inferior good and that as incomes grow consumers switch out of rice

Box 2.3. Has rice become an inferior good in Asia?

It has been argued that rice has already become an inferior good – i.e. the per capita rice consumption has started declining in Asia (Ito *et al.*, 1989). By estimating a complete demand system in two stages where the rate of urbanization is included as an additional explanatory variable beside per capita incomes, Huang and David (1992) showed that the downward trend in rice consumption in some countries in Asia is largely due to urbanization, and that the negative income effect cannot be generalized for Asia as a whole. They showed that the threshold level of income at which consumers start substituting rice for other foods has not yet been reached for the major rice-producing and consuming countries, such as China, India, Indonesia and Bangladesh. These four countries account for more than 70 percent of the total rice consumption, and they still dominate the growth in demand for rice in Asia.

The estimates of income and price elasticities of Huang and David (1992) show that for Asia as a whole, the income growth-induced demand for foodgrains is still positive. A 10 percent increase in per capita income will increase per capita consumption by 3.4 percent for rice, 3.2 percent for wheat and 1.8 percent for coarse grains. The own- and cross-price elasticities are high for wheat and coarse grains but relatively low for rice. A 10 percent increase in the price of the commodity will reduce its demand by 6 percent for wheat, 5 percent for coarse grains, and only 2 percent for rice. This result confirms the average Asian consumers' strong preference for rice so that when rice prices increase, they respond only marginally by substituting wheat or coarse grains for rice in their diet. Given a certain shortfall in supplies, rice prices (compared with prices of wheat or coarse grains) will then have to increase faster in order to clear the market.

Urbanization has a strong influence on the pattern of consumption of cereal grains. The demand elasticities of urbanization show that a 10 percent increase in urbanization will reduce per capita consumption by 1 percent for rice and by 5 percent for coarse grains, but it will increase the consumption of wheat by about 5 percent. Thus, the observed increase in the per capita consumption of wheat in countries with high incomes and rapid economic growth, in contrast to the pattern found for rice, is mainly the result of changes in food habits due to urbanization and only partially the outcome of income growth.

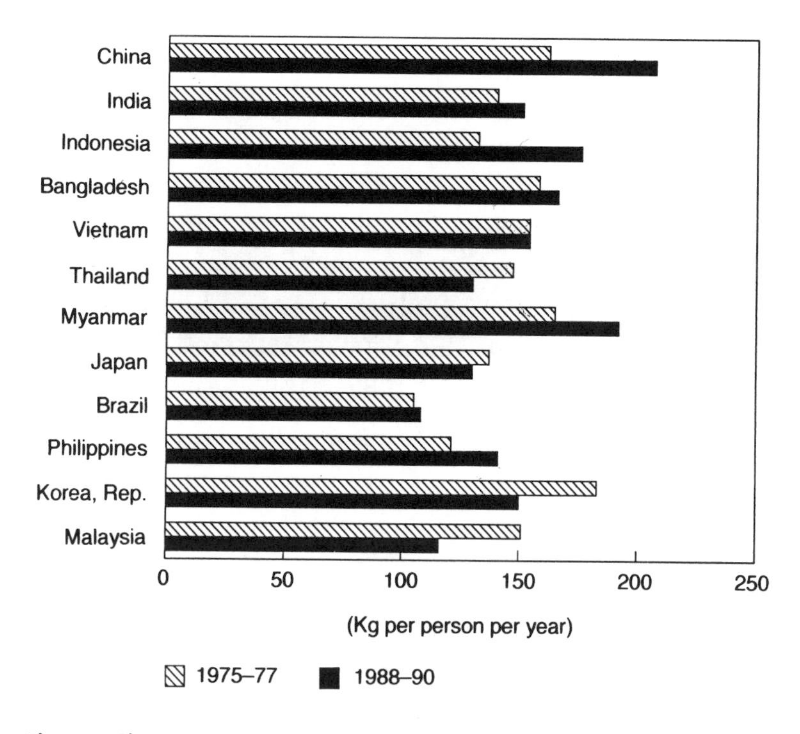

Fig. 2.6. Changes in per capita cereal consumption, 1975–1990. (Source of basic data: FAO *Agrostat Database*, various issues.)

into wheat, meat, fish and vegetables. While the argument is intuitively appealing, it is not at all clear that Asian rice consumers have reached the point of switching out of rice. In fact, with a few exceptions, rice consumption per capita is increasing and not decreasing.

Cereal grains account for two-thirds of the calorie intake in the average Asian diet. Except in Pakistan and the northwestern part of India and Northern China, rice continues to be the dominant staple grain in Asia, contributing 40–75 percent of the total calorie intake. Studies on consumption behavior show that per capita rice intake largely depends on income level, urbanization and changes in occupational structures. At low levels of income, rice is considered a luxury commodity and, with increases in income, people tend to substitute low-cost sources of energy, such as coarse grains and sweet potatoes, for rice. But at high levels of income, rice becomes an inferior good, as consumers substitute rice for high-cost quality food, such as vegetables, bread, fish and meat. A recent study argued that rice has already become an inferior good in Asia (Ito *et al.*, 1989, see Box 2.3 for a discussion and critique of this study). The FAO Food Balance Sheets,

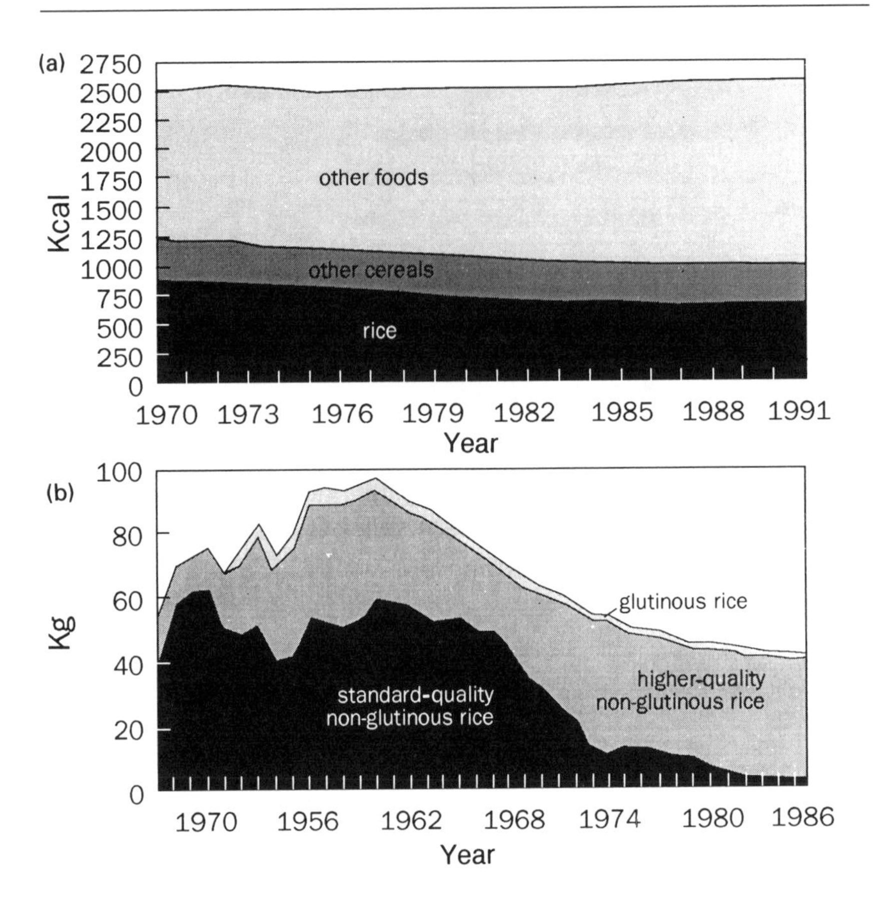

Fig. 2.7. (a) Availability of daily nutrients in Japan, 1970–1991 (per capita in Kcal). (Source: Food Balance Sheet 1991, Ministry of Agriculture, Forestry and Fisheries, Tokyo, Japan, 1993. (b) Changes in rice consumption in Japanese non-farm households (per capita annual data). (Source: Comprehensive Time Series Report on the Family Income and Expenditure Survey, 1947–1986.)

however, show that among Asian countries it is only in Japan, Republic of Korea, Malaysia and Thailand that per capita rice consumption has declined substantially since the mid-1970s – these are all middle- and high-income countries (Fig. 2.6). In Japan and Malaysia, rice consumption in 1990 was about 30 percent lower than in 1975 (Fig. 2.7(a) for Japan). But these countries jointly account for less than 3 percent of Asian rice consumption.

An examination of the changes in per capita rice demand in Japan from 1947 to 1986 shows a very different story than what is generally presented in the literature (Fig. 2.7(b)). The change in rice consumption at the two end years went from 55 kg per capita in 1947 to 45 kg per capita in 1986. However, the type of rice demanded changed from an almost exclusive reliance on standard-quality rice to an almost complete switch to high-quality rice. What is also interesting in this comparison is that per capita rice consumption rose from 55 kg to almost 95 kg by 1960. The substitution out of rice that most analysts have been tracking is from this peak consumption level.

Several important implications come out of these trends in Japanese rice consumption: (i) as incomes rise from low levels there is first a rise in rice consumption and then a fall at higher income levels; (ii) as the decline levels off at a certain amount of consumption, one should not expect a complete switch to other sources of calories; and (iii)

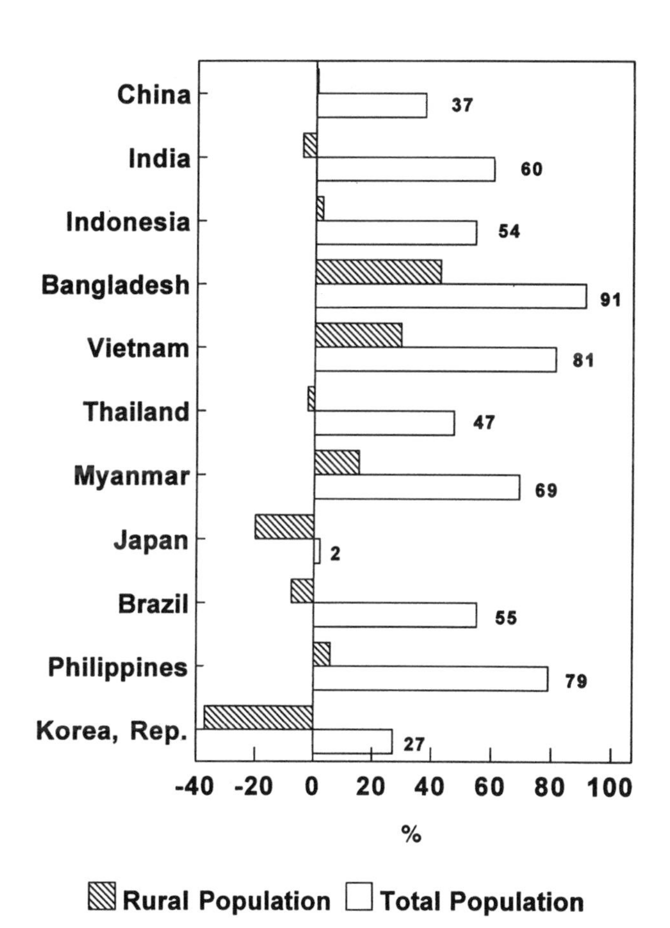

Fig. 2.8. Projected growth of population in major rice-growing countries, 1990–2025. (Source of basic data: IRRI, 1995.)

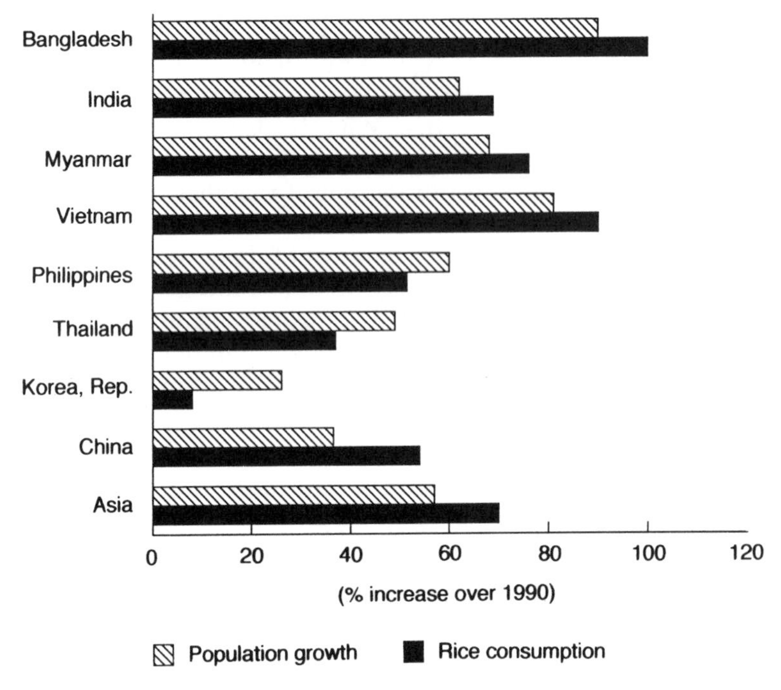

Fig. 2.9. Expected population growth and demand for rice in selected Asian countries by 2025. (Source: Bulatao *et al.*, 1990 – as cited in Hossain and Laborte, 1993.)

poorer-quality rice is almost completely replaced by high-quality rice as incomes rise.

The countries which have experienced a high rate of increase in per capita rice consumption are Indonesia (40 percent), Cambodia (36 percent), DPR of Korea (30 percent), Nepal (27 percent), China (21 percent), Myanmar (16 percent), and India (16 percent) (FAO Food Balance Sheets, various years). Huang and David (1992) showed that the threshold level of income at which consumers start substituting rice for other food has not yet been reached for the major rice-producing and consuming countries of Asia, such as China, India, Indonesia and Bangladesh. These four countries account for more than 70 percent of the total rice consumption, and dominate the growth in demand for rice in the world. The income-induced demand for rice is likely to be strong in large parts of Asia. Over the next several decades, this will be particularly true in South Asia.

Besides, population growth is still a major force behind the increasing demand for rice. The population in Asia whose main staple is rice is growing by nearly 40 million each year. Population growth is not expected to level off until the middle of the 21st century. Asian population is expected to increase by 18 percent during the 1990s and by 53 percent in the next 35 years (FAO, 1992). Most of the additional population will be located in urban areas and marketed surplus of rice has to increase to meet the urban demand (Fig. 2.8). In South Asia alone, where unmet demand for food is still large due to widespread poverty, another 800 million people will be added over the next 35 years. Recent projections (Fig. 2.9) show that, at prevailing price levels, the demand for rice may increase by 70 percent over the next three decades, most of it for feeding a larger population (Agcaoili and Rosegrant, 1994).

International Trade and Domestic Rice Supply

An important factor behind the prevailing complacency regarding the favorable food situation in Asia is the downward trend in world rice prices. In spite of the slowdown of production growth in recent years,

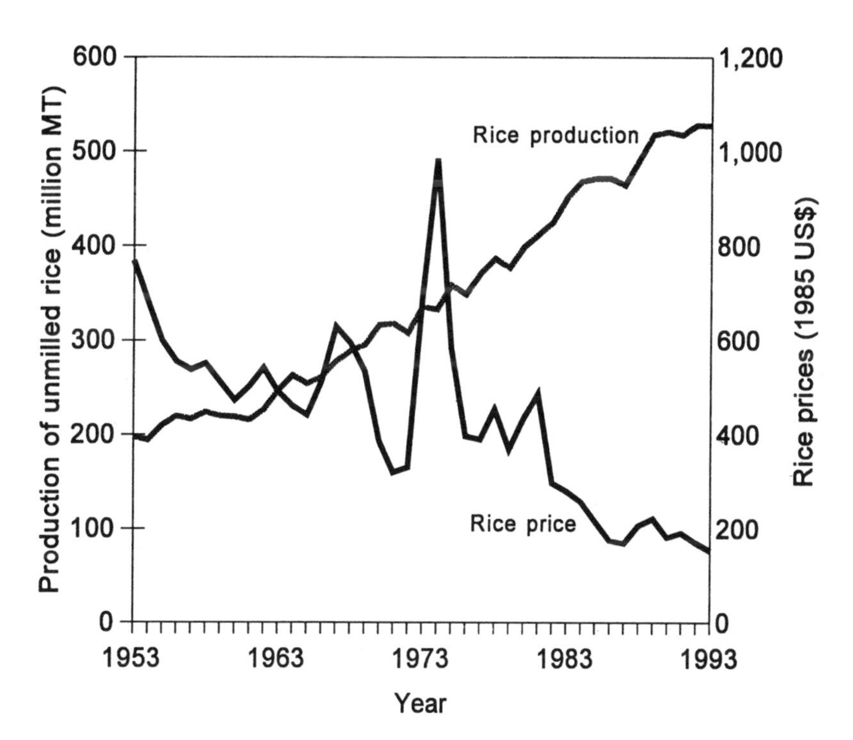

Fig. 2.10. Trends in world rice production and price, 1953–1993. (Source of basic data: IRRI, 1995.)

the downward trend in real rice prices is still continuing (Fig. 2.10), and policymakers in rice-exporting countries are facing political pressures to support domestic rice prices. Export prices, however, reflect the demand and supply situation in the world market which transacts only a very small proportion of global rice production. Less than 4 percent of the rice production in developing countries is currently traded internationally, in contrast to 30 percent for wheat and 15 percent for coarse grains.

In most of Asia, rice is grown on small family-based farms with an average size varying from less than 1 to 3 hectares, and hence the marketed ratio for rice is small. Variable natural conditions cause year-to-year shortages and surpluses, which make domestic prices highly unstable. An important political objective in most rice-growing countries is to achieve self-sufficiency in rice production and maintain price stability through domestic procurement and adjustment of stocks (Childs, 1990). In poorer countries, the objective of achieving self-sufficiency is dictated by limited availability of foreign exchange to finance major international purchases and the experience of unfavorable international deals, that prices are high in years of large deficits but low in years of surpluses. Even high-income East Asian countries with the capacity to import relatively low-cost rice from abroad have tried to maintain self-sufficiency by providing support to rice farmers and encouraging high cost domestic production (Anderson and Hayami, 1986). Consequently, the international trade in rice has been limited.

The continued decline in rice prices in the international market in recent years is primarily the result of increased competition for a stagnant import market. Of the major rice-consuming countries of Asia, few (except Malaysia and Sri Lanka) are major importers, and it is only during times of natural calamities that Asian countries go to the world market. The share of Asia in total rice imports was 62 percent in 1970; it came down to 37 percent in 1980 and further to 23 percent during the 1989–91 period (Table 2.7). This was brought about by large rice-importing countries, such as India, Indonesia, the Philippines and Bangladesh, achieving self-sufficiency in rice production. The international rice market is characterized by a large number of small importers and a small number of large exporters. Major rice importers are Sub-Saharan Africa, the EEC countries, Iran, Iraq and Saudi Arabia. The imports of milled rice which increased from 8.4 to 13.8 million tons during the 1965–81 period fluctuated around that level during the last decade. The share of imports in global rice production fell from 5 percent in 1981 to 3.7 percent in 1992. On the export side, only four countries – Thailand, USA, Myanmar and Pakistan – compete for two-thirds of the world market (Table 2.8). Myanmar and

China, major rice exporters in the 1960s, saw their share of the export market fall from 42 percent in 1960 to only 5 percent in 1989–1991. The most recent decline in rice prices is perhaps due to Vietnam's emergence as a major exporter, which increased supply from 97,000 tons in 1988 to 1.9 million tons in 1992. Vietnamese farmers have responded favorably to economic liberalization policies introduced in recent years, bringing the country from a net importer status to the third largest exporter in less than a decade (Pingali and Xuan, 1992). The availability of cheap labor enables Vietnam to outcompete Thailand and other exporters for low-quality rice.

Table 2.7. Changes in the share of rice imports, 1960–1990 (percent of total imports).

Importing region	1960	1970	1980	1985	1989–1991
Asia	61	62	37	26	23
South Asia	23	16	4	8	6
Southeast Asia	27	31	18	9	10
East Asia	11	15	15	9	7
Africa	8	10	20	24	25
Middle East	4	6	14	19	17
Latin America	6	5	8	7	11
Europe, North America and Oceania	21	17	21	24	24

Source of basic data: FAO Trade Yearbook, various issues.
Source: Hossain and Laborte (1993).

Table 2.8. Changes in the share of rice exports, by country, 1960–1990 (percent of total exports).

Exporting country	1960	1970	1980	1985	1989–1991
Thailand	19	15	23	35	36
United States	12	21	23	17	19
Myanmar	25	8	5	5	1
Pakistan	2	4	9	9	7
China	17	19	7	8	4
Others	25	33	33	26	33

Source of basic data: FAO Trade Yearbook, various issues.
Source: Hossain and Laborte (1993).

The trend in rice prices in the international market is a poor indicator of the overall supply–demand balance in individual rice-growing countries. While the long-term trend is steady there is substantial short-term variability in international prices that makes it very difficult for governments to meet unexpected deficits (Fig. 2.10). If consecutive natural disasters force China or India to import a small fraction of their demand, or if Japan and the Republic of Korea open up their domestic market due to international pressures for liberalization of trade in agricultural commodities, rice prices in international markets will increase substantially. Similarly if Myanmar, Thailand, Cambodia and Vietnam decide to exploit their excess production capacity and push for exports, prices will continue to fall irrespective of the supply–demand situation in other countries.

The slack that is available on the international market is very small and major rice consuming countries would not be able to rely on it to meet large production shortfalls. See Monke and Pearson (1991) for a very clear account of why the Indonesian government would be very cautious in relying on the world rice price for determining Indonesian rice strategy.

Moreover, the world rice market is segmented by variety and quality and prices are often determined by particular tastes and preferences which differ from country to country. The price elasticity of substitution among varieties may be quite low particularly at high income levels, where expenditure on rice accounts for a small fraction of total consumer expenditures[4].

Conclusions

Despite a slowdown in the demand for rice in some high-income countries of Asia, the continent as a whole continues to see increasing demand for rice. This is both due to population growth and due to income-induced demand for increased food consumption. For three-quarters of Asia's population the current mood of complacency about food supplies is not warranted. For the densely populated, low-income countries of Asia, the food crisis is still within the realm of possibility.

The success of the Green Revolution came about through a congruence of climatic, biophysical and economic factors working together. The matching of these factors was not accidental; it was the result of careful planning and orchestration. Future attempts at increasing and/or sustaining food productivity growth ought to keep the necessary balance among the factors that clearly contributed to the success of the recent past.

The international rice market is very thin and volatile. It has not been, and will not be, a reliable source of supplies to meet periodic deficits in the large rice-consuming countries, especially if two or more of them face large deficits at the same time. Moreover, with the liberalization of the East Asian market, one ought to expect an increased emphasis on high-quality rice in the international market. Adequate supplies of standard- and lower-quality rice at affordable prices for the poor will continue to be within the domain of national policy and production capabilities.

Notes

1. See Pingali and Binswanger (1987) for a detailed treatment on the relationship between demographic pressures, intensification and irrigation investments.
2. The Club of Rome's projections of food crisis, published in the famous book, *Limits to Growth,* were based on an almost exclusive reliance on area expansion for output growth. Their thesis was proven wrong by yield-increasing technical change induced by increasing population densities.
3. Of course, the agronomic yield potential could vary from experiment station to experiment station. Maximum yield trials on the same variety conducted, for instance, in the humid tropics (Philippines) and in the subtropics (China) could have very different yields because of plant growth differences due to temperature and daylength, as discussed in the first section of this chapter.
4. Production of high-quality rices is constrained by very specific agroclimatic conditions. Japan, for instance, consumes high-quality Japonica rices. Among potential exporters, only Northeastern China has a great potential for increasing and exporting this type of rice. But since the Chinese economy is also growing very fast, rice production for exports may not remain competitive in the future.

3

Sustaining Farm Profits through Technical Change

The primary reason for the rapid adoption and spread of modern rice varieties in Asia was that it was enormously profitable, compared to the use of traditional varieties. Once investments in land modification were made (such as irrigation, land leveling, etc. which were generally done with government support and/or subsidy), the move to modern rice technology resulted in a dramatic fall in the unit cost of production. With a fall in the unit costs of production and rice prices remaining high, there was a substantial shift in farm-level profits. The early adopters of the Green Revolution technology made significant rents from the use of this technology. Further expansion of modern variety use in the irrigated environments and in the favorable rainfed environments was motivated by the desire to capture the higher rents obtainable from the new technology. As the geographic area under the technology spread and as rice supplies increased, domestic prices dropped and reduced the profit differential. Despite the drop in profits, at current prices, the returns to modern variety cultivation continue to be significantly higher than the cultivation of traditional varieties.

The search, through research and technical change, for further yield growth and/or improved input efficiencies is motivated by the desire to recapture the lost profits of the Green Revolution. This task is becoming increasingly difficult with other contemporaneous changes in the factor markets that are driving up input prices. The rapid withdrawal of labor from the agricultural sector, the diversion of land to other agricultural and non-agricultural activities and the withdrawal of subsidies for agricultural inputs have all led to increasing costs of rice production, and the prospects are for continued increases (Chapter 9), The transformation of rural societies in Asia with increased urbanization and commercialization has substantial implications for the future viability of current farming systems (Chapter 8).

At the same time, there has been a significant reduction in government and donor support for irrigation infrastructure (Chapter 9), the implications of which are reduced productivity and profits. To add to all of the above trends, there has been a decline in yield and productivity due to intensive rice monoculture in the highly productive rice bowls of Asia (Chapter 4). This chapter discusses the following:

1. The technical change pathway that sustains farm-level profits from rice cultivation.
2. The reduction in cost per ton of paddy production during the Green Revolution period through land intensification and the adoption of modern rice varieties.
3. The shift towards labor-saving technical change in the post-Green-Revolution period.
4. The current need for sustaining farm profits through input intensification, with specific reference to knowledge-intensive technologies.

Phases of Technical Change in Rice Systems

The introduction of modern high-yielding rice varieties led to a sharp drop in the cost per unit of rice output in the irrigated lowlands of Asia. The downward shift in the cost function for rice production, holding output prices constant, provided farmers with the incentive to switch to high-yielding varieties. Table 3.1 provides examples from several locations across Asia comparing costs per unit of output for modern and traditional varieties in the early stages of the Green Revolution. At the time of adoption, unit costs of production were generally lower for modern varieties; at the very least, as in the case of Thailand, they were the same as the unit costs of producing traditional varieties, while the net returns were invariably higher. For a given location, unit costs are lower in the dry season and for higher nitrogen levels. It is also important to note that nitrogen use in the intensively cultivated rice bowl areas of Asia, such as those shown in Table 3.1, preceded rather than followed the introduction of modern varieties. Modern varieties being more responsive to nitrogen improved its use efficiency.

The adoption of modern rice technology is more than a switch in varieties; it involves a change in inputs, crop management practices and informed farmer decision making. While the switch in varieties

Table 3.1. Comparison of modern and traditional varieties in the early Green Revolution period (1971).

Location	Season	Rice type	Yield	Cost per ha	N (kg ha^{-1})	Net revenue per ha	Cost per ton	Price per ton
Tamil Nadu, India (rupees)	DS	TV	2.9	931	60	954	320	650
	DS	MV	5.7	1300	110	2126	230	601
Uttar Pradesh, India (rupees)	WS	TV	1.7	930	31	450	560	820
	WS	MV	3.7	1325	84	755	400	562
West Java, Indonesia	DS	TV	2.6	19.1	100	32	7.35	20
(000 rupiah)	DS	MV	3.4	21.5	82	41	6.32	18
Central Plains, Thailand (US$)	WS	TV	2.4	31	16	72	13	43
	WS	MV	3.1	40	17	93	13	43
Laguna, Philippines	WS	TV	3.1	1365	56	321	440	543
(pesos)	WS	MV	3.4	1056	51	481	310	458

TV, MV, DS, WS, Traditional variety, modern variety, dry season, and wet season, respectively.
Source: IRRI (1972).

and fertilizer use took place rapidly, the modernization of production systems was a more gradual and evolutionary process that responded to trends in factor and output markets. The continual and incremental modernization of rice production systems has allowed for the long-term reduction in unit production costs through the early 1990s, thus the private profitability of rice has been maintained despite the decline in domestic rice prices. Figure 3.1 shows unit costs in real terms declining and price-to-unit cost ratios remaining steady for Laguna and Central Luzon in the Philippines, for Punjab, India, and for West Java, Indonesia.

The transition of rice production systems can be categorized into three distinct phases: (i) the land augmentation phase; (ii) the labor substitution phase; and (iii) the knowledge and management intensity phase. Byerlee (1992) identifies similar stages in the transition of intensive wheat production systems in South Asia. The above phases of technical change were characterized in terms of the development and diffusion of technologies to substitute for emerging factor scarcities. Sustaining rice profits and incomes depends on the unconstrained transition through the above stages in response to market signals.

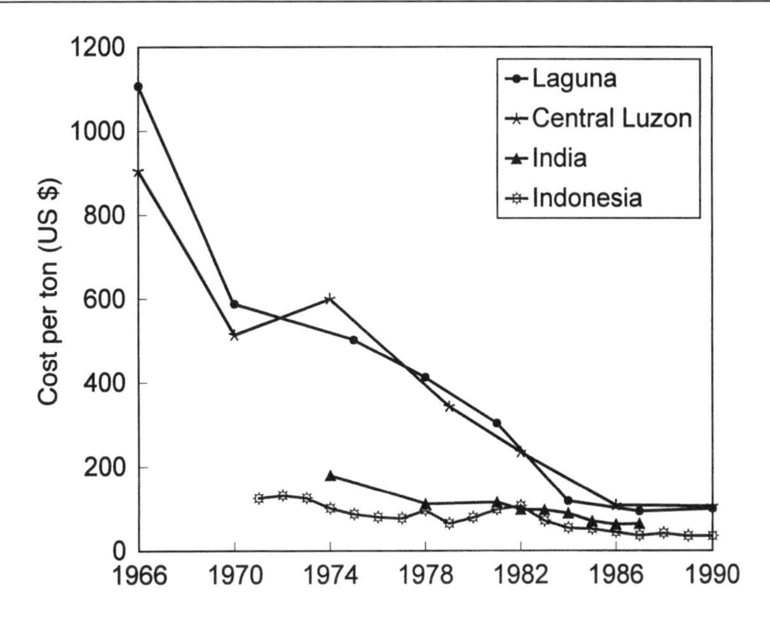

Fig. 3.1. Trends in unit cost of production for selected Asian countries (in US$ 1985 prices).

The land augmentation phase

The transition from low-yield, land-extensive cultivation systems to the land-intensive double- and triple-crop systems is only profitable in societies with an exhausted land frontier (Boserup, 1965; Hayami and Ruttan, 1985; Pingali and Binswanger, 1987). Investments in irrigation infrastructure and the availability of short-duration varieties and fertilizers are necessary for increasing cropping intensities. National and international research systems exploited a backlog of scientific knowledge to provide the genetic improvements that made this technical revolution possible (Evenson, 1974). Much of South and Southeast Asia made the transition to land-intensive production systems in the 1950s and 1960s. East Asia made this transition earlier (Barker and Herdt, 1985; Bray, 1986). By the 1980s, the land frontier had largely been exhausted across most Asian countries. Pockets still exist though where the transition to land-intensive production systems are not yet economically viable. The Irrawaddy Delta of Myanmar, the rainfed lowlands of Cambodia and Laos, and the outer islands of Indonesia are cases in point. Intensification would occur in the above areas as rising population densities and/or improved access to international markets led to an increase in land values.

The fall in the unit cost of rice production came about through the synergistic interaction among three factors: modern high-yielding varieties, chemical fertilizers, and the timely availability of irrigation water. The widespread adoption of modern varieties was followed by a 'post-Green-Revolution' phase of continued land-savings input intensification, especially fertilizers and water, and increased cropping intensities. In economic terms, intensification was equivalent to an improvement in allocative efficiency, resulting from improvements in input supply, capital accumulation and farmers' 'learning by doing'.

The labor substitution phase

The movement from single-crop cultivation systems to double- and triple-crop systems shifts out the demand for labor and also increases the demand for timely completion of operations. Farm surveys conducted by IRRI in 30 villages across five Asian countries in 1970–1971 indicate a shift in the demand for labor with modern variety adoption (Table 3.2). The demand for both family and hired labor increased. A more recent comparison of labor requirements in the irrigated and rainfed rice production systems in six Asian countries indicates that adoption of modern varieties in favorable environments significantly increases labor use per hectare by raising labor requirements for crop care, harvesting, and threshing (David and Otsuka, 1994) (Table 3.3).

Table 3.2. Farmers reporting a change in pre-harvest labor requirements following the introduction of modern varieties by location and by type of farming, 1971–1972.

		Farmers reporting (%)					
		Family labor		Hired labor from			
				Village		Outside village	
Location/ type of farming	No. of villages	More	Less	More	Less	More	Less
India	10	27	2	82	3	55	0
Indonesia	5	26	2	18	12	11	4
Pakistan	2	0	0	2	0	0	0
Malaysia	2	51	13	36	5	—	—
Philippines	9	62	15	65	7	20	7
Monoculture	12	42	13	55	4	24	3
Mixed farming	16	36	4	56	7	33	4

Source: IRRI (1973).

Table 3.3. Patterns of labor use in selected Asian countries across different rice production environments.

Rice production environment	Thailand	Bangla-desh	Nepal	India	Indo-nesia	Philip-pines
			(mandays ha^{-1})			
Irrigated	64	243	144	195	156	82
Favorable rainfed	—	209	132	229	174	71
Unfavorable rainfed	62	239	—	176	70	91

Source: David and Otsuka (1994).

The absolute quantity of labor required for transplanting, weeding, harvesting and threshing increased significantly with modern rice cultivation systems, because of the higher quantity of output and because of more systematic and intricate planting methods. Increased chemical fertilizer use resulted in higher crop output but also led to higher weed density and competition with the rice plant. Higher nitrogen use increased the plant's susceptibility to insect and disease infestations, hence requiring greater farmer effort in pest control. The period between harvesting and threshing one crop, and the land preparation and planting of the next crop becomes very short and time-bound by the availability of water. Hence, a significant labor peak occurs between harvest and transplanting. Significant labor peaks were also observed for handweeding during both the wet and the dry season.

Much of this additional labor requirement was met through hired labor but family labor requirements for supervision and management also rose significantly (see Table 3.2). Initially, the switch to modern varieties was profitable, even with higher labor requirements per hectare, because the labor requirements per ton of paddy were lower as compared to that of traditional varieties (Table 3.4; Barker and Cordova, 1978). Over time, however, increasing labor demand for peak period operations led to a rise in real wages for these operations, even in densely populated labor-surplus countries such as India (Bardhan, 1970, Lal, 1976) and Indonesia (Naylor, 1994). The rise in wages was further exacerbated by the concurrent growth in the rural non-farm sector and increasing employment opportunities in the urban sector.

Widespread adoption of labor-saving mechanical and chemical technologies has alleviated the growing labor constraints, and contributed substantially to overall productivity growth (Sidhu and Byerlee, 1991). A detailed discussion on the evolutionary pattern of mechanization of rice production systems is provided in the section on 'Increasing Profitability . . .' later in this chapter.

Table 3.4. Labor input for modern (MV) and traditional varieties (TV) of rice in rice-growing areas in Asia.

			Labor input					
			Mandays per ha		MV / TV	Mandays per ton		MV / TV
Study site[a]	Year	Season	MV	TV	(%)	MV	TV	(%)
Central Korea	1974	Summer	139	126	110	19	23	83
Laguna, Philippines	1975	WS	110	86	128	31	34	91
Central Thailand	1972	WS	117	81	144	40	39	103
Java, Indonesia[b]	1969/70	WS	262	235	111	57	71	78
Mymensingh, Bangladesh	1969/70	DS	194	137	142	57	62	92
Ferozepur, India[c]	1967/68–1969/70	WS	92	79	116	22	29	76
Hyderabad, Pakistan[d]	1972	WS	58	49	118	29	35	83

[a] In all sites except Laguna, labor input for area in MV is compared with labor requirement for area in TV during the same season. In Laguna, the same farms are compared with zero MV in 1966 and 95 percent MV in 1975.
[b] Pre-harvest labor only.
[c] Mehra (1976). Original source of information is 'Studies in economics of farm management in Ferozepur District (Punjab) and in Mazaffarnagar District (Uttar Pradesh)', combined reports.
[d] Khan (1975).
TV, MV, DS, WS, traditional variety, modern variety, dry season, and wet season, respectively.
Source: Barker and Cordova (1978).

Knowledge and management intensity phase

By the late 1980s, the most advanced 'post-Green-Revolution' areas of Asia, such as the Punjab of India and Central Luzon of the Philippines, had reached a point of sharply diminishing returns to further intensification and had entered a second 'post-Green-Revolution' phase characterized by the use of better knowledge and management skills to substitute for higher levels of input use (Byerlee and Pingali, 1994). Productivity gains accrue to farmers from differences in the way these inputs are used; that is, the timing and method of using inputs rather than their levels (Byerlee, 1987; Pingali *et al.*, 1990). For instance, two farmers in the same location with the same soil types and irrigation access using the same varieties and nitrogen levels may have different yields because of differences in the timing and method of nitrogen application. Over time, farmer technical knowledge and management skill become the primary determinants of differences in productivity and profits between farmers[1].

Improved knowledge and skills help farmers improve the technical and allocative efficiency of input use. Varietal selection, fertilizer timing and placement, water use, and pesticide application are some areas in which efficiency gains could lower unit costs of production. It should be recognized, however, that the adoption of knowledge-intensive technologies requires farmers to trade off reductions in input levels for higher time commitment to management and supervision. Farmer incentives for adopting techniques that increase input efficiency depend upon the price of inputs relative to the opportunity cost of time (which has been rising over time – see Chapter 9 for further details).

The movement along the above phases of technical change is guided by growing factor scarcities, first for land, then for labor, and finally for the other factors of production. The primary determinant for the movement along the technology change path is farm-level profits. The ability to reduce unit production costs through saving inputs as their relative scarcity increases has helped sustain the profitability of rice production, despite the temporal decline in rice prices. It is often not unusual to find farmers in the lower phases of technical change even when technologies for labor saving and/or increasing input efficiencies are available. Non-adopting societies may not have reached the level of factor scarcity for particular input to warrant technical change.

Land Intensification and the Adoption of the Seed–Fertilizer Technology

As mentioned above, increasing the productivity of ricelands has come about through an increase in cropping intensity and an increase in yield per hectare per crop. A detailed discussion on the adoption and spread of modern rice varieties has been provided in Chapter 2, along with a discussion of the constraints to further expansion in area under modern varieties. Constraints to further expansion in irrigated area are discussed in Chapter 9. The essential point to be made here is that the productivity and profitability gains that could be made through an expansion of area under modern varieties have already been achieved. Further gains in productivity would have to come from an intensification of input use, i.e. a movement along the production function in areas that are already using modern varieties. This section will therefore restrict itself to discussing the opportunities for further increasing irrigated rice productivity through enhancing input use.

Movement along the production function is through increased use of fertilizer, labor and pesticides. The opportunities for enhancing feritlizer use on modern rice varieties are discussed in this section. Discussion on increased labor use due to intensification and the opportunities for labor savings are discussed in the next section. The role of pesticides, especially insecticides and herbicides, is discussed in Chapter 11.

David and Barker (1978) provide a comprehensive assessment of the fertilizer responsiveness of modern varieties for a wide set of locations in Asia, for both irrigated and rainfed conditions. Figure 3.2 provides two examples derived from experiment station data of fertilizer response functions for modern and traditional varieties in India and the Philippines. Without the use of chemical fertilizers under irrigated conditions, the yields of traditional and modern varieties are about the same (David and Barker, 1978). Given that all other inputs and management practices are the same, unit costs of irrigated rice production would be similar for traditional and modern varieties at zero chemical fertilizer levels. Under irrigated conditions, the reductions in unit cost of rice production come about with the exploitation of fertilizer responsiveness of modern varieties. Holding other inputs constant, higher fertilizer responsiveness of a variety implies lower cost per ton of paddy produced.

Unit costs of modern rice production, at given nitrogen levels, are lower in the dry season than in the wet season due to higher solar radiation and lower disease pressure. The responsiveness of modern varieties to nitrogen is also positively related to the reliability of irrigation water supply. Wickham and Sen (1978) found that yields dropped by 30 percent under average irrigation as compared to ideal irrigation (Fig. 3.3).

In the case of rainfed agriculture, at zero or low nitrogen levels, traditional varieties would dominate modern varieties because the former are more adapted to climatic and hydrologic risks. Moreover, increased nitrogen application for traditional varieties leads to an increase in straw biomass rather than an increase in grain as in modern varieties.

With the widespread adoption of modern rice and wheat varieties, fertilizer demand increased dramatically from approximately 7 million tons in 1965/66 to around 60 million tons by the early 1990s, almost a ninefold increase (FAO, 1995). Asia accounts for 47 percent of global fertilizer intake and fertilizer demand continues to grow at the rate of 7.3 percent per annum. Cereal crops as a whole account for over two-thirds of the total fertilizer intake in Asia. In the 1990s, nitro-

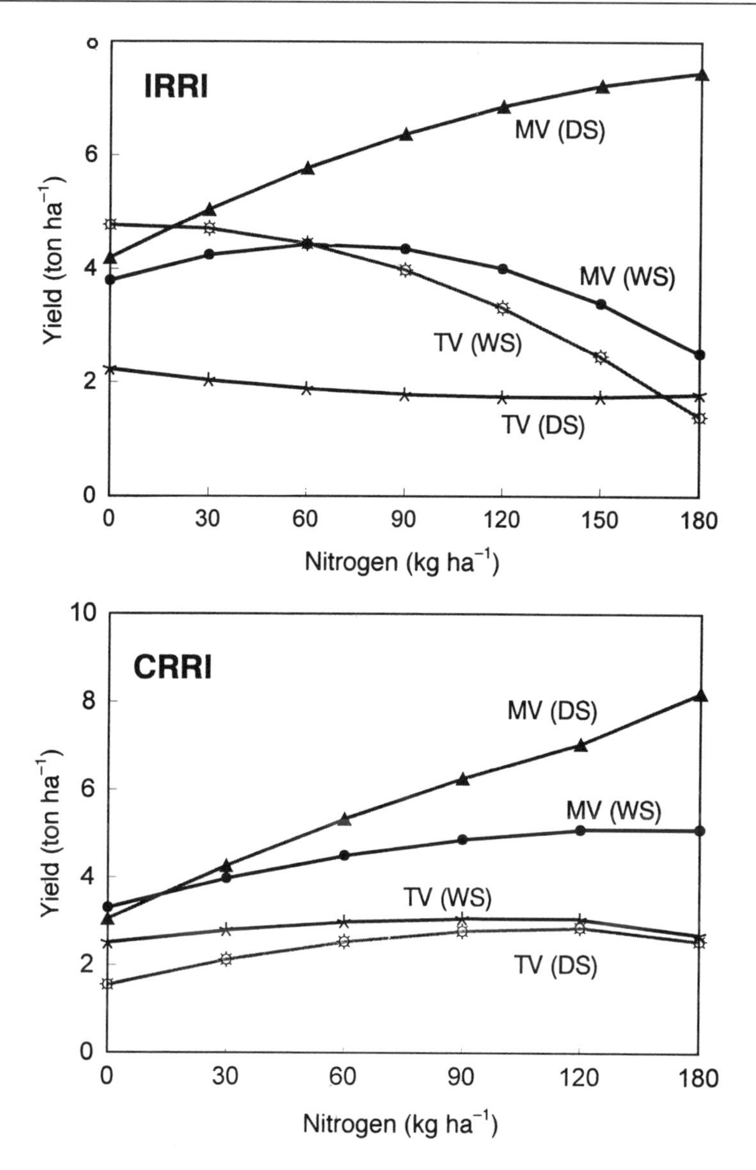

Fig. 3.2. Average yield response to N of modern varieties (MV) [IR-8 (dry season) and IR-20 (wet season)], and traditional variety (TV) Peta, IRRI, Philippines, 1968–1975, and at Cuttack Rice Research Institute (CRRI), India, 1967–1969.

gen application rates for the irrigated rice systems of South and Southeast Asia are around 80–150 kg ha^{-1} (Chapter 11). Fertilizers currently account for 20–25 percent of total production costs in irrigated rice systems (Pingali *et al.*, 1996b).

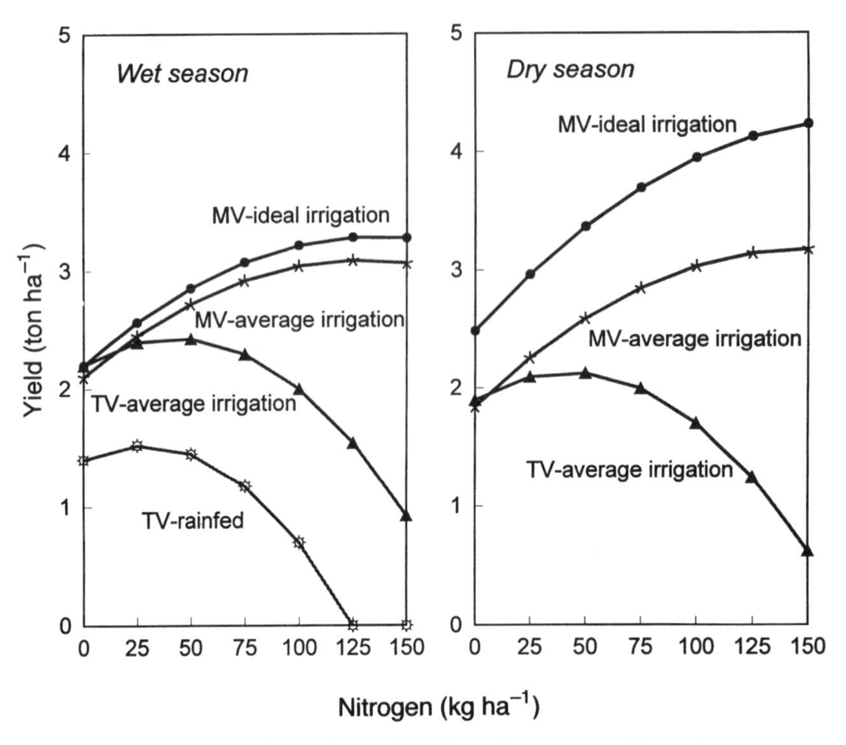

Fig. 3.3. Response of modern (MV) and traditional varieties (TV) to nitrogen at different irrigation levels, Philippines, 1976 wet and dry seasons. (Source: Wickham and Sen, 1978.)

Deliberate interventions by governments in the fertilizer market contributed to the rapid growth in fertilizer use across Asia during the 1970s and the 1980s. Establishment of government monopoly on procurement and distribution of fertilizer through parastatal institutions, direct fertilizer subsidy programs, regulation of private trade, and input and output price controls ensured favorable fertilizer prices for farmers, thus providing farmers the incentive to adopt modern varieties and increase fertilizer use (Table 3.5). In recent years, structural adjustment programs have begun to put pressure on governments to reduce public sector production and subsidized distribution of fertilizers. The long-term prospects are for less distortions in the fertilizer market and for a narrowing of the gap between domestic and world fertilizer prices.

Would the anticipated increase in fertilizer price relative to the rice price cause a slowdown in rice productivity growth? Pingali *et al.* (1996b) indicate that the opportunities for profitably expanding fertilizer use in the irrigated environments are limited for the intensively cultivated rice bowls of Asia. At current fertilizer-to-rice price ratios,

Table 3.5. Overview of policy changes in fertilizer sector, selected Asian countries.

Country	Year of initiating policy changes	Restrictions in fertilizer trade	Production capacity owned by the govt (%)	Fertilizer trade operated by the govt (%)	Pricing policy	Subsidies
Bangladesh	1978	None	100	nil	Ex-factory price fixed by government	None since 1973
China	1979	Import controlled by state-designated trading companies	100	50	Govt fix prices for certain portion of fertilizer	Yes
Indonesia	1986	None	100	100	Prices fixed by government	Yes
Myanmar	1990	Allocation of foreign exchange to importers	100	80	Fertilizer retail price fixed	None
Pakistan	1986	Import controlled by state-designated trading companies	30	50	Prices of potash and compound fertilizer fixed	Yes
Philippines	1986	5% excise duty on fertilizer manufactured in the country	77	50	No intervention	None since 1992
Thailand	1990	None	0	0	Export tax on rice before 1984 and export subsidy since 1993	None
Vietnam	1989	Import and trade controlled by State designated companies	100	12	Import duty abolished	Transport subsidy on phosphate and potash
India	1991	Import controlled by Mineral and Metal	53 for urea and 29 for phosphate	0	Ex-factory prices fixed	Yes

Source: FADINAP (1994). *Agrochemical News in Brief*, ESCAP/FAO/UNDP.

farmers at the high end of fertilizer use (farmers applying rates that are one standard deviation above the sample mean levels) are applying rates that are beyond the economic optimum in the rice bowls of the Philippines, Indonesia and India (Chapter 11). Higher fertilizer prices resulting from a removal of subsidies and other market distortions would reduce the incentive to increase fertilizer use per hectare. Higher relative price of fertilizer could, however, increase the incentives for farmers to increase the efficiency of fertilizer use. The conditions under which farmers would adopt efficiency-enhancing fertility management technologies are discussed in Chapter 11.

Further opportunities for unit cost reductions through the use of the seed–fertilizer technology are limited in the absence of a further shift in fertilizer responsiveness of rice cultivars, which would come about through a shift in the yield frontier for rice. A new 'super' high-yielding plant type and hybrid rice are examples of forthcoming rice technologies that could further reduce the unit cost of production (Chapter 10). Even with the current varieties, there are opportunities in some countries and some rice environments for a movement along the yield frontier through additional input use; these opportunities are discussed in Chapter 11.

Increasing Profitability of Labor-saving Technologies in the Post-Green-Revolution Rice Production Systems

Mechanical and chemical techniques have been employed to substitute for the increased human labor requirements with intensification and thereby reduced unit costs of production. Table 3.6 classifies rice societies in terms of their level of substitution out of labor by agricultural operation by the level of economic development. Substitution out of labor to mechanical power occurred first for power-intensive operations such as land preparation, threshing, pumping, milling and transport[2]. This transition occurred smoothly in the irrigated environments, even for low-wage countries such as India, Bangladesh and the Philippines, because of rising relative wages for these operations with intensification. In fact, the mechanization of transport and milling operations preceded the Green Revolution. Mechanization reduced the costs of power-intensive operations significantly as well as ensured their timely completion[3]. In high-wage countries such as Japan, Korea and Taiwan, mechanization has been carried through to the more control-intensive operations such as transplanting and harvesting.

In the case of labor-saving chemical technologies, consider the following two; chemical fertilizers and herbicides. One does not generally think about chemical fertilizer as a labor-saving technology but it does

Table 3.6. Adoption of labor-saving technologies by income growth, Asia.

	Low income	Medium income	High income
Tillage	Animal draft to power tillers	Power tillers to tractors	Tractors
Crop establishment	Manual transplanting	Direct seeding – manual transplanting	Direct seeding or mechanical transplanting
Weeding	Manual	Herbicides	Herbicides
Harvesting	Manual	Manual to combines	Combines
Threshing	Manual and animal treading	Portable threshers	Combine-harvesters
Milling	Small mills	Commercial mills	Commercial mills

lead to significant labor savings. The rice plant responds to nutrients without differentiating between the sources of nutrient supply. In essence, the nutrient requirements for modern varieties could just as well come from organic sources as from chemical fertilizers. In much of the humid tropics of Asia, a 5-ton harvest of rice can be obtained by providing the rice crop around 100 kg of nitrogen nutrients. The required nitrogen nutrients per hectare can be obtained from 250 kg of urea or from 8–10 tons of farmyard manure. The latter, a farm-produced input often called 'free', is highly labor intensive. When the opportunity costs of family labor are accounted for, its use at the above levels is economically non-viable. In fact without the availability of chemical fertilizers, modern rice production systems would not have been economical[4].

In high-wage countries, herbicides have been in use for several decades for controlling weeds (Chapter 11). Their use has been increasing in countries with recently rising wages such as Thailand, Malaysia and to a smaller extent in the Philippines. In these countries, the adoption of herbicide use has been in association with the adoption of direct seeding. Direct seeding eliminates the transplanting operation by broadcasting germinated seeds onto prepared (puddled) paddy soils. Unlike the transplanted rice system, there is no standing water in the direct seeded paddies when the seedlings are broadcast; water is re-introduced 7–10 days later. During the early period of crop establishment, a direct-seeded rice crop faces intense competition from germinating weeds. Early-season weed competition is prevented in transplanted systems by the presence of standing water which prevents the germination of weed seeds. Herbicide application a day before broadcasting the pre-germinated rice seeds reduces the threat of early-season weed competition. The joint adoption of direct seeding and herbicide use reduces labor requirements by 77 percent relative to transplanting and handweeding. Table 3.7 presents an example from Central Luzon, Philippines, of changes in unit production costs in the switch from transplanting to direct seeding.

Why has direct seeding/herbicide use not been more widely adopted? First, there could be a yield trade-off in areas without good water control; second, in the more temperate latitudes of East Asia, there could be a temperature constraint to seed germination for the winter rice crop; and lastly, in systems where a short duration non-rice crop is grown in between the wet and dry season rice crop, such

Table 3.7. Costs and returns of rice production by crop establishment method, Nueva Ecija, Philippines, 1986 dry season.

	Transplanted rice	Wet-seeded rice	Difference[a]
Revenue			
Yield (t ha^{-1})	4.3	4.1	0.2
Net yield (t ha^{-1})	3.8	3.5	0.3
Rough rice price (US$ kg^{-1})	0.14	0.14	0.0
Total revenue (US$ ha^{-1})	518	486	32
Costs paid (US$ ha^{-1})			
Material inputs:	120	125	5
Seeds	26	32	6***
Fertilizer	66	65	1
Herbicides	7	9	2**
Other pesticides	21	19	2
Hired labor:	25	8	27***
Land preparation	3	3	0
Crop establishment	28	1	27***
Crop care	5	4	1
Power:	67	57	10
Machine	34	57	7
Animal	2	0	2*
Total costs paid	222	191	31***
Gross margin (US$ ha^{-1})	295	296	1

[a] ***,**,* – Significant at the 1, 5 and 10 percent levels, respectively, based on t-tests of means.
[b] Net of harvester/thresher share. US$ 1.00 = PHP 20.39.
Source: Erguiza *et al.* (1990).

as in the Red River Delta area of Vietnam, direct seeding competes with high-value dry-season crop production, usually vegetables.

In the rice bowl provinces of Southeast Asia, reductions in unit costs that can be made with labor-saving mechanical and chemical technologies have been made. Further opportunities for using these avenues for unit cost reduction are minimal (see Table 3.7 for example from Central Luzon, Philippines). The one operation in Southeast Asia where further gains in productivity can be made through mechanization is harvesting (Chapter 12). Small harvesting machines have been in use in Malaysia since the mid-1980s; they began emerging in Thailand in the early 1990s and could be expected to be commercially available in other Southeast Asian countries by the late 1990s.

In the case of South Asia, the next shift in labor productivity could be expected from increased herbicide use and the shift to direct seeding (Chapter 12). Mechanization of harvest operations, as in Southeast Asia, ought to be expected initially only in the high-potential rice bowl areas, such as the Indian Punjab. In discussing unit cost reductions through the use of mechanization and herbicides, explicit consideration ought to be given to the equity and environmental costs associated with their widespread adoption. We postpone discussion on equity consideration of mechanization, especially in terms of female labor displacement, to Chapter 12. Environmental and health costs associated with pesticide use, both insecticides and herbicides, are discussed in Chapter 5.

Farmer Knowledge: The Next Scarce Resource

Post-Green-Revolution productivity differences between farmers are not explained by differences in access to technology or inputs but rather by differences in land quality and farmer technical knowledge (Pingali, 1994). Farmer technical knowledge determines the speed with which a farmer reaches the production frontier and his/her ability to move along the frontier[5]. For example, two farmers may use the same varieties and same levels of fertilizer but get very different yields because of differences in timing and method of fertilizer application. Box 3.1 summarizes for the rice bowl provinces of Southeast Asia the impact of human capital on input productivity, with emphasis on fertilizer productivity.

For both South and Southeast Asia, the only unexploited avenue for further productivity growth is in enhancing input use efficiency through the use of knowledge-based technologies. The importance of emphasizing more efficient use of inputs has been reinforced by recent evidence of significant problems in sustaining the quality of the resource base for intensive rice production systems in Asia (Pingali

Box 3.1. Productivity differences between farmers: the influence of human capital.

In Box 2.2 of Chapter 2, we argued that for the irrigated rice farms in Asia the gap was no longer between the technological yield potential and observed farm yields but rather between farmers themselves. This box argues that productivity differences are not explained by differences in access to technology or inputs but rather by differences in land quality, and differences in farmer technical knowledge (human capital).

During the early years of modern variety adoption, yield differences between farmers were related to differences in the levels of fertilizer applied (Herdt, 1987). In recent years, however, between-farmer differences in fertilizer use have been negligible. There are differences in the timing of fertilizer application and the method of application but generally not in the quantities applied. The same is true of other purchased inputs – seed, pesticides, and mechanical power. There is also a negligible between-farmer difference in total labor use.

Determinants of between-farmer productivity differences

Pingali *et al.* (1990) study productivity differences between irrigated rice farmers in the Philippines, Thailand and Indonesia. In order to identify the sources of productivity differences between farmers, they were ranked in terms of their performance. Farmer performance was defined as the ratio of farmer yield to the farmer-specific yield potential. The farmer performance index is a continuous variable that indicates the extent to which the farmer has been able to exploit the yield potential. If the farmer performance index is one, then the farmer's yield matches the yield potential; if the performance index is less than one, then the farmer faces an unexploited yield potential. The farmer performance index is a measure of the farmer's technical efficiency. Farmer-specific yield potential was extrapolated from production functions estimated using long-term trial data from neighboring experiment stations: IRRI for Laguna, PhilRice for Central Luzon, Sukamandi Rice Station for West Java, and Suphan Buri Rice Experiment Station for Central Plains, Thailand.

Farmer performance was empirically related to input levels, farmer characteristics and reliability of irrigation using a translog function. Farmer characteristics and reliability of access to irrigation water determine the quality of technology use and input use efficiency. Farmer characteristics are given by: family size per hectare; age of household head and schooling of household head. Family size per hectare represents the degree of specialization in farm tasks and the supply of supervision labor. Age (a proxy for experience) and schooling of the household head represent the stock of knowledge and the farmer's ability to effectively use increasingly knowledge-intensive technologies. Distance from the main irrigation canal is an indicator of the reliability of water access and is expected to be negatively associated with farmer performance.

Human capital, irrigation access and farmer performance

The study found that farm and farmer characteristics (family size per hectare; schooling; age of household head and distance from the main irrigation canal) have significant effects on farmer performance.

In most cases the importance of family size/hectare, farmer age and schooling as determinants of farmer performance has increased over time. Education and farming experience (age as a proxy) improves the farmer's ability to acquire and process information on particular technologies and to adapt them to their specific circumstances. Family size per hectare indicates the farmer's access to supervision labor. The increasing importance of human capital variables in explaining between farmer productivity differences attests to the growing complexity of rice production in Asia. Byerlee (1987) argues that the productivity gains in the post-Green-Revolution era will come from more efficient use of existing inputs to exploit the genetic potential of existing varieties. These 'second generation technologies' (such as better fertilizer incorporation technologies, integrated pest management, etc.) are more knowledge intensive and location specific than the modern seed-fertilizer technology that was characteristic of the Green Revolution. Productivity gains accrue to farmers who have the ability to learn about the new technologies and to use them effectively.

The impact of farm and farmer characteristics on the efficiency of fertilizers and pesticides was assessed. Results imply that marginal improvements in farmer performance with respect to fertilizer use will come more from improving the quality of fertilizer application (better timing and incorporation techniques) rather than increasing the quantity of fertilizer applied. Similarly, marginal improvements in farmer performance with respect to pesticide use will come from the judicious and discriminate use of pesticides rather than increasing the quantity of pesticides applied. Farmer education, experience and the availability of supervision labor were found to significantly improve the efficiency of fertilizer and pesticide use at the farm level. These results are consistent with the Pingali and Carlson (1985) study on the effects of human capital variables on pesticide use in the United States.

The importance of distance from the main irrigation canal as a determinant of farmer performance has also increased over time. For given input levels and farmer characteristics, farmers further away from the canal are less productive than farmers closer to the canal. The successful adoption of technologies that increase input efficiencies requires adequate and timely water supply which, of course, is inversely related to the distance from the irrigation canal. The productivity of fertilizers and pesticides decreases as the reliability of water control declines, this occurs with distance from the irrigation canal.

and Rosegrant, 1993; Cassman and Pingali, 1995). These sustainability problems are most evident in the rapid decline in partial factor productivities, especially for nitrogen fertilizer, and the leveling off and/or decline in the growth of total factor productivity (Chapter 4). In other words, the yield gains achieved in the 'post-Green-Revolution' period are being maintained by increasingly higher levels of inputs to compensate for degradation of the lowland resource base (Byerlee and Siddiq, 1994; Cassman and Pingali, 1995). Declining productivity trends can be directly associated with the ecological consequences of intensive monocultural systems, such as buildup of salinity and waterlogging, declining soil nutrient status, increased soil toxicities and increased pest buildup, especially of soil pests (Chapter 4).

Examples of knowledge-based technologies are: improved varietal selection; improved timing and application systems for fertilizers; integrated pest management; and judicious water management. Among the techniques available for enhancing input use efficiency, varietal selection is the most cost effective. Information on varietal characteristics is easily transmitted to farmers and switching seeds is a task easily accomplished. Farmers' adoption of varieties with multiple resistance to a wide spectrum of insects and diseases has helped contain the growth in insecticide use (Rola and Pingali, 1993) and virtually eliminated the need for fungicides in the humid tropics of Asia (Byerlee, 1993). Chapter 11 argues that the long-term benefits of using host plant resistance to rice pests are substantially higher when farmers have a better understanding of the pest–predator ecology and its dynamics with pesticide use.

With the exception of host-plant resistance, attempts to improve input use efficiency have generally failed, although they were scientifically impeccable. Integrated pest management (IPM), improved timing and application of fertilizers, and improved water management are examples of knowledge-intensive technologies that have had limited farmer interest although scientific knowhow about these techniques has been available for decades. Often, the time cost of acquiring and processing the information required for making a judicious choice at the farm level is substantially greater than the monetary value of the input saved (see Chapter 11 for an example of knowledge-intensive fertilizer management). The recent and highly acclaimed success of IPM in Indonesia was only possible because of the government decision to ban a large number of chemicals and to remove subsidies on the remainder. This policy switch resulted in an increase in the relative price of insecticides, and made it profitable for farmers to invest time in learning about IPM (Rola and Pingali, 1993). Similar success in improving fertilizer efficiency is unlikely given the long-term secular decline of fertilizer prices (Figure 3.4) relative to the cost of labor

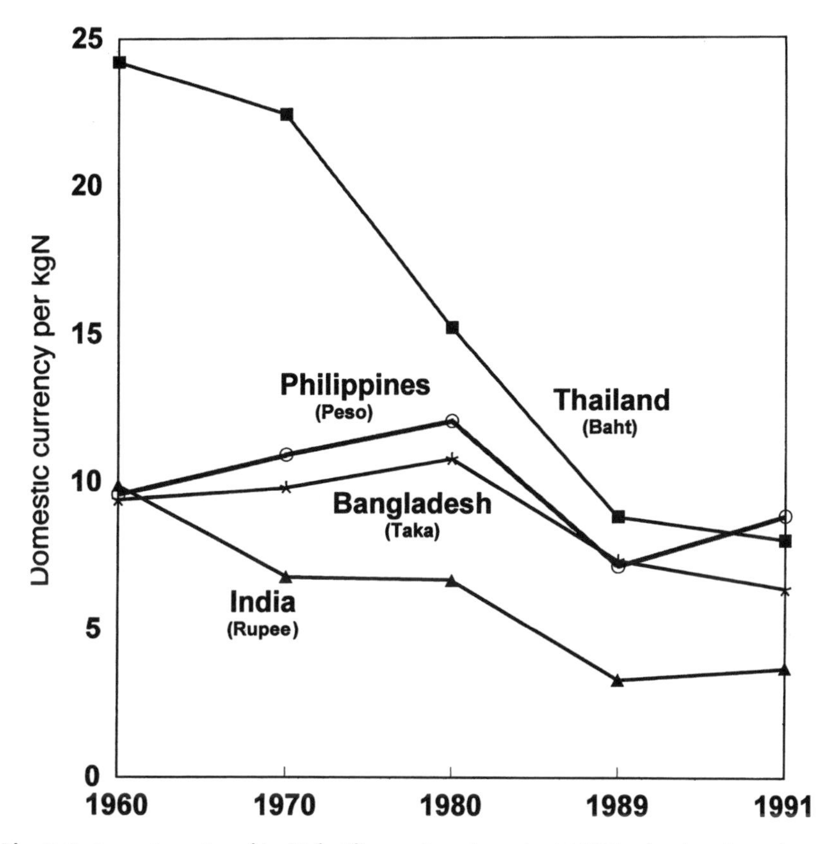

Fig. 3.4. Long-term trend in N-fertilizer prices (constant 1985 prices), selected Asian countries. (Source of basic data: IRRI, 1995.)

which has been stable throughout the years. Water use efficiency was similarly discouraged by water pricing policies. However, with the growing scarcity of irrigation water (see Chapter 9), the private costs of ensuring adequate water supply have been rising and could over time induce more rational use[6].

Chapter 9 argues that real agricultural wages are rising and can be expected to continue to rise due to the net withdrawal of labor from the agricultural sector with overall economic development. Rising agricultural wage rates also imply increasing opportunity cost of family labor. While further opportunities for unit cost reduction require improvements in input efficiencies, these will come only at the cost of increasingly expensive farmer time. If knowledge-intensive technologies are to be adopted, then one or both of the following conditions ought to be satisfied: (i) decision rules, although derived from complex scientific investigation, ought to be simple enough for the farmer to adopt with minimal additional time expenditure[7]; and (ii) contracting out the decision process itself ought to be possible[8].

Finally, it ought to be emphasized that the profitable adoption of knowledge-intensive technologies that enhance input efficiency will only occur when the price is right. Holding input prices low through the use of subsidies or output prices high through price support programs would reduce the likelihood of adopting input-efficient technologies.

Conclusions

Technological change has helped sustain the profitability and reduce the unit costs of rice production systems in Asia. Profitability of rice production has been sustained despite the long-term decline in rice prices due to technical change. Current evidence indicates that the opportunities for further productivity growth through technological change are limited, especially in the high-potential irrigated lowlands of Asia. A significant shift in the yield frontier for rice is needed in order to reduce the unit costs of production further. Technologies in the pipeline for shifting the yield frontier are discussed in Chapter 10.

In addition to the exhaustion of the opportunities for technological change, there are increasing factor scarcities due to commercialization and competitiveness (Chapter 9). The resulting rise in input prices – especially for land, labor and water – is leading to further decline in rice profitability. There is an urgent need for technologies that increase input use efficiencies in order to sustain rice profitability in an increasingly competitive world. Desire for increased input efficiencies is also driven by increasing environmental concerns due to intensive use of irrigation water, chemical inputs and the practice of rice monoculture over the long term (Chapter 4). Technologies for increasing the efficiency of input use are discussed in Chapters 11 and 12.

Notes

1. Byerlee (1992) identifies a similar phase of management and knowledge input for increasing wheat productivity in the post-Green-Revolution phase.
2. Agricultural operations can be broadly classified into power-intensive and control-intensive operations. Power-intensive operations, such as land preparation, threshing and milling, require high levels of energy and little human judgment. While control-intensive operations such as transplanting, weeding and harvesting require judgment and discrimination in addition to energy. Binswanger (1978) has shown that even in

low-wage societies power-intensive operations are the first to be mechanized; control-intensive operations are mechanized only under high-wage conditions.

3. Mechanization of an operation does not necessarily mean ownership of the machine. Contract services that emerged allowed farmers timely access to machines without having to own them.
4. In fact, Hayami and Ruttan (1985) argue that research investments in developing a fertilizer-responsive rice plant type were prompted by the long-term secular decline in fertilizer prices.
5. Farmer technical knowledge has played a crucial role in sustaining the productivity of traditional rainfed systems. Fujisaka (1994) has provided a comprehensive assessment of the importance of farmer knowledge in rainfed systems, in the choice of lands to intensify, in the matching of varieties to land forms and soil types, and in the understanding and averting of physical and pest stresses. The stock of farmer knowledge on the rainfed systems was built up over generations of farmers, in a particular location, adapting to physical, demographic and socioeconomic pressures. The introduction of modern technology in the irrigated environments made the farmers' traditional knowledge stock essentially obsolete. There has been a need for re-discovery and re-learning within a very short period.
6. The private costs of ensuring adequate water supply include water conflicts, supervision of water flows, farmer management of irrigation systems, and use of tubewells, among others.
7. Chapter 11 provides an example from insect pest management on how a very complex ecological study on pest–predator interactions led to a very simple rule on the timing of insecticide application.
8. A good example of contracting out the decision process is the use of professional scouting services in US agriculture for assessing pest infestation levels and for making judgments on the need for chemical control (Carlson, 1989).

4

Intensification-induced Degradation of the Paddy Resource Base

The Green Revolution Strategy for increasing food production in Asia was based on the intensification of the lowlands through massive investments in irrigation infrastructure and in crop research. It was presumed that the lowlands are resilient to intensification pressures and would sustain output growth over the long term. This strategy was meant to relieve pressures on the fragile uplands by creating employment opportunities in the lowlands. The strategy worked exceptionally well for rice, up to the mid-1980s (Herdt and Capule, 1983; Dalrymple, 1986). Since then rice productivity growth has slowed down in the intensively cultivated areas across Asia (Rosegrant and Pingali, 1994).

Aggregate rice output growth rate for Asia increased from 2.1 percent per annum during 1955–1965 to 2.9 percent per annum during 1965–1980. Rice output growth rate surpassed annual population growth rate of 2.3 percent. Area expansion contributed to nearly one-third of Asian rice output growth in the 1960s and one-fifth in the 1970s. Rapid yield growth from 1965 to 1980, due to the adoption of modern rice varieties, was the primary contributor to rice output growth. In the past decade, however, the growth in aggregate rice output has declined to 1.5 percent per annum. Rice yield growth in Asia also declined sharply in the 1980s, from an annual growth rate of 2.6 percent in the 1970s to 1.5 percent during the period beginning in 1981.

Virtually all future output growth must come from increased rice yield per unit of land since the opportunities for further area expansion are minimal. There is increasing evidence that the growth in rice yields has leveled off and there is a danger of future declines in yield growth, especially in the irrigated lowlands of Asia (Pingali *et al.*, 1990).

The slowdown in rice productivity growth in Asia since the 1980s has been caused by: (i) world rice price-induced factors; and (ii) intensification-induced factors. The world rice price has been on a declining trend in real terms since 1900, a decline which has sharpened in the 1980s (Mitchell, 1987). The declining price of rice has caused a direct shift of land out of rice and into more profitable cropping alternatives, and has slowed the growth in input use and yields. Probably more important in the long run, the declining world price has caused a slowdown in investment in rice research and irrigation infrastructure (these issues are discussed in detail in Rosegrant and Pingali, 1994).

Does intensification of irrigated land use, independent of world rice price effects, lead to a long-term decline in rice productivity? Intensification is defined here as the permanent movement from one rice crop per year followed by a dry season fallow, to two or three consecutive rice crops per year on the same land. This chapter argues that the practice of intensive rice monoculture itself contributes to the degradation of the paddy resource base and hence declining productivities. The consequences of intensification on the paddy resource base vary by agroclimatic and management factors and can be observed only over the long term.

Intensification and Rice Productivity Decline

Productivity decline on experiment stations

The best illustration of the consequences of intensification, independent of world rice price effects, can be found in the examination of yield trends from long-term trials conducted on experiment stations. The objective of long-term trials is to monitor maximum yields obtained over time, holding input levels and crop management practices constant. The long-term continuous cropping experiment conducted by the International Rice Research Institute (IRRI) in the Philippines is an excellent example of such trials. This experiment set up in 1963 has been monitoring the yield impact of rice monoculture (with three crops per year) and as of 1993 had completed 89 consecutive crops on the same plot (Cassman *et al.*, 1994). Table 4.1 presents yield trend from this as well as other long-term trials in the Philippines, India, Thailand and Bangladesh. The trends indicate that, even with the best available cultivars and scientific management, yields, holding input levels constant, decline over the long term. The declining yield trend in long-term experiments has been documented by

Flinn and De Datta (1984), Cassman *et al.* (1994), Pingali (1994) and Cassman and Pingali (1995).

IRRI started releasing modern rice varieties in the mid-1960s. IR-8 was the first of the modern varieties widely grown in Asia. At the time of its release in 1966, IR-8 yielded as much as 10 t ha^{-1} in the dry season and 6 t ha^{-1} in the wet season at IRRI's experimental farm in Laguna, Philippines. At that time, farmers in the neighborhood of IRRI growing traditional rice varieties were getting yields of 2.0–2.5 t ha^{-1} (IRRI, 1967). Since its initial release, the yields of IR-8 have been on a

Table 4.1. Growth in rice yield potential (percent) for selected Asian countries.

		Annual growth in yields	
		Wet season	Dry season
Philippines	IRRI	– 1.29	– 1.28
(1966–1968)	Maligaya Rice Research and Training Center	– 1.01	+ 0.15
	Visayas Rice Experiment Station	+ 0.18	+ 0.18[a]
	Bicol Rice and Corn Experiment Station	– 0.62	– 0.38
India	Coimbatore	– 0.27[a]	0.44[a]
(1969–1989)	Raipur	– 1.41	—
	Pantnagar	– 0.89	—
	Rajendranagar	0.12[a]	—
	Mandya	– 1.11	—
Bangladesh	Comilla	– 1.80	—
(1977–1988)	Joydebpur	– 0.13[a]	—
Thailand	Chiang Mai	2.30	—
(1977–1990)	Suphan Buri	– 2.50	—
	Chiang Rai	– 1.80	—
Vietnam (1977–1989)	Hanoi	– 3.35	—
Malaysia (1980–1987)	Tuaran, Sabah	– 3.37	—
Indonesia (1978–1990)	Sukamandi	—	0.99
Nepal	Bhairahawa (1977–1988)	– 1.77	—
	Paruwanipur (1977–1982)	– 2.99	—
Pakistan	Muingara, Swat (1985–1990)	5.09	—
	D.I. Khan (1979–1985)	– 4.57	—

[a] Not significantly different from 0%.
Source: Growth rates for the Philippines were estimated using data from the long-term fertility trials conducted at the above experiment stations. Growth rate estimates for the other countries used basic data from INGER.

declining trend even when grown under scientific management on the IRRI farm. Flinn *et al.* (1982) estimate that since 1966 the wet-season yields of IR-8 have declined by 0.2 t ha^{-1} yr^{-1} and the dry-season yields have declined by 0.26–0.47 t ha^{-1} yr^{-1}. The most commonly attributed cause of this decline is the greatly increased insect and disease pressure to which IR-8 is not resistant. (Insect and disease infestations have risen with the growth in intensive rice production across Asia, see below.)

Following IR-8, 40 modern rice varieties have been released in the Philippines. These varieties have better insect and disease resistance, shorter crop duration and to some extent better eating quality than IR-8. However, none of the later varieties has been able to match the initial yield potential of IR-8[1]. Indeed, De Datta *et al.* (1979) report that in recent years rice yields of over 9 t ha^{-1} are rarely recorded at IRRI. Perhaps more disturbing is the observation by Flinn and De Datta (1984) that the highest yields obtained from the nitrogen response trials have been exhibiting a long-term decline. They estimate the decline at an annual rate of 0.10–0.16 t ha^{-1} in both the wet and dry seasons. Figure 4.1 graphs the highest wet- and dry-season yields obtained in the nitrogen response experiment for the years 1966–1991.

Flinn and De Datta (1984) also provide evidence of yield stagnation or decline in both the wet and dry seasons in three other experiment stations in the Philippines in addition to IRRI. These experiment stations are in Nueva Ecija, Bicol and Visayas, covering a fairly representative cross-section of the irrigated rice areas of the Philippines. For Nueva Ecija, they find that the highest yields in the nitrogen response experiments at the Philippine Rice Research and Training Center (PhilRice) have remained stagnant in both the wet and dry seasons for the years 1968–1980. Figure 4.2 graphs the highest wet- and dry-season yields obtained in the nitrogen response experiments at PhilRice.

Pingali *et al.* (1990) updated the Flinn and De Datta analysis for all four stations by expanding the long-term yield response data set to include all years up to 1988. The highest yield for each nitrogen level was used as the dependent variable (in logs) and the independent variables in the regression were time trend, log nitrogen (kilograms per hectare), the square of log nitrogen and log rainfall. Rainfall was measured in terms of the actual amount from transplanting to two weeks before harvest. All other production and management inputs were held constant throughout the experiment. Generalized least squares regressions using the Parks method were run for identifying the long-term yield trend. The Parks method corrects for heteroscedastic and autocorrelated error terms and provides consistent coefficient estimates (Parks, 1967). The results are presented in Table 4.2. A

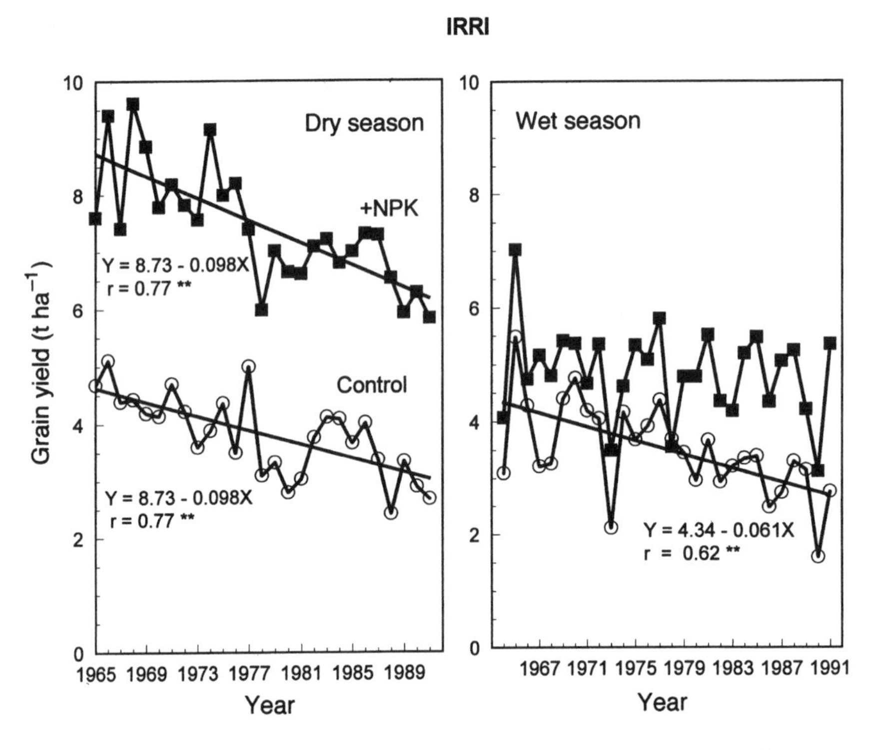

Fig. 4.1. Yield trends of the highest-yielding variety in the Long-Term Fertility Experiment conducted at the IRRI Research Farm in Laguna Province, Philippines, since 1964 in treatments that receive complete nitrogen (N), phosphorus (P), and potassium (K) inputs in each crop cycle (+NPK), and in the control treatment without fertilizer-nutrient inputs. (Source: Cassman and Pingali, 1995.)

significant negative yield trend was found for the wet season for three of the four locations and for two of the four locations for the dry season. The exceptions, VRES for both seasons and PhilRice for the dry season, exhibited stagnant yield trends. At IRRI, wet-season yields declined at the rate of 1.29 percent per year, while at PhilRice and BRES the yield decline was at the rate of 1.01 percent and 0.6 percent per year, respectively. VRES had a slight growth in wet-season yield at the rate of 0.18 percent per year. For the dry season the yield decline at IRRI was significant at the rate of 1.28 percent per year, while at BRES the yield decline was at the rate of 0.38 percent per year. PhilRice exhibited a slight yield growth at the rate of 0.15 percent per year, while at VRES the time trend was not significant.

Similar long-term yield declines have been observed in other experiment stations in India, Bangladesh, Thailand and Indonesia.

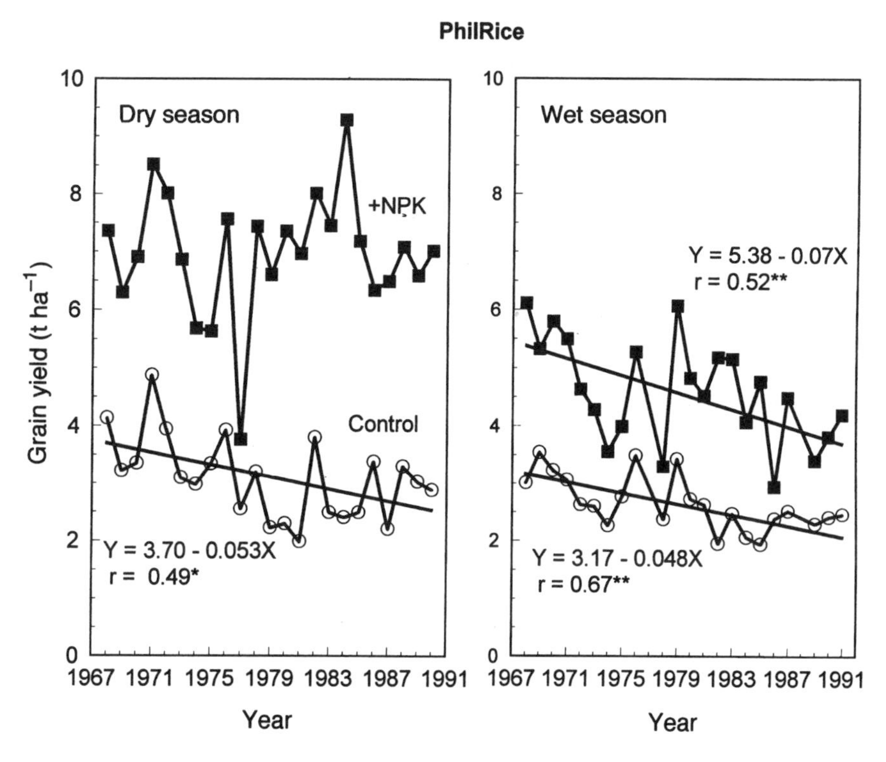

Fig. 4.2. Yield trends of the highest-yielding variety in the Long-Term Fertility Experiment conducted at the PhilRice Research Station in Central Luzon, Philippines, since 1968 in treatments that receive complete nitrogen (N), phosphorus (P), and potassium (K) inputs in each crop cycle (+NPK), and in the control treatment without fertilizer-nutrient inputs. (Source: Cassman and Pingali, 1995.)

Long-term fertilizer experiments conducted in four locations in India, from 1972 to 1982, found declining long-term yields in two and stagnant yields in two locations (Nambiar and Ghosh, 1984). The four locations were spread across the country, one each in the following states: Punjab, Uttar Pradesh, Orissa and Andhra Pradesh. Table 4.1 provides data on five additional experiment station yield trends from India over a 20-year period. Negative trends are observed in four of the five stations and the exception has a stagnant yield trend.

Long-term yield declines have been witnessed in 11 years of continuous rice cropping at Maros, Indonesia (INSURF, 1987) and 10 years of continuous rice cropping in Thailand (Gypmantasiri *et al.*, 1980). Table 4.1 presents data from three additional research stations in Thailand and two in Bangladesh. Negative trends are evident in all but one case.

Table 4.2. Experiment station yield growth rates (percent) with and without adjustment for environmental degradation.

	IR-8	Highest yielding entries	Highest yielding entries (adjusted)
Dry season			
BRES	–5.01***	–0.38***	4.63***
IRRI	–5.89***	–1.28***	4.61***
PhilRice	–0.58***	0.15**	0.72***
VRES	–1.30***	0.18	1.49***
Wet season			
BRES	–3.83***	–0.62***	3.22***
IRRI	–5.17***	–1.29***	3.88***
PhilRice	–1.67***	–1.01***	0.66***
VRES	0.93***	0.18*	0.75***

Adapted from Pingali *et al.* (1990).
Dependent variable: log yield ha^{-1}.
* Significant at 10% level; ** significant at 5% level; *** significant at 1% level.

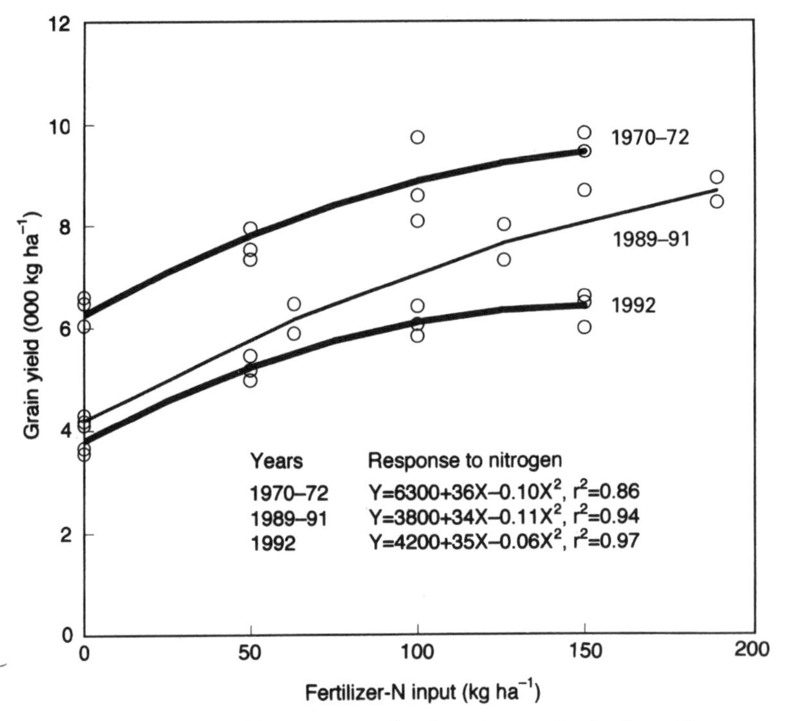

Fig. 4.3. Dry-season yield response to fertilizer-N rates in the Long-Term Continuous Cropping Experiment at IRRI based on the highest-yielding variety each year in three time periods. Bar intervals for the 1992 means represent ± SE. (Source: Cassman and Pingali, 1995.)

Declining experiment station yields holding constant input levels and crop management practices could indicate one or both of the following: (i) the genetic potential of later generation varieties is lower than earlier varieties; and (ii) the productive potential of the paddy resource base itself has been declining over time. Substantial evidence

Box 4.1. Experimental yield trends adjusted for environmental degradation.

The long-term decline in yield potential under intensive irrigated rice production can be attributed to either a degradation of the paddy environment or a decline in the genetic potential of the breeding materials used for generating cultivars.

The rate of environmental degradation can be measured by the rate of decline in yield over time, holding variety and input levels constant. In order to do this, yield time series on one particular variety are needed. At IRRI and at PhilRice the variety grown each year since the start of the long-term yield experiment is IR-8. Twenty years of data for IR-8 are available at IRRI and 18 years of data for PhilRice. Linear regression equations were estimated for IRRI and PhilRice using generalized least squares techniques for pooling time series and cross-section data. The results are provided in Table 4.2.

A significant decline in IR-8 yields was observed for all locations for the dry season and three of the four locations for the wet season. The long-term wet-season IR-8 yield decline at IRRI is at the rate of 5.17 percent per year, while at PhilRice it is at the rate of 1.67 percent per year. The corresponding dry-season yield declines per annum are 5.89 percent for IRRI and 5.17 percent for PhilRice. The long-term performance of IR-8 at BRES and VRES is found in Table 4.2.

To obtain the yield potential adjusted for the environmental degradation the highest yields in the long-term experiment for all four locations were compensated with an amount equal to the annual rate of decline in IR-8 yield. After adjusting yields for environmental degradation the highest yields were found to be increasing significantly in all locations for the dry and the wet season. The highest wet- and dry-season yields at IRRI have been increasing at an annual rate of 3.88 percent and 4.6 percent respectively. For PhilRice the wet-season time trend after adjustment for environmental degradation is 0.66 percent per year and the dry-season trend is 0.72 percent per year. These results imply that the rate of degradation of the paddy environment is greater than the rate of growth in the yield potential, hence the observed long-term declining trend in the highest experiment-station yields. These results also oppose the alternative view that the cause of the declining yield potential is an erosion of the genetic potential of the breeding materials.

Source: Pingali *et al.* (1990).

exists now to indicate that the genetic potential of latter-day varieties is not different from early generation varieties (see Khush, 1990; Cassman *et al.*, 1994). In fact, yields of over 10 tons have been recorded for IR-72, a variety released in the early 1990s. Current evidence seems to indicate clearly a long-term degradation of the paddy resource base (Cassman and Pingali, 1993; Cassman *et al.*, 1994). Estimates of nitrogen response functions using data from the long-term trials at IRRI show a significant downward shift of these functions over time while the slope remains the same (see Fig. 4.3). At zero nitrogen level (and at any nitrogen level), the long-term experiment plots at IRRI are giving significantly lower yields today than they were in the late 1960s and early 1970s.

Several hypotheses have been generated and are being tested for explaining the long-term yield decline under continuous cropping on experiment stations. These are: (i) long-term decline in soil nitrogen supply due to intensive cropping on water-saturated soils; (ii) increased incidence of diseases with high nitrogen use; and (iii) buildup of soil pests, especially nematodes, that hamper root growth. Do these hypotheses also explain productivity trends on farmer fields? Yes and no. No, because there could be several other factors operating on farmer fields – inefficient irrigation-water use, poor-quality irrigation water, lower human capital, and limited ability to diagnoze post-Green-Revolution problems, etc. On farmer fields, issues such as salinity buildup and micro-nutrient deficiencies could be just as important as soil nitrogen supplying capacity.

Farm-level yield stagnation and productivity decline

Experiment-station yield trends can be used as a means of forecasting future farm-level trends. The essential message from the experiment-station results is that under intensive rice monoculture systems, productivity over the long term is difficult to sustain, even with the best scientific management. For the intensively cultivated irrigated lowlands of Asia, that have been under rice monoculture systems over the past two decades, one should expect signs of declining productivity.

At the farm level, declining yield trends are usually not observed since input levels are not held constant over time. However, in areas where intensive rice monoculture has been practiced over the past two to three decades, one does observe stagnant yields and/or declining trends in partial factor productivities, especially for fertilizers, and declining trends in total factor productivities (Pingali *et al.*, 1990; Pingali, 1992; Cassman and Pingali, 1993).

Yield trends

In Chapter 2 the deceleration in national average yields was noted. It is, however, difficult to assess trends in farm performance and the technical efficiency of farmers over time using national-average yield time series because these data are based on the pooling of heterogeneous rice-growing environments (Herdt, 1988; Barker *et al.*, 1985). However, if countries are stratified by cropping intensities, it turns out that the rate of deceleration in yields is higher for countries with higher cropping intensities (Table 4.3). These are invariably countries with an exhausted land frontier that sought output growth through land-augmenting technical change; China, Korea and the Philippines are examples. Rapid yield growth in the decade of the 1980s has come from an increase in intensification in the low-intensity countries such as Laos, Nepal, and Cambodia (Table 4.3). India and Vietnam have been exceptions to the above trends; the former due to an increase in irrigation infrastructure in the 1980s (Rosegrant and Pingali, 1994) and the latter due to policy reforms of the mid-1980s (Pingali and Xuan, 1992).

Table 4.3. Comparison of annual growth rate in rice yield by cropping intensity for selected countries in Asia (percent).

	1970–1980	1980–1992
High-intensity areas		
Pakistan	0.58	– 0.65
Japan	0.20	1.04
China (incl. Taiwan)	2.41	2.31
Indonesia	2.93	2.17
Korea, Republic of	2.72	1.32
Malaysia	1.48	0.19
Philippines	3.72	1.84
Vietnam	– 1.18	3.35
Sri Lanka	1.73	0.96
India	1.44	2.92
Low-intensity areas		
Thailand	– 0.68	0.40
Laos	0.25	4.45
Nepal	– 0.74	2.43
Myanmar	4.32	– 0.23
Cambodia	– 3.44	2.97
Bangladesh	2.13	2.66

Source of basic data: World Rice Statistics 1990, IRRI (1991).

Within each country, a comparison of yield and productivity trends by provinces stratified by cropping intensities gives a better assessment of the state of intensive rice monoculture systems. Table 4.4 provides a province-level comparison of yield trends for Bangladesh, India and the Philippines. The essential point that comes out of this comparison is that in the irrigated, high-intensity rice provinces of these countries, growth rate in yields is decelerating. In several instances, yields in the late 1990s are not significantly different from those in the early 1980s. Yields have actually fallen in some instances. For provinces with lower levels of irrigation investments, yield growth has been stagnant through the entire Green Revolution period.

The long-term impact of intensive monoculture systems on productivity can be best assessed using farm-level panel data sets. Long-term, farm-level data that can be used to discern trends in productivity are rare throughout Asia. The exception are the long-term farm monitoring data from Laguna and Central Luzon provinces in the Philip-

Table 4.4. Comparison of yield trends in the high- and low-intensity provinces of selected Asian countries (percent).

High-intensity areas			Low-intensity areas		
Bangladesh	1970–1980	1980–1987		1970–1980	1980–1987
Dhaka	2.09	0.18	Faridpur	1.15	3.43
Comilla	1.12	1.87	Dinajpur	0.85	2.29
India	1970–1980	1980–1988		1970–1980	1980–1988
Andhra Pradesh	2.61	1.94	Madhya Pradesh	0.06	2.86
Haryana	4.46	-0.85	Orissa	3.63	2.34
Punjab	4.60	0.70	West Bengal	1.28	4.53
			Bihar	0.51	4.02
Philippines	1970–1980	1980–1991		1970–1980	1980–1991
Central Luzon	4.96	1.02	Northern Mindanao	2.24	2.84
Bicol	3.68	1.60			
Southern Tagalog	2.94	2.79	Western Mindanao	8.36	2.17

Sources of basic data:
BBS, Yearbook of Agricultural Statistics of Bangladesh (various years), Bangladesh Bureau of Statistics (BBS), Ministry of Planning, Government of the Peoples' Republic of Bangladesh, Dhaka.
1960: India Central Statistical Organization, Statistical Abstract. 1961–1984: India Directorate of Economics and Statistics, All India Estimate of Rice (various issues). 1985–1988: India Directorate of Economics and Statistics, Agricultural Situation in India (various issues).
Bureau of Agricultural Statistics (BAS), Department of Agriculture, Manila, Philippines.

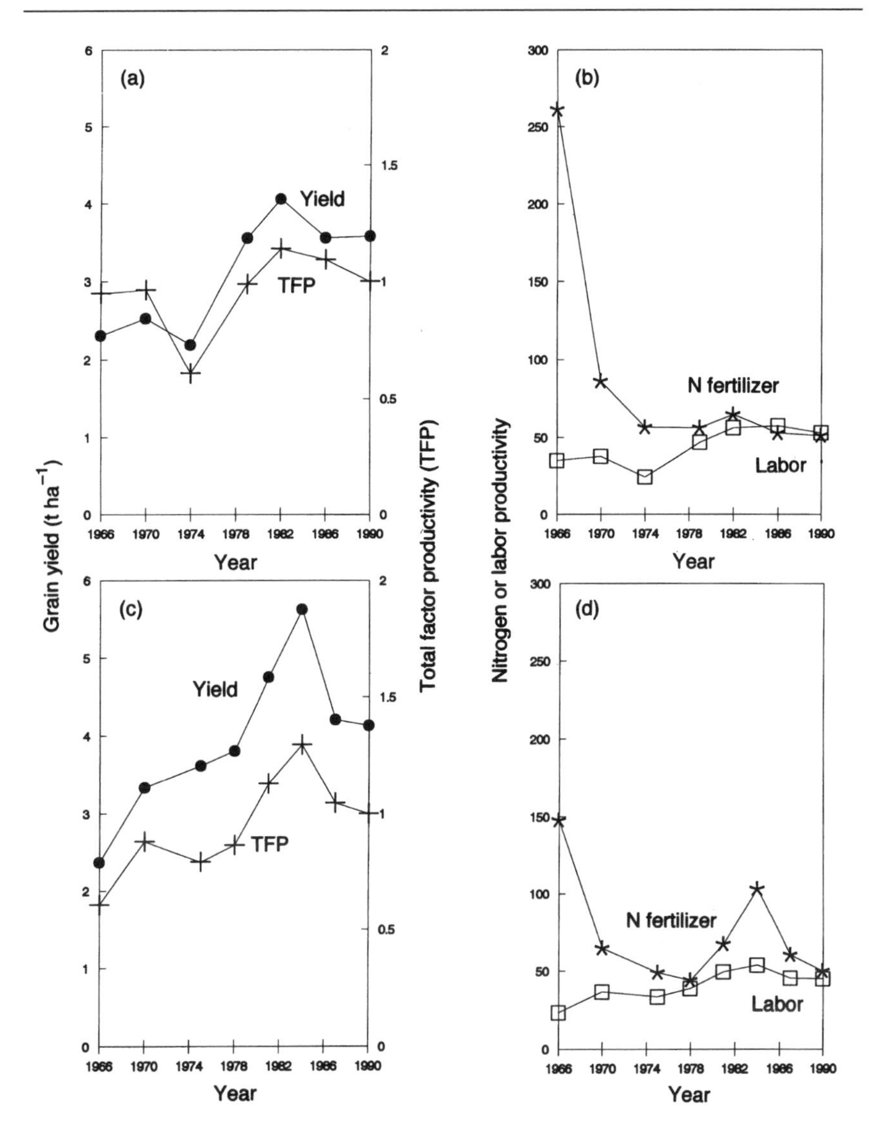

Fig. 4.4. Trends in rice yield and total factor productivity (TFP), and partial factor productivity for N fertilizer (kg paddy per kg applied N) and labor (kg paddy per person–day) achieved by farmers in Central Luzon (a and b) and Laguna (c and d) provinces, Philippines. (Source: Cassman and Pingali, 1995.)

pines which were collected by IRRI's Social Sciences Division from 1966 to 1990. The 'cost of cultivation' surveys conducted in India are a close approximation to long-term farm data sets. Two intensively cultivated locations were selected from these data sets for comparison

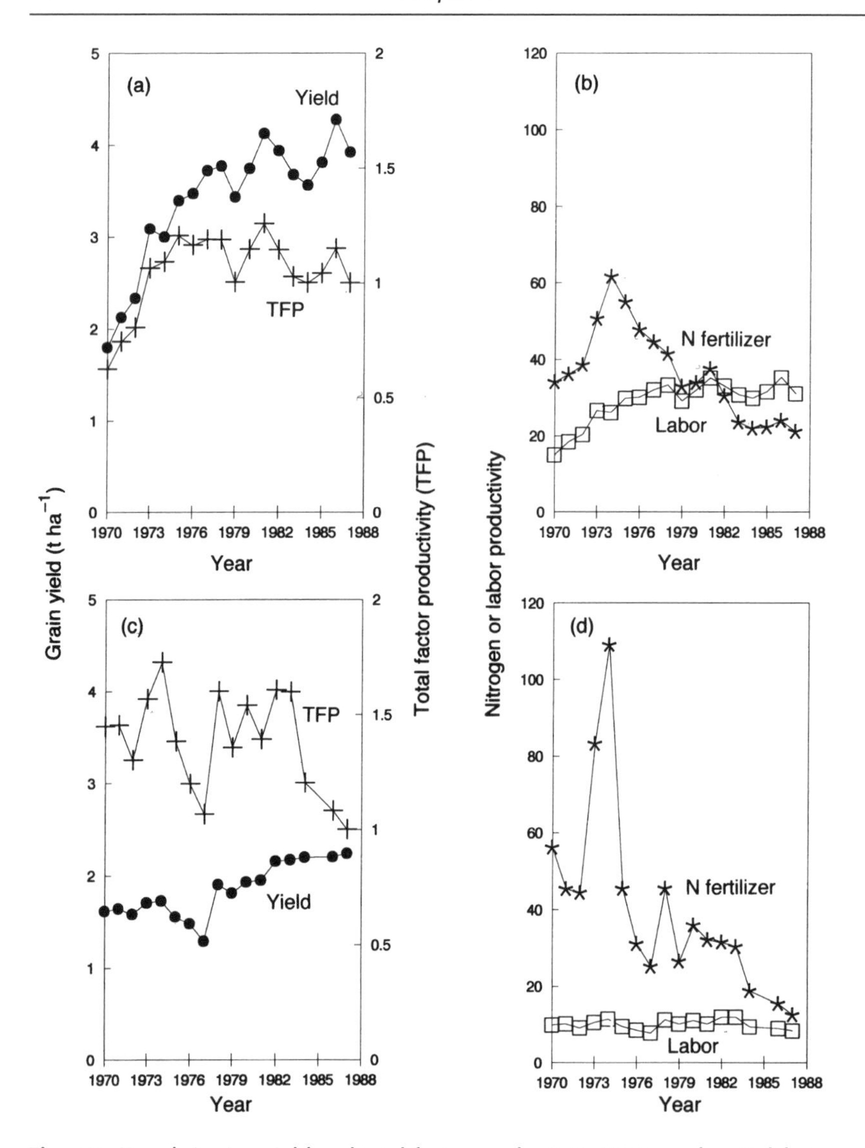

Fig. 4.5. Trends in rice yield and total factor productivity (TFP), and partial factor productivity for N fertilizer (kg paddy per kg applied N) and labor (kg paddy per person–day) achieved by farmers in Ludhiana, Punjab (a and b) and Krishna District, Andhra Pradesh (c and d), India. (Source: Cassman and Pingali, 1995.)

with the Philippines. The Indian locations are the Ludhiana District in the State of Punjab and the Krishna District in Andhra Pradesh.

Yield trends within these selected domains indicate patterns that are similar to the aggregate national and regional trends in rice output:

(i) there was a dramatic increase in yields following the rapid adoption of modern rice varieties and management practices in the 1960s; (ii) yields continued to increase in the second decade after adoption; and (iii) stagnant or declining yield trends are indicated in the most recent decade. Based on the farm monitoring data from the Philippines, average wet-season (WS) rice yields were approximately 2.5 t ha^{-1} in 1966, and increased to 4.2 t ha^{-1} in Central Luzon, and 4.7 t ha^{-1} in Laguna, by the early 1980s (Figs 4.4(a) and 4.4(c)). Since then, however, there has been a gradual decline in yields so that grain yield was 0.5 t ha^{-1} lower in 1990 than in the early 1980s in both domains.

In Ludhiana, Punjab, where an intensive rice–wheat double crop system is practiced, average rice yields rose from 1.8 t ha^{-1} in 1970 to 4.0 t ha^{-1} by 1980 and have remained relatively constant thereafter (Fig. 4.5(a)). Investment in irrigation and adoption of modern rice technology combined to produce similar trends in the delta areas of Southern India where intensive, continuous rice monoculture is practiced. Yield trends in Krishna District, a delta district of Andhra Pradesh, indicate yields remain on a positive trend but are growing slowly (Fig. 4.5(c)).

Growth in input use and factor productivity

Long-term, farm-level yield trends alone do not tell the complete story because farmers tend to maintain yields with increasing levels of inputs per hectare. Therefore, while an increase in yields may be observed at the farm level, it may be coming at the cost of a proportionately greater increase in input use. National data across Asia indicate that the rate of growth in yields has been substantially smaller than the rate of growth in fertilizer use over the past three decades (Table 4.5).

Consider the following intensively cultivated rice bowls of Southeast Asia: Central Luzon, Philippines, Central Plains, Thailand and West Java, Indonesia, for the period 1980–1989. Farm panel data sets for each of these locations indicate that in the 1980s the rate of growth in yields was lower than the rate of growth in input use (Pingali, 1992). In Central Luzon, a 13-percent yield increase over a 10-year period was achieved with a 21-percent increase in fertilizers, and a 34-percent increase in seeds. In the Central Plains, for the same period, yields increased by 6.5 percent, while fertilizer levels increased by 24 percent and pesticides by 53 percent. Similarly, for West Java, yields increased by 23 percent, while fertilizer use increased by 65 percent and pesticide use increased by 69 percent (Pingali, 1992). Table 4.5

Table 4.5. Comparison of growth in rice yield and growth in nitrogen input use (percent) for selected Asian countries.

	Annual growth rate	
Country	Rice yield	Nitrogen use
Bangladesh (1968–1989)	1.59	10.03
India (1961–1991)	2.04	9.76
Myanmar (1962–1992)	2.80	12.16
Nepal (1961–1991)	0.38	15.38
Vietnam (1976–1992)	3.52	10.62

Sources of basic data: World Rice Statistics 1990, IRRI (1991). AGROSTAT, IRRI, Los Baños, Laguna. Planning Management and Evaluation Division, Agriculture Corporation, Myanmar. Statistical Yearbooks, various years, Socialist Republic of Vietnam.

provides several additional examples of increased nitrogen input requirements for sustaining yields under intensive cultivation sytems.

Increasing input requirements for maintaining yield levels indicates declining partial factor productivities. Consider once again the two intensively cultivated locations in the Philippines and those in India described above. In the Philippines, grain yield per unit of applied N decreased markedly from 1966 to the early 1970s, then remained relatively constant at about 50 kg grain per kg N input (Figs 4.4(b) and 4.4(d)). A steady decrease in productivity from applied N is also apparent in the Indian districts since the mid-1970s (Figs 4.5(b) and 4.5(d)). Partial factor productivity for labor has risen slightly due to mechanization of land preparation in all cases except Andhra Pradesh where relatively little mechanization has occurred (Figs 4.4 and 4.5).

Where intensification is not associated with a change in the inherent productivity of the paddy resource base, declining factor productivity indicates a movement along a production function. Where intensification leads to reduced productivity of the resource base, declining factor productivities signify both a shift downward of the production function as well as a movement along the new production function.

Trends in total factor productivity (TFP)

Yield and other partial factor productivity measures discussed above do not provide a holistic picture of the long-term impact of technological change on rice farming systems. In the absence of externalities, the appropriate measure for assessing long-term system-level changes is the index of total factor productivity.

> Defined as the total value of all output produced by the system during one crop cycle divided by the total value of all inputs used by the system during one cycle of the system, a sustainable system has a non-negative trend in total factor productivity (TFP) over the period of concern
>
> (Lynam and Herdt, 1989, p. 209)

Changes in TFP over time may result from changes in the socioeconomic and biophysical environment. Evaluating TFP trends after specifying constant prices for inputs and outputs eliminates fluctuations due to changes in market conditions but does not account for changes

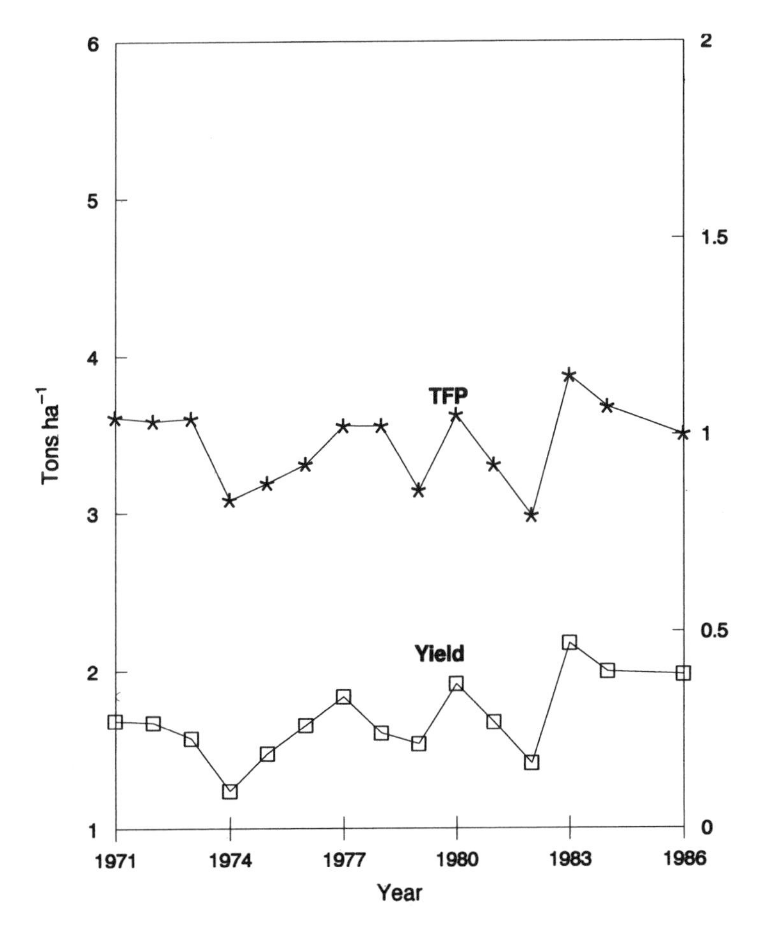

Fig. 4.6. Trends in yield and total factor productivity in Orissa, India. (Source of basic data: Social Sciences Division, IRRI.)

in the policy environment or changes in the biophysical resource base which may influence the quantity and efficiency of inputs used by farmers. Thus, monitoring of the TFP trend line based on constant prices indicates that the efficiency of input use has changed but does not explain why the change is occurring.

TFP trends were calculated for each of the locations in the Philippines and India described above. Based on 1990 prices for inputs and outputs in each year in the Philippines, trends in TFP indicate that productivity increased steadily until the early 1980s, followed by a slight decline through 1990 in both Central Luzon and Laguna (Figs 4.4(a) and 4.4(c)). TFP trends in Ludhiana follow a similar pattern to those in the Philippines (Fig. 4.5(a)), while there has been a sharp decline in TFP since 1984 in Andhra Pradesh (Fig. 4.5(c)).

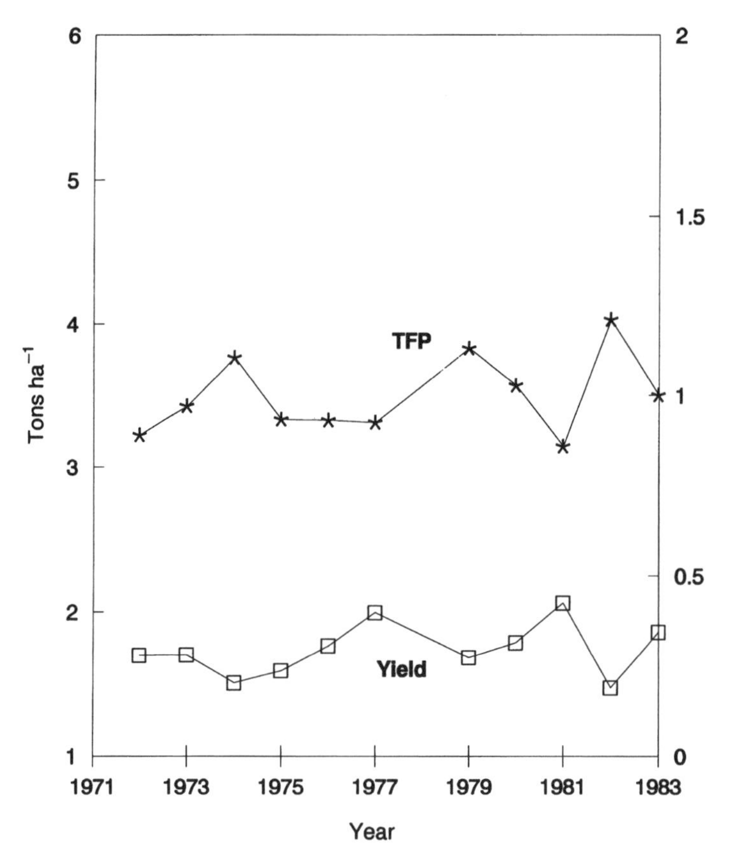

Fig. 4.7. Trends in yield and total factor productivity in Bihar, India. (Source of basic data: Social Sciences Division, IRRI.)

Contrast the above trends from the intensively cultivated rice areas with trends from two Indian states where only one crop of rainfed rice per year, using traditional cultivars, is the norm with little, if any, purchased inputs. The yield and TFP trends from the states of Orissa and Bihar are presented in Figs 4.6 and 4.7 respectively. Yield has remained relatively stagnant and so has TFP for both states.

Ecological Consequences of Intensification

Intensive rice monoculture on the lowlands results in the following changes in production systems: (i) rice paddies flooded for most of the year without adequate drying period; (ii) increased reliance on inorganic fertilizers; (iii) asymmetry of planting schedules; and (iv) greater uniformity in the varieties cultivated. Over the long term, these changes impose significant environmental costs due to negative biophysical impacts (Table 4.6).

Environmental consequences of rice monoculture systems vary by agroclimatic condition, soil type, and the source and quality of irrigation water. The following, however, are the most common consequences: (i) buildup of salinity and waterlogging; (ii) changes in soil nutrient status, nutrient deficiencies and increased incidence of soil toxicities; (iii) formation of a hardpan (subsoil compaction); and (iv) increased pest buildup and pest-related yield losses. A brief description of each of these problems and the possibilities for reversing them are discussed below. At the farm level, long-term changes in the biophysical environment are manifested in terms of declining total factor productivity, profitability and input efficiencies.

Salinity and waterlogging

> The long-term viability of irrigated agriculture (especially in the semi-arid and arid zones) becomes one of salt management – devising strategies to prevent salts from accumulating either in the irrigated area or downstream
>
> (NAS, 1989, p. 96)

Intensive use of irrigation water in areas with uneven toposequence and poor drainage can lead to a rise in the water table due to the continual recharge of groundwater. In the semi-arid and arid zones this leads to salinity buildup, and in the humid zone to waterlogging. Salinity is induced by an excess of evapotranspiration over rainfall causing a net upward movement of water through capillary action and the concentration of salts on the soil surface. The groundwater itself

Table 4.6. Intensification-induced degradation of the irrigated lowlands.

Resource base degradation	Possible/probable causes	Farm-level indicators of resource degradation	Economic impact
Build up of salinity/ waterlogging	■ Poor design of irrigation systems (poor drainage) ■ Intensive use of irrigation water	■ Reduced yields and/or reduced factor productivities ■ Reduced cropping intensities ■ Abandoned paddy lands in the extreme	
Increased incidence of soil toxicities and micro-nutrient deficiencies	■ Long-term flooding/ water saturation of paddy soils ■ Increased reliance on low-quality irrigation water ■ Depletion due to continuous rice monoculture	■ Reduced yields and/or reduced factor productivities	Declining trends in TFP Declining profitability of rice cultivation
Hardpan (subsoil compaction)	■ Increased frequency of puddling (wet tillage)	■ Reduced flexibility for non-rice crop production in the dry season	
Changes in soil nitrogen supplying capacity	■ Changes in organic matter quantity and quality due to slower rate of decomposition in continuously flooded soil	■ Declining efficiency of nitrogen fertilizer use	Increased social costs of negative externalities on the environment and human health
Increased pest buildup and pest-related yield losses	■ Continuous rice monoculture ■ Increased asymmetry of planting schedules ■ Greater uniformity in varieties cultivated	■ Increased pesticide use	

need not be saline for salinity to build up; it can occur due to the long-term evaporation of continuously recharged water of low salt content (Moorman and van Breemen, 1978). High water tables prevent the flushing of salts from the surface soil. Postel (1989) estimates that 24 percent of the irrigated land worldwide suffers from salinity problems, with India, China, United States, Pakistan, and the former Soviet Union being the most affected. In the short term, salinity buildup leads to reduced yields while in the long term it can lead to abandoning of paddy lands (Samad *et al.*, 1992; Postel, 1989; Mustafa, 1991)[2].

Induced salinity problems are caused by excessive irrigation and poor drainage (especially, seepage from unlined canals). Poor irrigation system design is a primary factor inducing salinity problems. For instance, in Pakistan's Sind Province, large areas became saline after the introduction of extensive irrigation, which led to a rise of the water table from a depth of 20–30 m to 1–2 m within 20 years (Moorman and Van Breeman, 1978); other examples from South Asia can be found in Dogra (1986), Abrol (1987), Chambers (1988). Dogra (1986) estimates that in India nearly 4.5 million hectares are affected by salinization and a further 6 million by waterlogging. 'Many irrigation projects in the drier rice-growing areas suffer from a lack of insight into the dynamics of salinity' (Moorman and Van Breeman, 1978)[3].

In the humid zone, induced salinity buildup is not as much of a problem because the higher rainfall levels help flush out the accumulated salts. However, excessive water use and poor drainage cause problems of waterlogging in this zone. Waterlogged fields have lower productivity levels because of lower decomposition rates of organic matter, lower nitrogen availability, and accumulation of soil toxins. These issues are discussed further along in this section.

Can salinity problems be reversed?

Once salinity has set in, it becomes very difficult and expensive to reverse it. Salts have to be flushed out of the soil and drained out of the area. This process, which can be very expensive, requires large quantities of fresh water and a drainage system in place. Often, retiring saline lands from irrigated agriculture may be more cost effective than trying to fix them. The opportunity costs of fresh water are very high, both for agricultural and non-agricultural purposes; hence de-salinization may not be the socially optimal use of this water. Similarly, where scarce national resources are to be diverted to the construction of drainage systems, the returns are higher for such investments on systems that are productive, rather than 'dead' systems.

The problem of induced salinity ought to be managed by addressing the causes of the problem – these are poor system design and inefficient use of irrigation water. For systems that are already in place, improving irrigation-water use efficiency would lead to a significant slowdown in salinity buildup. An essential component in improving water use efficiency is pricing irrigation water at its 'true' cost. The opportunities for improving water use efficiency are discussed further in Chapter 9. For new systems, drainage systems have to be planned and costed out right from the start, even if this means that many new systems may not be cost effective.

Changes in soil nutrient status

The most commonly observed effect of intensive rice monoculture systems is a decline in the partial factor productivity of nitrogen fertilizer. Recent work at IRRI (Cassman *et al.*, 1994) indicates that this is due to a decline in the nitrogen supplying capacity of intensively cultivated wetland soils. In addition, increased incidence of phosphorus, potassium and micro-nutrient deficiency is also observed due to a lack of nutrient balance in fertilizers applied (De Datta *et al.*, 1988).

Dynamics of soil nitrogen supply

Fertilized rice obtains 50–80 percent of its nitrogen requirement from the soil; unfertilized rice obtains an even larger portion, mainly through the mineralization of organic matter (De Datta, 1981). The nitrogen-supplying capacity of the soil depends on the previous cropping history and residue management, the quantity and quality of soil organic matter, and the moisture regime which affects the composition and activity of the microflora and fauna that govern the decomposition of soil organic matter and crop residues. In continuous cropping of flooded soils with two and sometimes three crops each year, organic matter is conserved or increased even when all straw is removed at harvest. This is due to the large carbon inputs from the aquatic biomass, such as the green and blue-green algae, and to a rate of organic matter decomposition slower than for dry soils. Despite this conservation, the soil nitrogen-supplying capacity decreases due to chemical changes in the organic matter and effects of flooded soils on microbial activity (Cassman *et al.*, 1994). The soil's capacity to provide nitrogen to the plant declines with continuous (two to three crops per year) flooded rice cultivation systems.

Declining soil N supply results in declining factor productivity of chemical nitrogen, since soil N is a natural substitute for chemical nitrogen. Holding fertilizer levels constant, intensive rice culture can over time reduce the soil's ability to meet the plant's nitrogen requirement, especially in the later period of crop growth, thereby affecting crop yields. Farmers have been increasing the amount of chemical fertilizers applied in order to maintain their yield levels. Declining soil nitrogen supply is not caused by the prolonged use of chemical fertilizers, but rather because of intensive rice monoculture systems themselves. This is evident when one looks at the yield trends on long-term continuous cropping experiments without chemical nitrogen use; the yields on these plots are also declining over time (Cassman *et al.*, 1994).

The magnitude of yields foregone due to declining soil nitrogen supply are estimated by Cassman and Pingali (1995). Using long-term experiment data from the IRRI farm, Cassman and Pingali estimate the decline in yields to be around 30 percent, over a 20-year period, at all nitrogen levels. In other words, the response function to nitrogen shifts downwards due to declining nitrogen-supplying capacity of the soil. How can these trends in nitrogen productivity be reversed? The primary leverage point is in the cropping system, a break in the continuous flooded rice cycle with a dry season non-rice crop (a crop that does not require standing water). In the southern coastal plains of China, a cropping system consisting of two crops of rice followed by a crop of barley or soybean, practiced over 18 years, has maintained a high and stable crop yield (Li, 1993, as cited in Kundu and Ladha, 1995).

The choice of the crop grown in the dry season determines the level of contribution of the crop rotation to enhanced soil N supplying capacity. In the double rice cropped areas of India, a short-duration legume such as cowpea and mungbean grown as part of the rotation (after the second crop of rice) increased yields by 1 t ha^{-1} and saved 30 kg N ha^{-1}.

Macro-nutrient deficiencies

In addition to nitrogen, phosphorus and potassium are the two other macro-nutrients demanded by the rice plant. Phosphorus and potassium deficiencies are becoming widespread across Asia in areas not previously considered to be deficient. These deficiencies are directly related to the increase in cropping intensity and the predominance of year-round irrigated rice production systems. For example, in China, it is estimated that about two-thirds of agricultural land is now deficient in phosphorus, while in India nearly one-half of the districts have been classified as low in available phosphorus (Stone, 1986; Tandon, 1987; Desai and Gandhi, 1989). Desai and Gandhi note that this is due to the emphasis on nitrogen rather than a balanced application of all macro-nutrients required for sustaining soil fertility. The result of unbalanced application of fertilizers has been a decline in the efficiency of fertilizer use over time (Ahmed, 1985; Stone, 1986; Desai and Gandhi, 1989).

Micro-nutrient deficiencies and soil toxicities

Perennial flooding of ricelands and continuous rice monoculture lead to increased incidence of micro-nutrient deficiencies and soil toxicities. Zinc deficiency and iron toxicity are the ones most commonly observed in the tropics. Waterlogging and salinity buildup aggravate these problems. In Asia, zinc deficiency is regarded as a major limiting factor for wetland rice on about 2 million hectares (Ponnamperuma, 1974). These are mainly soils of low zinc content. Soils that are not initially of low zinc content also show signs of induced zinc deficiency due to rice monoculture. Drainage, even if temporary, helps alleviate this deficiency by increasing zinc availability (Moormann and van Breemen, 1978; Lopes, 1980).

Most ricelands do not start off with any soil toxicities; however, toxicities may build up in some soils due to continuous flooding, increased reliance on poor-quality irrigation water and impeded drainage. Iron toxicity is the most commonly observed soil toxicity due to intensive rice cultivation. Other toxicities are related directly to chemical content and pollutants in the incoming irrigation water. For instance, boron toxicity has been observed on the IRRI farm due to high levels of boron in the irrigation water.

Rice researchers in Asia and US-based rice researchers familiar with Asia scored the rice production problems that they identified as significant for South and Southeast Asia (Barker and Duff, 1986; Herdt and Riely, 1987). A computation was made of the value of rice production forgone to each problem. Among the soil-related problems, phosphorus, zinc and sulfur deficiencies were ranked as the top three problems, accounting for forgone production of 5 million, 3.6 million and 2.5 million tons of rice per year, respectively. Salinity was ranked the fourth and accounted for forgone production of 2 million tons, while iron toxicity accounted for forgone production of 1.6 million tons of rice annually. Jointly these soil-related problems account for approximately 14 million tons of rice production forgone annually, around 10 percent of the total rice produced in Asia.

Once diagnosed at the farm level, micro-nutrient deficiencies are relatively straightforward to correct. Zinc deficiencies can be corrected by adding zinc, for instance. Diagnosis is itself not easy though; quite often micro-nutrient deficiencies are misdiagnosed as pest-related damage. In the case of soil toxicities, farm-level diagnosis is equally complicated and corrective actions are not as straightforward. In both cases, however, the problem ought to be attacked at the cause rather than the cure stage. Periodic breaks in rice monoculture systems and improved water use efficiency go a long way towards reducing the incidence and magnitude of the above problems.

Water quality

Rice productivity can be negatively affected by waterborne pollutants, both physical pollutants (such as silt and mine tailings) or dissolved chemicals that cause soil toxicities. Pollutant concentrations in irrigation water have been increasing due to: the degradation of the watersheds that replenish the irrigation systems (Castañeda and Bhuiyan, 1988; De Vera, 1992); industrial pollutants discharged into the river system; and increased pumping of brackish groundwater. The factors driving these negative externalities and their impact on agricultural productivity will be discussed in Chapter 5.

Long-term changes in soil physical characteristics

Seasonal cycles of puddling (wet tillage) and drying, over the long term, lead to the formation of hardpans in paddy soils. The hardpan refers to compacted subsoil that is 5–10 cm thick at depths of 10–40 cm from the soil surface. Compared to the surface soil, a plow pan has higher bulk density and less medium to large-sized pores. Their permeability is generally lower than that of the overlying and deeper horizons. The formation of hardpans makes it difficult to grow non-rice crops after rice in a cropping system; for the rice crop, it contributes to impeded root growth and ability to extract nutrients from the subsoil, and leads to the buildup of soil toxicities due to the perennial waterlogged condition of the soil layer above it.

A striking example of the problem of hardpans is seen in the rice–wheat cropping system in South Asia. With the advent of short-duration rice and wheat varieties, over 12 million hectares of paddy lands are grown to wet-season rice followed by a dry-season wheat crop (Hobbs and Morris, 1996). The productivity of the wheat crop is affected by the poor establishment of wheat after puddled rice. If the hardpan is broken through deep tillage and soil structures are improved through the incorporation of organic matter, it affects the productivity of the subsequent rice crop by reducing water-holding capacity of the soil. Intensification has reduced the flexibility of dry-season crop choice by changing the soil physical structure.

Increasing losses due to pests

The use of purchased inputs for plant protection was unimportant for rice prior to the mass introduction of modern varieties. Farmers had traditionally relied on host-plant resistance, natural enemies, cultural methods, and mechanical methods such as handweeding. Relatively minor pests – leaffolder, caseworm, armyworm and cutworm – started to cause noticeable losses in farmers' fields as area planted to modern varieties increased. The green leafhopper and brown planthopper (BPH) became major problems, the former as a vector of the rice tungro virus (RTV) and the latter as a direct result of insecticides killing its natural enemies (Teng, 1990). Soil pests, especially root nematodes, have also increased with intensification (Prot *et al.*, 1992). Root nematodes hamper root growth and thereby affect rice yields. Pest buildup in irrigated rice systems is related to: continuous rice cultivation; increased asymmetry of planting; uniformity of varieties cultivated; and injudicious pesticide use. Heong *et al.* (1992) argue that prophylactic pesticide application has led to the disruption of the pest–predator balance and a resurgence of pest populations later in the crop season.

Injudicious and indiscriminate pesticide application is related to policies that have made these chemicals easily and cheaply accessible. Rola and Pingali (1993) have argued that pesticide use has been promoted by policymakers' misperceptions of pests and pest damage. Policymakers commonly perceive that intensification of rice cultivation and modern variety use necessarily lead to increased pest-related crop losses and that modern rice production is therefore not possible without high levels of chemical pest control.

Perceptions on high crop losses are based on experiences from early modern varieties that were susceptible to pest damage. Most varieties released since the mid-1970s are highly resistant to a broad spectrum of pest infestations, yet the perception of a close link between modern varieties and pesticides persists. This perception has led to the promotion of pesticide use through subsidies and credit programs. Pesticide mismanagement, disruption of the pest–predator balance, and increased pest losses are the result.

Even with appropriate policies, there are two areas of growing concern with respect to intensification and pesticide use. The first is the increasing incidence of plant diseases, especially in association with high fertilizer use. The second is the growing incidence of pest resistance to chemicals, especially in the case of weeds. Farmers targeting high yields usually apply high levels of chemical fertilizers. The resulting luxuriant and dense rice canopy is more susceptible to disease infestations, such as blast, sheath blight and bacterial blight (Teng, 1990). High disease incidences have been observed in Japan, Korea, Vietnam and China. Fungicide use has been increasing in these areas and is expected to follow suit in other parts of Asia as higher yields are sought.

The increasing importance of herbicides in Asian rice production with the shift from transplanting to direct seeding brings with it a concern for changes in weed ecology and possible emergence of herbicide resistance. Moody (1994) has reported the increased incidence of 'red rices' in direct-seeded areas in the Philippines. These weedy forms of rice are taller than cultivated rice, lodge, and have different grain characteristics. Rice seeds carried on the farm from season to season become the main reservoir of weed seeds and the main cause of farm-level weed buildup over time. In countries with high rice herbicide use, such as the United States, there are signs that some of the weed species are becoming resistant to herbicides. As herbicide use increases in Asia such trends may also emerge.

Intensification Policies: Where Did They Go Wrong?

Ironically, the very policies that encouraged increased food supply through intensive monoculture systems also contributed to the declining sustainability of these systems. Rice policies operated under two presumptions: (i) that the lowlands are resilient to intensification pressures, and that they could sustain productivity growth indefinitely; and (ii) that modern technology provided a 'silver bullet' solution to food supply problems of Asia. Traditional farming systems were sustainable because of lower intensities of cultivation and because they benefited from a stock of farmer technical knowledge, about the crop and paddy resource base, built over millennia. Neither science nor farmer knowledge was able to predict the changes imposed by intensification and modern technology use on the biophysical resource base.

Massive investments in irrigation infrastructure were essential for the success of the Green Revolution. Without these investments rapid growth in productivity would not have been possible. However, in retrospect, it is clear that many of the systems were poorly designed. Drainage investments were deliberately left out of irrigation projects to keep the cost down (NAS, 1989). The long-term environmental costs of irrigation investments were rarely accounted for and hence the social returns from these investments were over-estimated. Timmer (1991, p. 9) states that 'the track record of incorporating environmental and public health costs into the design and evaluation of irrigation projects is dismal, whether the projects were funded by external donors such as the World Bank or came directly from the country's own budget'. What is surprising, retrospectively, is that the consequences of poorly designed irrigation systems could be predicted prior to the Green Revolution period (USDA, 1954).

It was not just the design of irrigation systems that was problematic; system management, water allocation and water pricing also contributed to environmental costs of irrigated agriculture. Water was, and still is, essentially 'free', and therefore used beyond the social optimum. Increasing water use efficiency has substantial environmental benefits (as discussed in the last section) and does not adversely affect yields, yet this leverage for improving the sustainability of the resource base was not used anywhere in Asia. Constraints to volumetric pricing of water are discussed in detail by Rosegrant and Binswanger (1994).

The dual goals of food self-sufficiency and sustainable resource management are often mutually incompatible. Policies designed for achieving food self-sufficiency tend to undervalue goods not traded internationally, especially land and labor resources. As a result, food self-sufficiency in countries with an exhausted land frontier could come at a high environmental cost.

Government intervention in the rice market, especially through output price support and input subsidies, provided farmers with incentives for increasing rice productivity. In addition to highly subsidized irrigation water, Asian farmers benefited from 'cheap' fertilizers, pesticides and credit (see Monke and Pearson, 1991, for an example of Indonesian price policy). The net result was that rice monoculture systems were extremely profitable through the decades of the 1970s and the 1980s, despite a long-term decline in the real world rice price through this period[4].

Input subsidies directly affect crop management practices at the farm level; they reduce farmer incentives for improving input use efficiency. Improving farm-level technical efficiencies requires farmer investment in learning about the technology and how best to use it. Where input prices are kept low through government intervention, farmers would not have the incentive to spend the time to learn about methods of increasing technical efficiency. The injudicious use of fertilizers, pesticides and water is a case in point. The persistence of inefficient input use leads to several of the ecological problems discussed earlier in this chapter.

Rice production policies of the Green Revolution era were enormously successful in moving the food deficit countries of Asia to food self-sufficiency and in some cases food surplus. These policies were often unidimensional and formulated in a crisis mind-set. The need today is for a more holistic approach to food production policies that explicitly account for externalities and long-term system sustainability.

Conclusions

The current problems of sustaining productivity growth in the Asian lowlands indicate that they are as susceptible to environmental degradation as the more fragile uplands. Environmental impacts of intensification on the lowlands, although not easily observed, tend to have a significant long-term impact on food production and food supply. For instance, if current rice yields on the irrigated lowlands of Asia dropped by 5 percent, the impact on total rice production would be 36 million tons per year.

The problem of sustainability of the lowlands does not emerge because of the technology being used but rather because of the intensity of land use itself and the choice of crops. Intensive rice monoculture systems over the long term are not sustainable without adequate changes in current technologies and management practices. The argument that rice has been sustainably cultivated in China, Japan, etc. for centuries is not an argument against the current problems being faced in Asia. In these countries, until recently, one crop of rice was grown per year, and the fallow period during the dry season allowed the land to be rejuvenated.

The problem of sustaining productivity growth comes about because of inadequate attention to understanding and responding to the physical, biological and ecological consequences of agricultural intensification. The focus of research resources ought to shift from a fixation on yield improvements to a holistic approach to the long-term management of the agricultural resource base. When yield per hectare is used as the only measure of productivity growth and the 'true' costs of production are not considered then research resource allocation will be biased away from understanding the systemic problems causing productivity stagnation or decline. It is unlikely that there will be quick answers for reversing the current negative trends in productivity growth, and sustained research investments are essential.

Sustaining productivity gains in the post-Green-Revolution era will have to come from more efficient use of inputs, including land and labor. Technologies for enhancing input efficiencies (such as better fertilizer management, integrated pest management, etc.) are more knowledge intensive and location specific than the modern seed–fertilizer technology that was characteristic of the Green Revolution. Productivity gains accrue to farmers who: have the ability to learn about the new technologies; discriminate among technologies offered to them by the research system; adapt the technologies to their particular environmental conditions; and provide supervision input to ensure the appropriate application of the technology.

The Green Revolution strategy of increasing lowland productivity has to a large extent relieved the pressure on the uplands by providing employment opportunities for migrant labor. However, if the current trend towards stagnation/decline in lowland productivity persists then one could expect a decline in employment opportunities in the lowlands and hence increased pressure on the uplands.

Notes

1. There is no evidence to indicate that, for irrigated rice, adding more plant traits has resulted in any yield sacrifice.
2. Parallel incidences of salinity buildup in US irrigated lands can be found in NAS (1989).
3. In a three-decade period since 1953, about 33,000 hectares of the service area of the Tungabhadra irrigation project in Karnataka State of India have been affected by waterlogging and salinity buildup. In these areas, farmers have switched from upland crops to rice, which is more tolerant, but their yields are much lower compared to unaffected areas. About 20,000 hectares had to be abandoned and the annual rate of increase of affected area is estimated to be 6,000 hectares. Similarly, in the Nagarjunasagar irrigation project area in Andhra Pradesh, India, nearly 18 percent of the 140,000 hectares of service area have been affected by waterlogging and salinity in 14 years.
4. In addition to private profitability, input subsidies could also increase social profitability. Consider the case of Indonesia, during the 1970s, when oil prices were high and there was a real danger of agriculture being affected by the Dutch Disease. Indonesian planners deliberately avoided the Dutch Disease by subsidizing agriculture to maintain its private profitability. If this was not done, Indonesia would have gone the way of Nigeria and Venezuela, in drawing people out of the agricultural sector during the oil boom and then not having a viable agricultural sector when the oil market busted.

5

Erosion, Pollution and Poison: Externalities and Rice

Introduction

While unit costs of production fall with agricultural intensification and modern technology use (Chapter 3), environmental costs tend to rise. Intensification-induced environmental externalities are soil erosion for the uplands, water and air pollution and pesticide poisoning for the lowlands. Upland degradation also imposes a negative externality on the lowlands through increased sediment buildup in irrigation infrastructures. Environmental externalities generated within a watershed transcend sectors of economic activity; for example, intensive utilization of the upper watershed for agriculture and mining affects fish production in lowland lakes because of increased lake water turbidity due to high silt content.

While intensification of the lowlands could have significant negative externalities associated with it, it could also have significant positive environmental benefits, the most important of which is the release of the fragile uplands from intensification pressures. Because of increasing productivity of the lowlands and the consequent employment generation, subsistence production systems in the uplands, especially upland rice production, have declined over time in most Asian countries.

In evaluating the returns to intensification and modern technology use, 'true' economic costs ought to be considered, including production and environmental costs. The rate of return to agricultural intensification and the adoption of modern yield-increasing technology will be over-estimated when these costs are not considered. This chapter offers several examples in which the environmental consequences of land and technology use were quantified.

Specifically, this chapter provides the following:

1. A watershed-level overview of the sources of environmental externalities by human activity, and highlights of the interlinkages across the transect and across sectors.
2. A discussion of the determinants of upland degradation and the consequent incidence of soil erosion.
3. A quantification of the relationship between upper watershed exploitation and the degradation of lowland irrigation infrastructure.
4. An assessment of the impact of watershed utilization, for agricultural and non-agricultural purposes, on freshwater fish production.
5. An economic evaluation of the health impact of pesticide use and a re-assessment of the costs and benefits of using pesticides for intensive rice production.

Environmental Consequences of Watershed Utilization

Environmental consequences of watershed utilization ought to be discussed separately for subsistence systems and commercial systems. In the case of subsistence systems, the driving force behind watershed exploitation is population growth and rising population densities in a particular watershed. Subsistence utilization of the watershed is mainly restricted to food production and human habitation. The driving force behind the commercial utilization of a watershed is enhanced trade opportunities, which come about mainly through improved transport infrastructure. In addition to food and agricultural production, commercial uses of watershed resources include logging (both legal and illegal), mining and mineral extraction, industrial and residential uses. This section describes the various economic activities along a watershed transect and points out the negative externalities created. The exact process by which the externalities are generated is discussed in the next three sections.

For millennia, forest farmers in Southeast Asia had been practicing a low-technology, yet sustainable, system of shifting cultivation. Small patches of forest were felled and the woody material burned if it was dry enough. Crops were planted on the open land, and a series of harvests and replantings were carried out until yields declined substantially or weed and pest invasions become too great (2–3 years). The area was then abandoned and allowed to lie fallow, and forest growth took over as a new plot was felled and planted. If the period of fallow was long enough for the forest to restore soil fertility and develop a tall, closed canopy to eliminate weeds and brush (at least 7 years), the forest farmer could return and initiate the process again. This had been a stable system throughout Asia and the Pacific, with farming areas of only 1–2 hectares (Hamilton, 1984).

Boserup (1965) described the process of intensification of shifting cultivation systems as population densities increased. Progressive reductions in fallow periods, induced by population growth, resulted in the evolution of land utilization from shifting cultivation to permanent agriculture. Pingali and Binswanger (1988) have argued that the long-term returns to intensification vary by slope and soil type, for a given agroclimatic zone[1]. Intensive agricultural systems are least profitable on the upper slopes due to relatively shallower soils and higher risk of drought and soil erosion. Consequently, intensive production systems in subsistence societies have generally been observed in the mid- and lower slopes of a watershed (Pingali *et al.*, 1987). Upper watershed areas were generally reserved for forestry and long fallow shifting cultivation systems. The process by which this equilibrium in watershed land management breaks down is described in the second section of this chapter.

Improved transport and market infrastructure results in increased commercial utilization of watershed resources. The holistic management of resources by local institutions gives way to centralized management and regulation by the state of individual sectors such as forestry, agriculture, mining, etc. Communication across sectors becomes sparse and understanding of inter-sectoral linkages is limited. The consequence is high levels of resource exploitation and environmental externalities. A stylized version of watershed utilization with improved transport infrastructure is shown in Table 5.1.

Environmental consequences of watershed exploitation are most easily observed in terms of soil erosion/sedimentation, water pollution, air pollution and pesticide-related health impairments. Environmental risks of human exploitation of the watershed vary by location along the watershed transect (Table 5.1). With the exception of air pollution effects, all other consequences of watershed management are felt mainly within the particular watershed. Air pollution effects, specifically methane emissions from rice paddies, and their impact on climate change are discussed in detail in Box 5.1.

The predominant environmental risk associated with the intensive exploitation of the upper slopes of the watershed is soil erosion. Shifting cultivation has most often been cited as the cause of soil erosion from the upper slopes. The next section of this chapter argues that it is not shifting cultivation *per se* that is the problem, but rather the occupation and exploitation by landless households of public forest land cleared and abandoned by illegal loggers. Lack of tenurial security has prevented farmers from making land investments necessary for

Table 5.1. Watershed utilization and externalities.

	Upper slopes	Mid-slopes	Lower slopes	Streams, rivers, lakes
Determinants of land use change				
Increasing population	Shifting cultivation	Intensive farming; human habitation	Intensive wetland rice systems; human habitation	Captured fisheries; water for domestic uses; water for irrigation
Improved transport infrastructure	Vegetable/fruit production; commercial logging	Mining; industrial estates	Aquaculture; industries	Cultured fisheries; water for domestic and industrial uses; water for irrigation
Externality risks				
Soil erosion	***	**	*	n.a.
Sediment buildup	n.a.	**	***	***
Polluted surface water	*	*	**	***
Contaminated groundwater	*	**	***	n.a.
Pesticides and human health	**	***	***	n.a.
Methane emissions	*	**	***	n.a.

*, **, ***, Relative importance of the particular externality by slope.
n.a., Not available.

reducing soil loss. Erosion problems have been aggravated by the increased production of high-value vegetable crops on the upper slopes, especially during the cooler months of the year in the rapidly growing countries of East and Southeast Asia.

Soil erosion from the upper watershed results in increased silt and sediment flow through the water systems of the mid- and lower slopes. The consequence for irrigated agriculture is reduced reservoir capacity and poorer water delivery due to canal siltation (see Downstream Effects of Soil Erosion, p. 99). Sediment flow also affects the quality of

Box 5.1. Global climate change: trade-off between rice production and methane emissions.

Significant global warming is predicted for the coming decades due to increases in anthropogenic atmospheric greenhouse gases (such as carbon dioxide (CO_2), methane (CH_4), chlorofluorocarbons (CFCs), among others) (IPCC, 1990). Of immediate concern to policymakers and scientists worldwide is the potential effect of climate change on the world's agriculture, and its major impact on food production (Rosenzweig *et al.*, 1992). In order to meet the demands of a growing human population, agricultural productivity must continue to increase at the rate of 4 percent per annum. If food production is adversely affected by global environmental changes, serious long-term food shortages could result.

Most of the environmental consequences of intensive watershed use as discussed above are felt locally, i.e. within that particular watershed. Concerns about global climate change, however, suggest global effects of local land use decisions, particularly the intensive use of lowlands for paddy production (Ranganathan *et al.*, 1994). By absorbing infrared radiation, greenhouse gases – carbon dioxide, methane, nitrous oxide and chlorofluorocarbons – can increase global mean temperatures. Climate change also threatens the irrigated lowlands with predictions of increased levels of ultraviolet-B (UV-B) and carbon dioxide. If atmospheric methane continues to increase at the current rate of 1–2 percent per year, it is expected to contribute more to future climate change than any other gas, with the exception of carbon dioxide.

Rice paddies, especially the irrigated rice ecosystem, have been identified as one of the main anthropogenic sources of methane. As in natural wetlands, flooded ricefields result in an oxygen-depleted soil wherein methane is a major end-product of organic matter fermentation. In wetland rice soils, vertical transport of methane to the atmosphere is enhanced by well-developed intercellular air spaces (aerenchyma) in the leaf blades, leaf sheath, culm and roots of the rice plant, making rice paddies a major source of methane emissions (Neue *et al.*, 1995).

The potential methane emissions from rice paddies have long been noted, but comprehensive measurements and studies on the factors controlling such emissions are recent (IRRI, 1992; 1993). Current research suggests that factors contributing to increases in rice production, such as incorporation of crop residues, inorganic fertilizer use and crop establishment practices, also enhance methane emission (Neue *et al.*, 1994; Neue and Sass, 1996). Concerns about long-term sustainability of rice production systems are being addressed by mitigation options such as water and fertilizer use efficiency. The potential trade-off between rice production and methane mitigation is therefore high, as mitigation options require changes in management practices that may affect rice production and productivity. In the end, the impact of climate change on rice production can change the latter's profitability, and add to the existing concerns of maintaining global food security.

The warm humid tropics of Asia are expected to be most adversely affected by global climate change, in terms of rice supplies. On the other hand, temperate rice-growing regions such as northern China, Japan and the Korean Peninsula could benefit from global warming, which would lengthen the season over which rice could be grown without risk of cold damage. However, even those potential benefits are not guaranteed. Much of the interaction of rice with climate change remains unresolved, and needs to be well understood. Predictions of future climate, food production and water supply at acceptable levels of accuracy and scale of resolution can only be achieved with integrated and global change research. Predicting future methane emission trends from rice agriculture, and developing mitigation technologies in accord with the needed increase in rice production, requires more information on processes affecting methane fluxes, and the geographic distribution of factors controlling these factors (Neue *et al.*, 1995). The growing demand for rice, and the real possibility of unreliable supplies, makes it imperative to assess the impact of enhanced UV-B and climate change on rice production and productivity, so as to set in place policies that respond to the threat of changing climate by minimizing its negative effects.

surface water resources and thereby the productivity of local fisheries (see Erosion, Effluents and Fish Productivity, p. 104). Where mining and industrial activities are dominant in a watershed, waterborne pollutants from these activities also affect the quality of freshwater resources and irrigation water (see Castañeda and Bhuiyan, 1988, for a Philippine case study; and Gleick, 1993, for a global survey).

An important but sparsely studied concern in Asia is one of pesticide contamination of groundwater due to intensive cultivation on the mid- and lower slopes. Bhuiyan and Castañeda (1995) have found evidence of pesticide residues in the groundwater of Laguna and Central Luzon, Philippines. Rural Asian farmers obtain water for drinking and household uses from shallow aquifers under agricultural lands, such that pesticides in the groundwater are of particular concern. The herbicide butachlor was most commonly found among the samples studied by Bhuiyan and Castañeda. Although the concentrations of herbicides detected were well below the maximum tolerance levels, the increasing use of herbicide in association with rising wage rates (Chapter 9) could lead to higher contamination levels in the future.

Indiscriminate and injudicious pesticide use, particularly insecticide use, could have serious human health consequences (see Pesticides and Human Health, p. 110). The health consequences of pesticide exposure of farm families is documented in Rola and Pingali (1993), Pingali *et al.* (1994), and Pingali and Roger (1995). The loss in farmer productivity resulting from impaired health was quantified for a sample of Laguna and Central Luzon respondents by Antle and Pingali (1994). Pesticide-related health impairments are generally concentrated in areas of high-intensity farming – these are the mid- and the lower slope areas. However, high-value vegetable crop production on the upper slopes could also lead to increased and indiscriminate use of insecticides, and thereby result in detrimental health consequences.

Determinants of Upland Degradation in the Humid Tropics

Soil degradation, defined here as soil erosion and/or a decline in soil fertility, is generally associated with deforestation caused mainly by commercial logging, intensive cultivation and overgrazing induced by population growth. Soil degradation is not a universal problem. First the threat of soil degradation varies by soil type, temperature, rainfall regimes and slope. It tends to be higher in the humid tropics on shallow soils with slope greater than 18 percent. Second, appropriate land use and land investments can prevent the problem, even on high-risk soils. Therefore, degradation problems are mainly restricted to areas where the rate of return to preventive land investments is low. Incentives for corrective land investments depend on the relative land endowments of the region and on the security of land tenure.

In Southeast Asia, rapid deforestation caused by unchecked logging and poor conservation efforts contribute more to soil degradation than population-induced agricultural intensification. Studies in the hill forests of Malaysia show that sediment production rates increase from less than 100 m^3 km^{-2} $year^{-1}$ in undisturbed tropical rainforests to between 388 m^3 km^{-2} $year^{-1}$ and 694 m^3 km^{-2} $year^{-1}$ in recently logged hill forests (O'Laughlin, 1985).

In the Philippines, approximately half the land is classified as 'alienable and disposable', which may be privately owned. The other half, which is mostly 18 percent slope, is public land, classified as 'public forest land'. Of the 15 million hectares of public forest land, only 6 million hectares have significant tree cover, and currently less than one million hectares of old-growth forest remains. Cultivated area in the uplands is approximately 3.9 million hectares or 43 percent of cleared land (Garrity *et al.*, 1991).

Upland cultivation *per se* does not necessarily lead to land degradation; it would set in if proper safeguards for land and soil conservation were not incorporated into the farming system. Conservation investments require the farmer to have a long-term decision horizon, implying long-term and secure access to land.

Without secure long-term rights to land, farmers would not have the incentive to plant trees and perennials for protecting the soil against leaching. As Sajise *et al.* (1988) put it, 'shifting cultivators threatened by insecure land tenure take an opportunistic short-term view of land development'. Noronha (1985) meanwhile points out that, in societies where the ownership of trees is tantamount to the ownership of the land the trees are on, farmers temporarily occupying the land would be prevented from planting trees or perennial crops. Raintree (1987) documents case study evidences from across the developing world on the importance of tenure security for investment in trees. In the humid tropics, where uncertain ownership or tenure prevents the planting of trees, degradation is most likely to occur.

In discussing the impact of land tenure insecurity on the degradation of the upland environment, we ought to separate out tenure security in privately held agricultural lands from tenure security in public forest lands. In agricultural societies with steadily growing populations private rights to land evolve over time. The transition from general cultivation rights to specific land rights with population growth is documented in Miracle (1967), Hopkins (1973), Noronha (1985), Binswanger and Pingali (1987), Dove (1987) and Pingali (1990). Hayami and Ruttan (1985) review the evolution of property rights to land in Japan, Thailand and the Philippines. They find population growth to be instrumental in the process, although expanded possibilities for trade also play a role. The transition to private land rights widens incentives to undertake investments required for the intensification of production and soil fertility maintenance. Formal land ownership as characterized by the possession of title also helps the farmer in acquiring credit for making the necessary investments in the land. Feder and Onchan (1987) and Chalamwong and Feder (1986) provide evidence on land ownership and investment for Thailand.

In agricultural societies, when does the evolution in land rights break down? The normal process of evolution can break down when unusually high population growth rates are experienced, as caused by a sudden increase in migration into the area, and hence an unprecedented increase in the opportunity cost of fallow land. Government policies, especially the movement to socialist economic systems and collectivized agriculture, can circumvent the evolution of land rights (see Hung, 1977, for Vietnam). Where farmers do not have long-term rights to land, the private rate of discount is higher than the social rate of discount, such that private investments made in sustaining land productivity will be lower than those society would consider optimal.

Farmers contemplating soil conservation investments do not expect to reap the full benefits and therefore under-invest.

In addition to inadequate rights to agricultural land, the uncontrolled access to public forest lands is a major contributor to upper watershed degradation. Commercial loggers (both legal and illegal) first pioneer into public lands and clear primary forests. They are then followed by migrant farmers who rent parcels of cleared timberland from the nominal 'owners' (the logging companies) or the 'pseudo-landlords' from nearby towns. Since the migrants do not have a secure long-term tenure on the land, they tend to maximize short-term returns and follow the logging companies to the next site once degradation has set in. For case study evidences from Southeast Asia see ACIAR (1984) and Fujisaka and Capistrano (1985). Fujisaka and Capistrano (1985) report on the increased incidence of erosion caused by the 'get it while you can' attitude of Philippine farmers in Southern Luzon, who cultivated upland areas without a secure long-term tenure to the land.

Governments in the region have made several attempts to improve the tenure security of the migrant upland farmers but with limited success in controlling degradation. In the Philippines, Certificate of Stewardship Contracts (CSC) have been issued to migrant farmers since the early 1980s. However, only 2.5 percent of the upland area has so far been included in the stewardship leases. There are several limitations to this contract arrangement. In general, migrant farmers have not been able to assert their CSC claim against that of the absentee pseudo-landlords. The lease is non-transferable, and hence cannot be used as collateral for loans to invest in farm improvements, and it is only inheritable within the 25-year lease period (DENR, 1990).

Despite secure long-term rights to land, one would expect erosion and degradation to persist in areas where collective action is required for watershed-level investments. The free-rider problems associated with such collective action are self-evident. Group action for making watershed-level investments in erosion control would be possible only where farmers have an incentive to cooperate or where they are coerced to do so. Incentives for cooperation will be higher in relatively closed communities under severe land pressure. Terraces in West Java are an example of such cooperation (Soemarwoto and Soemarwoto, 1984).

Downstream Effects of Soil Erosion

Literature on upper watershed degradation has generally mentioned the externalities imposed on lowland productivity especially in terms of sediment flow affecting irrigation infrastructure (see, for example,

Cruz *et al.*, 1988). The externalities generally cited are soil erosion, sediment buildup in reservoirs, and the increased incidence of flooding. There is no general consensus on the importance of human activity in the uplands as a determinant of the problems created for the lowlands. Some soil scientists argue that soil erosion is caused by natural processes and human activity has little to do with it (Johnson, 1988; Burton *et al.*, 1989) while others implicate human activity more definitely (Sanchez, 1976; Lal, 1980). In any case, it is safe to say that susceptibility to erosion is dependent on soil properties and land management practices (Sanchez, 1976).

Human activity in exploiting the upper watershed would have an impact on soil erosion and therefore would impose negative externalities on the lowlands. The negative impact of upper watershed degradation on irrigation infrastructure through an increase in sediment buildup is easier to observe and quantify.

There are several case studies in Asia on the continued reduction in the storage capacities of reservoirs because of excessive sedimentation rates. The Korangkates reservoir in East Java, Indonesia, which has a storage capacity of 0.343 km^3 of water, has a sediment inflow 18 times more than the design rate, shortening the economic life of the reservoir by two percent each year (Brabben, 1979). In a study of eight large reservoirs in India, the actual sedimentation rates were found to vary between 1.4 and 16.5 times the expected rates (Dogra, 1986).

The Pantabangan watershed mirrors the ravages of soil erosion. It has a total watershed area of 82,900 hectares, 44 percent of which is forested. Grasslands which have taken over formerly forested lands now occupy 44 percent. Croplands take up a mere 12 percent. Prior to dam construction in the late 1960s, sediment yield was estimated to be around 20 tons ha^{-1} yr^{-1}. With widespread forest denudation, this increased to 81 tons ha^{-1} yr^{-1} (Cruz *et al.*, 1988), implying a loss in the dam's service life by 61 years. Similarly, the Magat watershed has some 83 percent of its 414,300 hectares already under severe to excessive soil erosion. Sediment yield at the dam site was estimated at 20 tons ha^{-1} yr^{-1} during planning stage (Coloma, 1984). This has been exceeded by present-day rates of up to 34 tons ha^{-1} yr^{-1} (Cruz *et al.*, 1988). As with other dams similarly affected, Magat dam has lost some 40 years of its service life. There was a proposal to extend service life by constructing a larger sediment storage. This, however, could not be rationalized, the obvious consequence being that the potentially irrigable areas downstream could no longer be serviced.

A study in 1988 by Cruz *et al.* on the on-site and downstream costs of soil erosion in the Magat and Pantabangan watersheds estimated annual erosion rate per hectare for open grasslands at 88 tons, and for other land uses at 28 tons. Since top-soil sheet erosion is about 40 percent of the gross erosion rate (MADECOR, 1985), gross annual erosion per hectare was estimated to be 219 tons for the open grasslands and 71 tons for other land uses.

On the downstream cost of soil erosion, sedimentation is the process that should be taken into account. Where the watershed drains into a major dam and reservoir system providing various services, the heavy impact of sedimentation becomes all the more evident. From an ex-post project perspective, sedimentation reduces project benefits by shortening reservoir and dam service life, and reducing useful storage capacity, as in the cases of Ambuklao Dam and Pantabangan Dam watershed projects. Massive sedimentation of Ambuklao Dam in Luzon, Philippines, resulted in the reduction of its useful lifespan from 60 to 32 years, yielding an estimated loss of 500 million pesos in 1985. To these costs should also be added benefits forgone from lost biodiversity with potential economic value as food, medicine, fuel, fiber and material of industrial usage (EMB, 1990).

Castañeda and Bhuiyan (1988) have shown, for a Philippine watershed, the impact of mining upstream on the quality and quantity of water provided to lowland farmers downstream. The upstream discharge of mine tailings caused abandonment of 24 percent of the system's water conveyance network due to sedimentation and a consequent reduction of about 40 percent of the system's service area. Similarly, Sajise (1986) cited that after 20 years the Agno River Irrigation System (ARIS) in the Philippines now operates at only 25 percent capacity due to siltation of the intake and canal system. ARIS can now irrigate only 9,546 hectares instead of the original 17,500 hectares intended. In 1976, losses from desilting costs and crop failure amounted to US$ 4.4 million (Coloma, 1984). Siltation is largely due to mine tailings discharged by nine mining companies in the watershed.

In India, UNDP/FAO estimated that 2400 million tons of silt are carried through the rivers every year. Vital reservoirs have seen their useful life significantly shortened because of sediment yield from mountain watersheds under human pressure (ADB, 1987). Bed levels of the Terai rivers are rising 15–30 cm annually. Over the 1972–1979 period, Himalayan torrents and streams have more than doubled in width while rivers have widened by more than one-third through erosion and siltation, reducing the capacity to carry floodwater. Many dams are silting up at twice their design inflow rates, reducing their planned life by decades.

The impact of bare soil, bare fallow or fallowing on surface runoff, erosion and sedimentation in four Asian countries are summarized in Appendix Table 5.A1. Depending on slope and land use, the practice of bare fallow increased surface runoff, sediment yield and erosion rate. In these studies, fallowing was compared to other upland farming conditions, namely grassed area, unmulched traditional cultivation, clean cultivation, and to a swidden area (see Appendix 5.A on sustainable farming systems for the densely populated humid tropics).

De Vera (1992) econometrically related upper watershed degradation to irrigation infrastructure for each region of the Philippines. Specifically, this study estimated the impact of forest production, mineral production, and upland rice production on irrigated area and rice area for both the wet and the dry seasons (Table 5.2). The extent to which real expenditures on system maintenance can sustain irrigation

Table 5.2. OLS results on the determinants of irrigation system degradation.

	Wet season		Dry season	
Determinants	Irrigated area	Rice area	Irrigated area	Rice area
Intercept	−2039.60 (9657.47)	−17532.91 (15263.71)	−11918.94** (4575.81)	−40126.59** (17064.19)
Wet-season irrigated area	—	—	0.620*** (0.115)	0.512*** (0.116)
Forest products	−0.010*** (0.003)	−0.011*** (0.003)	—	—
Upland rice	−0.033 (0.091)	−0.012 (0.079)	—	—
Value of mineral production	−0.002* (0.001)	−0.002** (0.001)	—	—
Lag of real rice price	—	29.005* (15.937)	—	29.120 (16.748)
Population density	224.732*** (62.723)	107.371** (54.468)	122.734*** (45.899)	139.263*** (46.995)
Real MOE	1.769*** (0.256)	1.653*** (0.220)	1.340*** (0.320)	1.586*** (0.324)
Regional dummy 1	−2774.807 (5196.079)	5831.208 (4465.894)	−20962.14*** (3293.276)	−24795.03*** (3330.396)
Regional dummy 2	56220.256*** (5197.079)	49806.621*** (4472.520)	14390.242** (6135.634)	15410.270** (6199.948)
Regional dummy 11	4814.971 (5246.165)	7507.559* (4486.266)	−910.916* (1739.077)	−2000.983** (1736.166)

***, **, *, Significant at 1, 5 and 10 percent respectively.
Figures in parentheses are the standard error of estimates.
Source: De Vera (1992).

infrastructure, given the externalities imposed by upper watershed degradation, was also assessed.

A time series cross-section data set consisting of regional aggregates for each of the 12 regions over the 1979–1988 10-year period was used for the analysis. Forest production and mineral production had significant negative effects on wet-season irrigated area. Upland rice area had a negative but insignificant effect. The effect would have been stronger if data for all upland crops were available. Real maintenance expenditures and regional population density had significant positive effects on wet-season irrigated area. If population density reflects regional land pressure, irrigation infrastructure is higher in more densely populated provinces. In essence, results showed that the degradation of national irrigation systems was significantly influenced by upper watershed destruction in addition to reduced maintenance

Table 5.3. Determinants of irrigation infrastructure degradation: elasticity estimates.

		Wet season		Dry season	
	Service area	Irrigated area	Rice area	Irrigated area	Rice area
	(565,058)	(409,277)	(382,424)	(329,974)	(310,740)
Wet-season irrigated area (409277)	—	—	—	0.7686	0.6739
Forest products (5599630)	–0.1102	–0.1364	–0.1576	–0.1048	–0.0992
Upland rice area (157121)	–0.0163	–0.0127	–0.0051	–0.0098	–0.0032
Value of mineral production (13050370)	–0.0637	–0.0695	–0.0710	–0.0534	–0.0447
Population density (143.57)	0.0934	0.0788	0.0403	0.0534	0.0643
Real MOE (62149)	0.2465	0.2686	0.2687	0.4589	0.4865
Lag of rice price (919.12)	—	—	0.0697	—	0.0861

Figures in parentheses are mean values.
Source: De Vera (1992).

investments. Increased production of logs, upland agriculture and mining were among the watershed activities identified to cause irrigation system degradation, the adverse effect of which is a loss in total rice production.

A 1 percent increase in forest product production, mineral production and upland rice area leads to declines in wet-season irrigated area of 0.14 percent, 0.07 percent and 0.01 percent respectively (Table 5.3). The corresponding declines in wet-season rice area are 0.16 percent, 0.07 percent and 0.01 percent respectively. To the extent that upland rice area is a small proportion of all upland crop area, the elasticity is under-estimated. Meanwhile, a 1 percent increase in real expenditures for maintenance of irrigation infrastructure leads to a 0.27 percent increase in wet-season irrigated area and a 0.27 percent increase in wet-season rice area.

Since there is no runoff from the upper watershed during the dry season, irrigation infrastructure in the dry season is a function of the wet-season infrastructure. A 1 percent increase in wet-season irrigated area leads to a 0.77 percent increase in dry-season irrigated area and a 0.67 percent increase in dry-season rice area. The effects of upper watershed degradation on dry-season infrastructure are evaluated through their wet-season effects.

The net effects of upper watershed degradation on irrigation infrastructure were evaluated by accounting for the positive effect of real maintenance expenditures. Extrapolation of the results indicates that, for the Philippines as a whole, upper watershed degradation leads to an incremental loss of 4200 ha yr^{-1} of wet-season irrigated land and 2700 ha yr^{-1} of dry-season irrigated land.

The above results on negative externalities on irrigation infrastructure imply that increases from irrigation maintenance investments are overwhelmed by the externality caused by forest degradation, due to forest, upland rice and mineral production. These results also imply that the long-term productivity of the lowlands cannot be viewed independently of upland degradation. On the contrary, increased upland clearing could diminish lowland productivity because of impaired hydrologic conditions, erosion and siltation problems.

Erosion, Effluents and Fish Productivity

The impact of watershed management is seen ultimately in the state of its water and aquatic resources. In addition to soil erosion, resulting from the poor management of the upper watershed, agricultural and industrial effluents also lead to adverse impact on the water system. Common agricultural effluents are livestock wastes and agricultural drainage water concentrated with chemical fertilizer and pesticide residues. Polluted water systems, whether from sediment or other sources, have lower levels of oxygen and nutrients essential for aquaculture, and therefore lead to a reduction of the quantity and quality of these resources. Laguna de Bay, Philippines, has become a classic example of the consequences of indiscriminate management and disposal of watershed resources. The lake, which has an area of 90,000 hectares, was a major source of fish supply for Metro Manila (see map, Fig. 5.1). Fish production in the lake dropped from 350,000 metric tons in 1963 to 202,000 metric tons in 1980 and further declined to 147,000 metric tons in 1988 (Valerio, 1990).

Valmonte (1993) quantified the impact of watershed utilization, for agricultural and industrial purposes, on fish production in Laguna de Bay. Private decisions of farmers, loggers, homeowners and industrialists along the lake's watershed have resulted in high social costs in

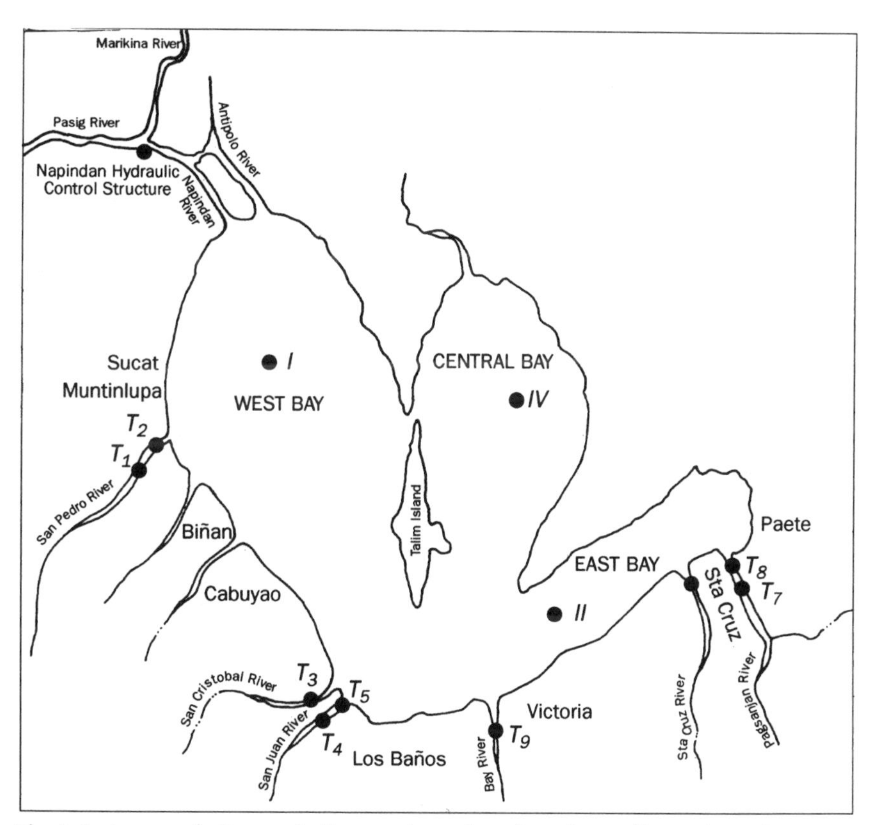

Fig. 5.1. Laguna de Bay and tributary rivers: location of sampling points. (Source: Valmonte, 1993.)

terms of water quality and sustainability of fishery resources. The following is a summary of the Valmonte study.

The transect of the Laguna de Bay watershed (Fig. 5.2) shows that it is used for agricultural, urban, industrial and forestry activities. Agricultural intensification and industrial development in the watershed have resulted in: (i) increased erosion and sediment flow due to land clearance and poor agricultural practices, especially on marginal lands in the upper parts of the watershed; (ii) accelerated eutrophication (nutrient enrichment) of lake water due to increased loads of nitrogen and phosphorus in the runoff water from paddy fields and poultry and hog enterprises; and (iii) increased pollution of heavy metals and pesticides from the disposal of industrial wastes and pesticide concentrations in the agricultural drainage water.

The biological carrying capacity of a water body is measured by net primary productivity (NPP) which indicates the amount of food (algae) available to fish at a given point in time (Payne, 1986). Fish production is positively related to NPP, which in turn is influenced by the nutrient content of the water and its turbidity. While phosphorus and nitrogen nutrients are required for algal growth, concentrations

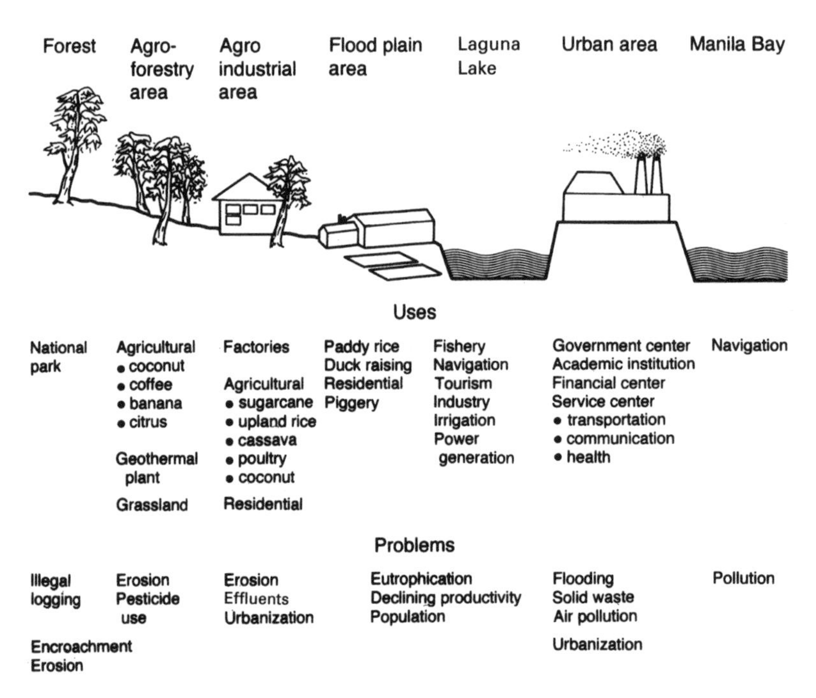

Fig. 5.2. General transect of the Laguna de Bay watershed. (Source: Valmonte, 1993.)

above a very small level lead to algal blooms. Excessive algal growth reduces light penetration and deters the replenishment of dissolved oxygen in the water, causing fishkill due to low oxygen supply. Nitrates are contributed by agricultural runoff including pig and poultry wastes, domestic wastes including human excreta, and industrial effluents. Phosphorus contributions to the lake come from agricultural drainage including livestock and poultry operations; domestic sewage including phosphates from household detergents; and industrial effluents. Time trend estimates presented in Valmonte (1993) indicate that the level of nitrates and phosphorus in the lake water has increased over the 11-year period from 1978 to 1988.

Water turbidity is caused by high levels of suspended particles in the water, which could occur because of high algal growth, high sediment inflow and/or effluent inflows. An increase in water turbidity hinders light penetration leading to, as discussed above, poor photosynthetic activity and low net primary productivity. Lake water turbidity has been increasing over the 1978–1988 time period, especially for the dry season (Valmonte, 1993). There was a clear positive relationship between the levels of nutrients and water turbidity in the rivers draining into the lake and the levels in the lake water. The river–lake relationship was particularly strong for the dry season. Lake water quality and lake productivity are directly related to the quality of water in the rivers, indicating the strong influence of watershed activities on the lake.

Using the monthly water quality data over the 1978–1988 period provided by the Laguna Lake Development Authority (LLDA), the effect of dissolved nitrates and phosphates (NPHOS) and turbidity (TURBI) levels on net primary productivity (NPP) of the lake water was estimated. The estimate shown below indicates a significant negative relationship between the level of pollutants in the water and its biological productivity.

$$NPP = 1.2628^{***} - 0.00068^{**}\ NPHOS\ (lag) - 0.01092^{***}\ TURBI$$

(degrees of freedom = 130 and F-value= 14.58^{***})

***, **, significant at the 1 and 5 percent levels respectively

Net primary productivity is an indicator of water quality and its conduciveness to fish production. There is a direct relationship between the biological fish yield of a water body and its net primary productivity (Payne, 1986). However, data on biological fish yield was not available from the LLDA; available data was on the actual amount of fish caught. Quantity of fish caught is a function of the biological yield (or net primary productivity) and fish price expectations. In other words, NPHOS and turbidity act as shifters of the supply function for fish. Since the levels of lake NPHOS and turbidity are related to the inflow of sediment and effluents from the watershed, the supply

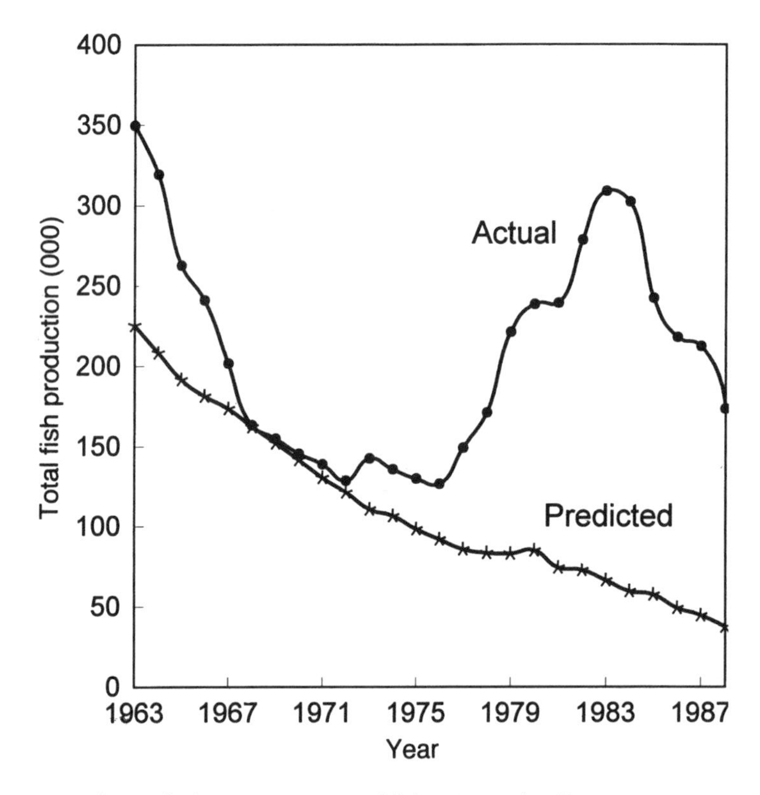

Fig. 5.3. Fish supply function in terms of fish price and pollutant sources, Laguna de Bay. (Source: Valmonte, 1993.)

function can be specified in terms of fish price and the sources of pollutants (Fig. 5.3). Included among the water quality determinants are the NPHOS equivalent of manure annually produced by the total livestock and poultry population in the system, chemical fertilizer use, number of industries, forest hectarage, and lag of fish prices[2]. An ordinary least squares (OLS) regression equation in logarithmic form of annual fish output for the period 1964–1988 as related to watershed activities affecting lake water quality and lagged fish price is presented in Table 5.4.

NPHOS contributions from chicken and duck and hog farms around the lake had a significant negative relationship with fish production. NPHOS contributions are the result of waste disposal from these operations.

Cropping intensification deteriorates water quality primarily through leaching excess agricultural chemicals into streams and tributaries via irrigation waters and runoff. In the model, chemical fertilizer use (CFERT) showed the expected negative relation. An increase in chemical fertilizer use in the lake area depresses lake water quality, and decreases fish supply.

The significant positive influence of forest area (FORHA) on fish supply indicates the importance of reduced water turbidity on NPP. Lower lake water turbidity improves light penetration, enhances algal photosynthetic activity and primary production, encouraging fish growth. The estimate implies that, for the period covered, a 1 percent increase in forest cover improved total fish production by 2.46 percent.

Industrial development within Laguna de Bay's watershed is a known deterrent factor of the lake's water quality. No less than 424 industries, 90 percent of which are highly pollutive, stood around the lake and its tributaries in 1986. Concentrated in the northern region and taking up around 63 percent of the total land area, most of these

Table 5.4. Fish production related to watershed variables, Laguna de Bay, 1963–1988

Independent variables	Variable names	Logarithmic model	
		Open	Total
Intercept		–7.5508	–15.7001*
Chicken and ducks NPHOS	CDNP	–1.0401*** (0.3498)	–0.6956** (0.3094)
Hog NPHOS	HOGNP	–0.1523 (0.2802)	–0.2576 (0.2478)
Fertilizer use	CFERT	–0.7972** (0.3528)	–0.6349* (0.3120)
Annual total number of industries	INDUS	2.7069*** (0.7169)	2.4272*** (0.6341)
Forest area	FORHA	1.9011** (0.7255)	2.4552*** (0.6417)
Lagged fish price (P kg^{-1})	LPRICE	–0.2392 (0.5171)	0.1154 (0.4574)
Degrees of freedom		24	24
R-square		73.77	74.44
F-value		8.44***	8.74***

***, **, *, Significant at the 1, 5 and 10 percent levels, respectively.
Source: Valmonte (1993).

have no fully operational pollution control devices. Toxic chemicals, poisonous heavy metals, solid wastes, enriching mineral nutrients, thermal pollution and their interactions have caused abnormalities in the lake ecosystem. These include the lake's weakened natural capacity to cleanse itself, food chain perturbations, and weakened fisheries (Sajise *et al.*, 1988).

Contrary to the above however, the model's industry variable (INDUS) gave an unexpected positive coefficient. The logarithmic estimates indicated an increase in fish supply given an increase in the number of industries around the lake.

The industry variable may have captured the consumers' demand for fish in Laguna de Bay, rather than the deleterious effects of industrial wastes on water quality. That is, as the number of industries in the lake area increases, so does the number of people employed in these industries, and the overall income of the area, leading to an increase in fish demand.

The relationship between lagged fish price and total fish supply was positive, although the coefficient value was not significant. A 1-percent increase in lagged fish prices increased total fish supply by 0.12 percent, indicating an inelastic supply function.

The above example shows that agricultural and non-agricultural developments in the aggregate generate externalities that influence the productivity of the watershed as a whole. Decisions on the utilization and management of resources in one subsystem when done in isolation could lead to significant costs, in terms of loss of productivity, on the other subsystems of the watershed.

Pesticides and Human Health

The last two sections discussed the off-site effects of agricultural intensification and watershed exploitation, with particular reference to water and aquaculture resources. This section provides an example of an on-site externality created by intensive lowland production systems, the case of injudicious pesticide use on human health. The negative impact of pesticides on farmer and farm-household health is an externality because farmer decisions on pesticide use are made without an awareness of health consequences (Warburton *et al.*, 1995).

Farmers and agricultural workers face acute and chronic health effects due to prolonged and unsafe exposure to pesticides. Eye, skin, pulmonary, neurologic and gastrointestinal problems are significantly associated with long-term pesticide exposure. Pesticides which might be linked with these impairments include certain organophosphates, organochlorines, organotins and phenoxy herbicides. Most of these chemicals, commonly available across Asia, are classified by the World Health Organization (WHO) as extremely hazardous chemicals (Category I and II chemicals) and are either banned or severely restricted for use in the developed world (Pingali and Rola, 1994).

Pingali *et al.* (1994) found that explicitly accounting for health costs associated with pesticide exposure substantially raises the cost of using pesticides[3]. The value of the rice crop lost to pests is invariably lower than the cost of treating pesticide-caused diseases. For rice production, when health costs are factored in, the natural control ('do nothing') option is the most profitable and useful pest control strategy. This section summarizes the Pingali *et al.* (1994) study on the impact of pesticide use on Philippine farmer health. In a complementary study on the same sample, Antle and Pingali (1994) found pesticide-related health impairments to cause significant reductions in labor productivity. The environmental and ecological consequences of pesticide use were also assessed and are discussed in detail in Pingali and Roger (1995).

The medical literature provides a set of indicators for assessing long-term health effects due to pesticide exposure (Morgan, 1977; Hock, 1987; Nemery, 1987). Of these, the impact of chemicals on the eye, the respiratory system, neurologic system, dermal effects and gastrointestinal problems are most discernible in a cross-section analysis. These effects are described below along with empirical evidence from the farm sample. Logit regressions were used to relate the positive incidence of these ailments to pesticide exposure.

Eye effects

The eye is very vulnerable to physical and chemical hazards in the agricultural setting. A chronically irritated eye can lead to the formation of a pterygium, a vascular membrane over the cornea, usually affecting older people and people exposed to dust and wind. With increasing severity, the vascular membrane may encroach on the pupil, diminishing visual acuity and requiring surgical removal to improve eyesight. Pterygium can therefore reduce farmers' productivity, initially because of bothersome symptoms, and later because of diminished vision.

Logit regression estimates presented in Table 5.5 indicate that the incidence of eye abnormalities increase significantly with age and with exposure to insecticides. Exposure to herbicides has the expected positive sign though it is not significant. General health status, as measured by the ratio of weight to height, has the expected negative sign on eye abnormalities, although not significantly. The probability of eye abnormalities among the sample households is 0.36; this was determined from the logit function at the mean levels of all variables.

An increase in the application of insecticides, from the mean level of one per season to two, will increase the probability of eye problems by 22 percent. Farmers applying three recommended doses of insecticides face a probability of 0.53 of having chronic eye problems.

Table 5.5. Logit regressions on pesticides and health impairments.

	Eye	Pulmonary	Polyneuro-pathy	Skin	Gastro-intestinal	Multiple impair-ments
Intercept						
alpha 1	-3.2373** (1.391)	–2.8914** (1.464)	–5.8633* (3.4546)	–6.7327*** (1.6778)	4.3077 (1.95)	0.0493 (1.2267)
alpha 2						–1.0807 (1.2287)
alpha 3						–2.5382** (1.2462)
alpha 4						–4.8817*** (1.3183)
Age	0.0455*** (0.0144)	0.0501*** (0.015)	0.0508 (0.0323)	0.0218 (0.0162)	–0.0443*** (0.0159)	0.0237* (0.0125)
Weight/height	–0.00174 (0.049)	–0.0263 (0.0488)	–0.0708 (0.1186)	0.151*** (0.057)	–0.1851** (0.0780)	–0.0354 (0.0419)
Smoking	—	0.6350* (0.3972)	—	—	—	0.6116* (0.3533)
Drinking	—	—	1.9852* (1.1394)	—	—	–0.7123 (0.3575)
Total dosage category I and II pesticides	0.3497** (0.1714)	0.259* (0.1566)	–0.1025 (0.2967)	0.2180 (0.1546)	0.0919 (0.1695)	0.3551*** (0.1363)
Total dosage of category III and IV pesticides	0.4986 (0.3942)	0.0291 (0.3918)	1.3815* (0.7131)	1.0414** (0.4267)	0.9849** (0.4348)	0.9616*** (0.3421)
Cloth cover over mouth	—	0.3833 (0.4106)	—	—	—	0.6309* (0.3487)
n	148	145	148	148	149	145
Chi-square	19.849	15.7	8.885	22.516	24.157	36.293

***, **, *, Significant at 1, 5 and 10 percent, respectively.
Figures in parentheses are standard errors of estimate.
Source: Pingali *et al.* (1995).

Skin effects

Skin problems are commonly observed for farmers frequently exposed to pesticides. Mixing, handling and applying pesticides could cause dermal contamination. Dermal contamination is greater when spraying with a knapsack sprayer than with a spinning disc applicator or an electrodyn sprayer (Durand *et al.*, 1984). The hands and forearms have the highest potential for pesticide contamination. Eczema, a chronic allergic dermatitis characterized by lichenification and fissuring, is a dermatologic health indicator of pesticide exposure. The skin appears thickened with accentuated markings.

The incidence of skin problems is positively related to the use of both insecticides and herbicides, although only the latter are significant. This is because most herbicides used in the Philippines are acetamides, and unprotected use of these chemicals is known to cause skin problems. Farmers at the sample average for age and nutritional status who do not apply any herbicides have a probability of 0.12 of having skin problems. The probability of skin problems rises to 0.30 for farmers with one herbicide application and 0.50 for farmers with two applications.

Respiratory tract effects

Long-term exposure to chemical irritants like pesticides can cause respiratory symptoms such as cough, cold, sputum formation, wheezing, rales, tenderness and decreased chest expansion (Hock, 1987; Nemery, 1987). Bronchial asthma and other abnormal lung findings are the two respiratory-tract indicators of pesticide exposure.

The incidence of respiratory abnormalities is significantly related to age, smoking and exposure to insecticides. Herbicides had the expected positive sign although not significantly. Nutritional status (weight/height) had the expected negative sign but not significantly. Evaluated at the sample mean, the probability of abnormal respiratory findings for farmers who do not smoke is 0.30. These farmers apply one recommended dose each of insecticides and herbicides. At this level of pesticide use, farmers who smoke have a 50 percent higher probability of abnormal respiratory findings. The probability of respiratory problems increases by 16 percent for farmers applying two doses of insecticides, and by 30 percent for farmers applying three doses, irrespective of their smoking habits.

Polyneuropathy

Organophosphorous compounds and 2,4-D are known neurotoxicants (Morgan, 1977). Both have been implicated as causative agents for polyneuropathy, a neurologic disorder that manifests typically as motor weakness in the distal muscles and sensory deficit with a 'glove-and-stocking' distribution. Absence of deep tendon reflexes in the early stages may be the only sign, but neuropathy may be purely motor or purely sensory.

The incidence of polyneuropathy is significantly associated with drinking and with pesticide use. Herbicides had a significant positive effect while the effect of insecticides was positive but not significant. Age had the expected positive sign and nutritional status had the expected negative sign on the incidence of polyneuropathy, although both coefficients were not significantly different from zero.

Farmers who do not drink alcohol, evaluated at the sample mean with respect to age, nutritional status and application of pesticides, were found to have a probability of 0.02 for positive findings of polyneuropathy. The probability rises by 550 percent for farmers who regularly consume alcohol. Farmers applying herbicides at the rate of three recommended doses per season were found to have a probability of polyneuropathy findings of 0.24 if non-drinkers. The same probability for farmers who drink is 0.70.

Gastrointestinal effects

Pesticides usually enter the gastrointestinal tract accidentally through the mouth. For example, a farmer applying pesticides who smokes or wipes sweat off near his mouth may unknowingly ingest pesticide particles. Carbamate insecticides formulated in methyl alcohol and ingested may cause severe gastroenteritic irritation (Morgan, 1977). Organophosphates and copper salts also irritate the gastrointestinal tract, manifested as intense nausea, vomiting and diarrhea. The health indicator chronic gastritis is clinically characterized by epigastric tenderness and pain associated with nausea and vomiting.

The incidence of gastrointestinal problems is positively related to pesticide exposure, of which exposure to herbicides was significant. Gastrointestinal problems have a significant negative relationship with nutritional status. The significant negative effect of age was unexpected. Farmers evaluated at the sample mean applying one recommended dose of herbicides have a probability of 0.27 of an abnormal gastrointestinal finding. Two recommended doses of the same chemicals increases the probability by 85 percent, and three doses by 167 percent.

Incidence of multiple health impairments

The analysis above attempted to isolate the impact of pesticides on each specific illness. However, farmers exposed to pesticides over the long term may face several of these illnesses at the same time. Pesticides may also cause other non-specific illnesses in addition to those mentioned above. Multinomial logit regressions indicate the incidence of multiple health impairments to be significantly and positively related to age, smoking habits, and to the use of pesticides, both insecticides and herbicides. Nutritional status and drinking habits had the expected negative signs, although both were, contrary to expectations, insignificant. The use of a cloth cover over the mouth and nose, a common practice during spraying, had an unexpected positive significant sign. This probably indicates that, while farmers believe it provides them protection, the use of such a cover actually creates more problems, since it absorbs chemicals which the farmers can inhale.

For non-smoking farmers in the sample, applying one recommended dose each of insecticides and herbicides, the probability of being affected by three or more illnesses at the same time is 0.19. An additional dose of insecticides increases this probability by 32 percent. Farmers who smoke face an additional 63-percent increase in the probability of three or more health impairments.

Valuation of the health cost of pesticide exposure

The costs faced by farmers due to health impairments were computed based on the medical tests conducted. The medical tests provided an assessment of each farmer-respondent's ailments and their seriousness, which may or may not be related to pesticide exposure. The treatment required to restore the farmer's health was assessed. Treatment costs (including medication and physicians' fees) plus the opportunity cost of farmers' time lost in recuperation formed a measure of the health cost per farmer.

Farmer health costs are associated with: pesticide exposure (the number of times the farmer comes into contact with pesticides); and 'other' farmer characteristics (weight over height, age, smoking, and alcohol consumption). Log-linear regression estimates for pesticide users identified the significance of each of the above variables in determining the cost of health impairment.

$$
\begin{aligned}
\log(\text{Health Cost}) = \; & \underset{(1.39)}{4.366^{***}} + \underset{(0.3130)}{1.192^{***}} \log(\text{age}) \\
& - \underset{(0.0316)}{0.0756^{**}} \text{ Ratio of weight to height} \\
& + \underset{(0.2360)}{0.9160^{***}} \text{ Smoking dummy} \\
& - \underset{(0.24)}{0.53^{***}} \text{ Drinking dummy} \\
& + \underset{(0.232)}{0.486^{**}} \text{ Log (insecticide dose)}
\end{aligned}
$$

$$- \ 0.042 \log \text{(Herbicide dose)}$$
$$(0.365)$$

(degrees of freedom = 100; $r^2 = 0.30$)
*, **, *** = significant at the 10, 5 and 1 percent levels respectively

According to the regression estimates, insecticides significantly influence farmer health costs, while herbicides have no insignificant effect. Costs increase by 0.49 percent for every 1-percent increase in insecticide dose. The insignificance of herbicides in the health cost equation could be due to the overwhelming number of insecticide-related illnesses relative to herbicide-related illnesses.

Weight-by-height ratio had a significant negative effect on health costs, while age and smoking habits increased health costs significantly. The coefficient of the drinking variable, although significant, had a negative sign. Some measurement deficiencies may influence this result; that is, some farmers might have stopped drinking because they already have a disease or ailment. Also, younger and healthier individuals tend to be more candid about drinking habits. This kind of information would reflect a high health cost for a non-drinker respondent in the data set.

The health-cost regression results were used to estimate expected health-cost values per recommended dose of insecticides. The mean cost per insecticide dose is based on estimates for a non-smoking, non-drinking farmer population. It assumes an average age of 44 years, a weight–height ratio of 23.5, and an average herbicide dose of 0.50 kg ai ha^{-1}.

When insecticides are not applied the health costs are on the average 1084 pesos; with the application of one recommended dose of insecticides, costs rise to 1519 pesos. One dose of insecticides per season is generally the economic threshold level (ETL) of insecticide application, the decision rule of integrated pest management (IPM). Most farmers usually apply two doses of insecticides and, for them, the health costs are 1849 pesos. A complete prophylactic application package, consisting of calendar spraying, requires approximately six recommended doses of insecticides per season. For these farmers, the health costs work out to be 2792 pesos on the average.

Valuation of the net benefits of pesticides

The expected effects of pesticides on the mean yields and variance of yields were determined (see Rola and Pingali, 1993; and Pingali and Roger, 1995, for detailed results). Six crop seasons of farm production data, for each of the sample households using pesticides, were used to estimate the moments of the yield distribution (Table 5.6). Insecticides were found to have a significant positive effect on the mean and a significant negative effect on the variance of the yield distribution. Herbicides had a small negative effect on the mean and did not affect the variance of the yield distribution.

Four insect pest management strategies were evaluated: natural control (no insecticide application); IPM (one recommended dose of insecticides); farmers' practice (two recommended doses of insecticides); and prophylactic control (six recommended doses of insecticides). Herbicide use was held constant at one recommended dose across the four pest-control strategies. Productivity benefits of insecticide use were evaluated, first assuming risk neutrality, and then assuming risk aversion. The latter analysis was done in an expected utility maximization framework. The coefficients of risk aversion were taken from Sillers (1980) who measured them in the same area of the Philippines using an experimental approach.

For a risk-neutral farmer, the net benefits of insecticides applied are negative when health costs are explicitly considered (Table 5.7). In

Table 5.6. Impact of pesticides on the mean and the variance of the yield distribution for irrigated rice production (in kilogram per hectare) in Laguna and Nueva Ecija, Philippines (in logarithm).

Independent variable	Laguna		Nueva Ecija	
	First moment	Second moment	First moment	Second moment
Constant	8.12 *** (0.34)	0.77 (0.85)	5.81 *** (0.26)	0.51 ** (0.16)
Nitrogen	0.12 * (0.08)	0.03 (0.17)	0.49 *** (0.06)	–0.06 (0.04)
Insecticide dose	0.06 *** (0.01)	–0.12 *** (0.02)	0.06 *** (0.01)	–0.40 * (0.02)
Herbicide dose	–0.12 *** (0.03)	0.18 *** (0.05)	–0.04 ** (0.02)	–0.03 (0.04)
Pre-harvest labor	–0.12 (0.09)	–0.16 ** (0.07)	0.09 *** (0.03)	–0.01 (0.02)
Season dummy[a]	0.29 *** (0.05)	–0.33 *** (0.08)	—	—

Figures in parentheses are standard errors of estimate.
***, **, * = significant at the 1, 5 and 10 percent levels, respectively.
[a] Season dummy = wet season, dry season.
Source: Rola and Pingali (1993).

Table 5.7. Net benefits and certainty equivalents of alternative insecticide use strategies considering health costs

Pest management strategy	No. of recommended doses	Estimated health costs (pesos)	Net benefits[a] (pesos)	Certainty equivalents[b] (pesos)
Laguna				
Complete protection	6	2792	11336	13862
Farmers' practice	2	1849	12385	15706
IPM	1	1519	12208	15791
Natural control	0	1084	12844	16664
Nueva Ecija				
Complete protection	6	2792	14058	17539
Farmers' practice	2	1849	14284	18078
IPM	1	1519	14227	18109
Natural control	0	1084	14743	18688

[a]Yields were estimated from Table 5.6; a price of 4.60 pesos per kilogram was used.
[b] Risk aversion coefficient of 1.37 was used.
Source: Rola and Pingali (1993).

other words, the positive production benefits of applying insecticides are overwhelmed by the increased health costs. This is due to the small positive effect of insecticides on the mean of the yield distribution relative to the large negative health effect. A farmer applying two recommended doses of insecticides increases his/her net profits by 492 pesos relative to a farmer who applies none; however, his/her health costs go up by 765 pesos, a net loss of 273 pesos. Pest-control strategies ranked in terms of net benefits, including health costs, indicate that the natural control option (no insecticide application) is the dominant strategy, followed by the current farmers' practice of two applications and then the IPM strategy. IPM strategy in its current form ranks lower than the farmers' practice because of the high labor requirement for monitoring pest populations.

Does the dominance of natural control hold when risk aversion is considered? To consider this, the pest control strategies were ranked in terms of the certainty equivalent level of return. Again, when health costs are explicitly considered, the natural control option is the dominant pest-control strategy for both locations. IPM and farmers' practice are not significantly different and complete protection comes last.

Insecticides shift the net benefit function to the left, as health costs increase the total cost of production. Once farmers are aware of the costs incurred due to pesticide exposure, the threshold levels that they use as decision rules to spray increase.

Synthesis and Policy Implications

The above examples indicate the difficulty of quantifying the impact of externalities generated by intensive watershed exploitation. Nevertheless, it is important to use case studies such as the ones presented above to derive guiding principles for resource utilization, research resource allocation, and policy formulation.

Private choices of individuals at one point along a watershed have consequences, both good and bad, on the well-being of individuals at other points of the watershed. The interdependency between the uplands and the lowlands is particularly strong. There are also strong interdependencies between sectors of economic activity; for example, poor forest conservation policies could have detrimental consequences on upland rice production and on fish production. Similarly, indiscriminate pesticide use can affect drinking water supplies and human health.

Upland resource conservation policies ought to take a holistic view of agriculture, forestry, mining and human habitation. Agriculture policy alone cannot prevent or reduce soil erosion. Clearly defined access rights to public forest lands are essential for selective clearing and re-planting of forests. For land under agricultural production, secured long-term tenure provides the incentive for farmers to invest in soil conservation technologies. Where the returns to private investment in soil conservation are low, government-encouraged collective action is required for watershed-level investments.

The Green Revolution strategy of increasing lowland productivity has to a large extent relieved the pressure on the uplands by providing employment opportunities for migrant labor. However, if the current trend towards stagnation/decline in lowland productivity persists, then one could expect a decline in employment opportunities in the lowlands and hence increased pressure on the uplands. Where upper watershed externalities are not explicitly accounted for, there is an under-investment of research resources in upland conservation.

Long-term productivity of the lowlands, especially the state of irrigation infrastructure, is not independent of upper watershed degradation. In the design of irrigation systems, siltation rates generally assumed tend to be independent of upper watershed degradation. One therefore tends to overestimate the life of the system and consequently the returns to irrigation investment.

Given the results of lowland degradation due to upper watershed externalities and due to lowland intensification, the steady decline of investments in real terms for the maintenance of irrigation infrastructure is a cause for concern (Chapter 9). For the Southeast Asian countries, donor lending for irrigation infrastructure has declined from US$

630 million in 1977–1979 to US$ 202 million in 1986/87 (Rosegrant and Pingali, 1994). Consequently, these countries reduced sharply their total spending on irrigation, both for new construction and for maintenance (Rosegrant and Pasandaran, 1990). The full effects of these declines are just beginning to be felt due to the lags in irrigation construction.

Where multiple uses of the watershed are common, proper management of waste disposal is imperative for reducing externality generation. Mining and industrial waste disposal are the two important leverage points for reducing the level of water pollution.

Pollution from agricultural inputs is closely related to input pricing policies and farmer knowledge of efficient use and negative consequences. In the case of pesticides, pricing and regulatory structure, plus inadequate storage, unsafe handling and improper sprayer maintenance taken together account for a risky public-health environment. Regulatory policies that discourage the use of highly hazardous chemicals ought to be examined, especially in terms of their impact on farmer health and crop losses due to pest damage. Investments in farmer training and information campaigns on proper pesticide management could help reduce some of the health and environmental risks.

Decentralization of watershed management decisions, allowing greater community participation in the decision-making and management process, could have long-term pay-offs in terms of sustainable utilization of watershed resources. Community participation would necessarily move the decision-making process away from sector-specific concerns and towards a more holistic watershed-level approach.

Notes

1. Upland areas in the humid tropical zone face special problems of agricultural intensification. Continuous field crop production on most soils of this zone leads to rapid leaching, soil acidification and/or erosion resulting in soil-fertility and yield declines (Allan, 1965; Ruthenberg, 1980; Kang and Juo, 1981; NRC, 1982; Sanchez *et al.*, 1982; and FAO, 1984). Ruddle and Manshard (1980) provide a comprehensive survey of the human impact on the humid tropical forests across the world.
2. The model was modified from the study of Valerio (1990) which described the effects of various watershed resource development activities on Laguna de Bay's fishery resources, and determined the major factors influencing the long-term trends in fishery output.
3. This study was conducted on a random sample of 152 rice-farming households in the Philippines, during the years 1989–1991 (see Pingali and

Roger, 1995, for a description of the sample and the methodology). These households varied in terms of the level of pesticide exposure, from none to high levels of exposure. The sample households were monitored over several crop seasons and detailed records were kept for input use, pest management practices, and pesticide storage and handling practices. A detailed medical assessment was conducted on the entire sample, including an interview, physical examination, a battery of laboratory tests and exposure history. A set of medical indicators of pesticide exposure were defined and related econometrically (using logit regressions) to a set of farmer characteristics, such as age, nutritional status, history of tobacco and alcohol consumption, and occupational exposure to insecticides and herbicides. Probabilities of health risk were assessed relative to these farmer characteristics. Health costs associated with pesticide exposure were quantified and explicitly accounted for in the benefit–cost ranking of alternative pest-control strategies. Pest-control strategies were evaluated for both the risk-neutral and risk-averse farmer.

Appendix 5.A: Sustainable Upland Farming Systems for the Densely Populated Humid Tropics

What are the economically viable options for increased food crop production in the uplands of the humid tropics in the face of increasing population densities? In general, a system of farming that closely mimics the dense natural vegetation of the humid forests is what will work in the long run. This is exactly what the shifting cultivators tried to achieve in their multi-storied cultivation of cleared tropical forest lands (Ruddle and Manshard, 1980). For instance, a system of intercropping short- and medium-term crops under a protective cover of trees is feasible since the ground would be covered for most of the year and therefore the detrimental effects to the soil would be minimized (Okigbo, 1974; Sanchez *et al.*, 1983; Sanchez and Salinas, 1981).

Technological options for permanent cultivation on the upland soils of the humid tropics vary by the extent of markets for tree crops in a particular area. The humid and subhumid tropics are well suited to perennial crops such as bananas and to tree crops such as rubber, cocoa, and oil palm. Once these crops have been established, their canopy reduces the growth of weeds and they demand less of the soil than annual crops. The effects on the soil resemble those of forest and bush vegetation in establishing a closed nutrient cycle (Sanchez, 1976; Ruthenberg, 1980). The tree cover moreover protects the soil from erosion. Food crops can be intercropped during the establishment phase of a tree crop and sometimes on the mature plantation. Ruthenberg (1980) lists the following four phases in the establishment of smallholder tree crop plantations:

1. Perennial crops are planted around the house as the shifting cultivator begins to maintain a permanent residence.
2. The area devoted to perennial crops is extended by interplanting perennials with annuals in newly cleared plots.

3. In time, the perennial crops predominate in the mixed cropping system, and young stands of tree crops are interplanted with arable crops.
4. As perennial crops create more shade, intercropping becomes less important.

In discussions of the importance of tree crops as a choice for the humid tropics it should be borne in mind that the returns on tree crops are determined by farmgate prices. The establishment of tree crops is not feasible where well-developed infrastructure for exporting the crop is not available. Tree crops are also not a wise choice in the face of falling or widely fluctuating export prices and a wavering government policy towards tree-crop exports.

The principles of sustainable small-farmer tree-crop plantations have been applied with some success to agrosilviculture. Agrosilviculture is a regulated system of plantation forestry, in which a public authority allocates forest land to shifting cultivators to clear. The cultivators are allowed to plant for a limited time, after which the cultivators are required to plant tree seedlings and to tend them until the trees are large enough to control weed regrowth. This is essentially a form of regulated shifting cultivation, in which the farmer is a short-term tenant or licensee of an area of land owned by a state authority (Norman, 1979; Budowski, 1980).

In the Philippines, the Integrated Social Forestry Program (ISFP) was launched in 1983 to 'maximize land productivity and enhance ecological stability'. Under the program, a 'kainginero' (shifting cultivator) family of five members is given a 25-year stewardship certificate of contract for continued occupancy of a forest clearing of 3–7 hectares. Specific requirements include protecting the forest ecosystem by maintaining a certain portion of the land under trees. These stewardship contracts can be renewed for another 25 years (Sajise, 1987; EMB, 1990). Other examples of such social forestry programs can be found in Pragtong (1987) for Thailand.

As population densities rise, agrosilviculture systems are economically feasible only in areas where other, more profitable, agricultural systems are unlikely to be established. The marginal lands on the upper slopes and the steeper mid-slopes are the most likely areas for agrosilviculture. These areas are already under forest cover, and permanent cultivation of field crops is not possible because of high levels of erosion and low soil fertility. The deeper soils of the mid- and lower slopes can be cultivated with a combination of perennial and annual food crops and are less likely candidates for agrosilviculture.

In the agroforestry system, annual crops are grown, usually in mixtures, under perennial forest or plantation trees. The home gardens of South and Southeast Asia and the compound farms of West Africa are the best known examples of agroforestry. Permanent compound farming represents an intensive management system that approximates the ecosystem of the forest. Lal (1983) points out that such a system can sustain and prolong the productivity of land and minimize degradation of the soil in the humid tropics even at high population densities. In densely populated Central Java, for instance, home gardens account for 22 percent of the total area cultivated (Ruthenberg, 1980). Fujisaka and Wollenberg (1991) report on a case from Southern Luzon, Philippines, where the switch from annual cropping on cleared uplands to an agroforestry system led to a reversal of degradation trends. Fujisaka and Wollenberg found that: the migrants adapted their techniques to local circumstances; they learned about and tried to obtain better local lands and soils; they experimented with crops and necessary farm practices; and they continually observed the successes and failures of their neighbors. This study suggests that a sustainable dynamic equilibrium can emerge, with farmer innovation overcoming environmental constraints.

Where markets for tree crops and forest products are limited and where rising population density leads to a reduction in fallow periods, the problem of finding a viable farming system for the humid tropical uplands appears. The problem is essentially that arable cropping patterns similar to those used in the temperate environment have adverse effects on the soil and therefore cause a rapid decline in yields (Poulsen, 1978; Kang and Juo, 1981; Lal, 1983). The problem is aggravated in sloping uplands which are highly susceptible to soil erosion. In the mid- and lower slopes of the uplands of Southeast Asia, permanent cultivation of rice, maize and vegetables is common and so are the associated soil degradation problems. Indigenous and modern options for arresting and perhaps reversing the soil degradation problem in the humid tropics are discussed below.

Terraces of West Java, Indonesia and Banaue, Philippines, are famous examples of indigenous techniques for sustaining production on sloping uplands. There are also examples for the Philippines of indigenous alley cropping, use of natural vegetative barriers (Fujisaka, 1990) and improved fallow management (MacDicken, 1990). Garrity *et al.* (1991) provide a comprehensive assessment of the technological options for sustaining permanent farming systems on the sloping uplands of the Philippines. Options commonly proposed for controlling soil erosion and/or sustaining fertility are: hedgerows; vegetative or grass strips; minimum tillage, and live mulches.

Table 5.A1 provides synthesis results from 78 studies on the impact of alternative soil conservation technologies on surface runoff, erosion and sedimentation and on upland crop yield. Each of these studies was summarized in Doolette and Macgrath (1990). Six studies found that the removal of forest cover for shifting cultivation purposes led to an increase in surface runoff. The increase in runoff and erosion was over 100 percent in all but one case. Terracing, plantation trees, and vegetative and grass strips decreased surface runoff by up to 100 percent in 17 studies. Five studies gave no significant change in surface runoff using the same options. Two cases indicated that agroforestry and terracing increased surface runoff by more than 100 percent.

Of the 11 instances that examined the impact of the switch from conventional tillage using a moldboard plow to zero or minimum tillage, five studies found that erosion and sedimentation decreased by 75 to 100 percent. In contrast, two studies indicated that zero or minimum tillage increased erosion and sedimentation by more than 100 percent. It should be noted that such impacts are influenced by other factors such as slope of the area, climate, soil type, land use and/or species grown in the area.

From the same synthesis, terracing was found to effectively control erosion and sedimentation, as confirmed by results of 16 studies indicating a

Table 5.A1. Impacts of soil conservation technologies (no. of studies), various slopes.

Soil conservation technology	Percentage changes –100 to 75	–75 to –50	–50 to –25	–25 to 0	NSC	0 to 25	25 to 50	50 to 75	75 to 100	>100	Total
I. On surface runoff											
A. Bare soil/bare fallow/ fallowing	0	0	0	0	0	0	0	1	0	4	5
B. No till/zero tillage/ minimum tillage	1	0	3	1	0	1	0	0	0	0	6
C. Agroforestry	0	0	0	0	0	0	0	0	0	1	1
D. Plantation trees	1	0	1	0	1	0	0	0	0	0	3
E. Terracing	4	2	4	3	4	0	0	0	0	1	18
F. Vegetative and grass strips	0	0	1	1	0	0	0	0	0	0	2
II. On erosion and sedimentation											
A. Bare soil/bare fallow/ fallowing	0	0	0	0	0	0	0	1	0	7	8
B. No till/zero tillage/ minimum tillage	5	2	0	0	0	0	1	1	0	2	11
C. Agroforestry	1	0	0	0	0	1	0	0	0	1	3
D. Plantation trees	8	1	0	0	2	2	0	0	2	4	19
E. Terracing	16	3	3	3	1	0	0	1	0	1	28
F. Vegetative and grass strips	5	2	0	0	0	0	0	0	0	0	7
III. On upland crop yield											
A. No till/zero tillage/ minimum tillage	0	0	1	2	3	1	2	1	0	0	10
B. Terracing	0	0	1	2	6	5	4	1	2	0	21
C. Use of animal manure	0	0	0	0	0	0	1	0	1	0	2
D. Mulching	0	0	0	1	0	0	5	3	2	0	11

NSC, no significant change

The above table was summarized from results of studies, as numbered below, cited in Doolette and Macgrath (1990):

I. Surface Runoff
- A - 20, 122, 157
- B - 73, 90, 122, 157, 212
- C - 208
- D - 20, 173
- E - 10, 11, 74, 75, 76, 84, 89, 102, 103, 167, 169, 177, 191, 219
- F - 3, 102

II. Erosion and Sedimentation
- A - 20, 41, 122, 139, 157
- B - 67, 73, 90, 122, 139, 157, 212
- C - 141, 208, 215
- D - 20, 30, 48, 72, 74, 90, 165, 176, 215, 216
- E - 33, 56, 68, 74, 102, 103, 104, 129, 143, 144, 167, 169, 177, 182, 198, 205, 219
- F - 3, 31, 102, 217

III. Upland Crop Yield
- A - 97, 122, 190, 212
- B - 17, 18, 34, 42, 56, 68, 69, 74, 75, 76, 104, 105, 124, 135, 182
- C - 21, 153
- D - 34, 37, 91, 105, 122, 149, 153, 185, 193, 197

decrease of 75 to 100 percent. Only one study stated that terracing led to an increase in erosion and sedimentation by more than 100 percent. These studies compared terracing to unimproved local farming/soil conservation practices such as clean cultivation, up-and-down slope cultivation, contour farming, or simple unterraced cultivation.

Vegetative and grass strips decreased erosion and sedimentation by 50 to 100 percent in seven studies. These were either in comparison to monthly tilled bare soil, tree plantations, or to conventional cultivation. The impact of agroforestry and plantation trees on soil erosion and sedimentation, however, gave varied results. Ten studies found that these options can decrease erosion by up to 100 percent, while five studies showed they increased it by more than 100 percent. These were compared to either bare soil, shifting cultivation, natural cover, undisturbed natural forest, continuous fallow, crop monoculture, or to intercropping.

In terms of upland crop yield and productivity, three studies showed no significant difference when a farmer shifts to farming with zero or minimum tillage. Two studies indicated a possible 25 percent decrease in upland crop yield while two other studies indicated an increase of 25 to 50 percent in the same. These studies compared zero or minimum tillage to either unmulched traditional cultivation, traditional mixed cropping, or to conventional tillage.

Terracing was found not to significantly affect upland crop yield in six studies. Twelve studies, on the other hand, found that it increased crop yield up to 75 percent. The impact of terracing was compared either against crop yield of unterraced cultivation or that of traditional mixed cropping. One study noted that 'highest yields were attributed mainly to minimized soil disturbance on the vegetative terrace'.

Two studies on the use of animal manure (specifically cow/poultry manure applied at four to five tons per hectare) showed an increase in upland crop yield of 25 to 100 percent. This was compared to applying manure at two tons per hectare, or to applying commercial fertilizers. Similarly, mulching was found to increase upland crop yield by up to 100 percent in ten studies. One study showed that it decreased upland crop yield by 25 percent. These studies compared mulching against unmulched traditional cultivation, and to clean cultivation.

While several Asian countries have extension efforts for promoting conservation technologies for the sloping uplands, the incidence of sustained adoption has been small. The lack of secure land tenure was implicated as a constraint to the implementation of any long-term land improvement among tenant farmers or occupants of public lands.

Asian Rice Market: Demand and Supply Prospects

6

Introduction

Asia has had an impressive record of feeding an evergrowing population with limited land resources. The production of rice, the most important staple grain in Asia, has increased faster than population over the last three decades. Most of the increases in rice production have come from growth in yields per hectare (Chapter 2). The growth in demand for rice has started slowing down due to urbanization, rapid increases in per capita incomes, and the high levels of rice consumption already reached in many countries. The growth in rice supply may, however, slow down faster than demand because of the decline in rice area, the closing of the yield gap in irrigated systems, decline in relative profitability of rice production, increasing concerns about environmental protection, and the non-availability of productivity-enhancing technology for the rainfed environments. Also, as Asia prospers, rice farmers particularly in East Asia are finding it difficult to compete with non-farm sectors for factors of production, and hence their ability to maintain production growth is questionable.

Two contrasting developments may substantially affect the rice market in Asia in the near future. First, the prosperous Asian countries will increasingly find it difficult to sustain producers' interest in rice farming. The move toward free trade in agricultural production, initiated by the Uruguay Round of the General Agreement on Tariffs and Trade (GATT), will have an impact on the sustainability of rice farming in these countries. Second, the potential for increased productivity created by the dramatic technological breakthrough in the late 1960s has almost been exploited, particularly for the irrigated and favorable rainfed environments. But population continues to grow at a high rate. As rice production is losing the race against population, sustainability of self-sufficiency in rice production will become a major issue for the land-scarce, low-income countries. This chapter looks into the emerging demand and supply trends, and assesses their impact on sustaining food security in low-income countries and its interface with the international trade in rice.

Emerging Trends in Demand

The growth in demand for rice depends on: (i) the level of per capita income; and (ii) the rate of growth of population. In addition to understanding the factors that affect long-term changes in total quantity of rice demanded, we also ought to be looking at changes in rice quality preferences. As incomes grow, rice consumers tend to switch from inferior to superior quality rices.

At low levels of income, when meeting energy needs is a serious concern, rice is considered a luxury commodity. With increases in

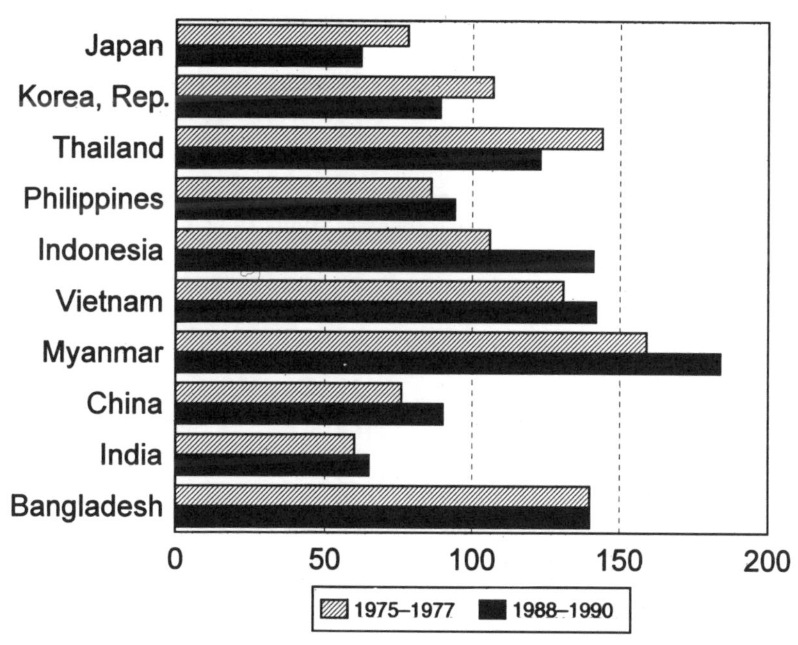

Fig. 6.1. Changes in per capita rice consumption, selected Asian countries 1975–90. (Source: Hossain, 1994.)

incomes, people tend to substitute low-cost sources of energy – such as coarse grains, cassava and sweet potato – for rice. But at high levels of income, rice becomes an inferior good (Ito *et al.*, 1989). As incomes rise further, consumers go for a diversified diet and substitute rice for high-cost quality food with more protein and vitamins, such as vegetables, bread, fish and meat. Growing urbanization which accompanies economic growth leads to changes in food habits and the practice of eating outside the home further reduces per capita rice consumption. Japan and South Korea have already passed through these phases and have experienced a decline in per capita rice consumption after reaching a high level several decades earlier (Fig. 6.1). Recently, Malaysia and Thailand are facing the same experience. But these high- and middle-income countries – where per capita rice consumption has been declining – account for less than 10 percent of total Asian rice consumption. The income threshold at which consumers start substituting rice for higher-quality and more varied foods has not yet been reached in large countries such as India, Indonesia, Bangladesh, Philippines and Vietnam. These countries account for more than 40 percent of total rice consumption and dominate the growth in consumption of rice. The per capita grain consumption in many of these countries is still lower than the peak level reached in Korea and Japan during their early phase of development (Table 6.1). With increased incomes and alleviation of poverty, per capita rice consumption may increase further in these countries.

But the major boost in increasing demand for rice over the next few decades will come from population growth. The Asian population

Table 6.1. Per capita consumption of cereal grains and rice in major rice-growing countries, 1990–1992 (in kg person^{-1} year^{-1}).

	Cereal grains	Rice
China	200	91
India	133	73
Indonesia	169	134
Bangladesh	163	158
Thailand	135	126
Myanmar	199	194
Philippines	137	88
Korea, Rep.	169	105

Source: FAO (1993).

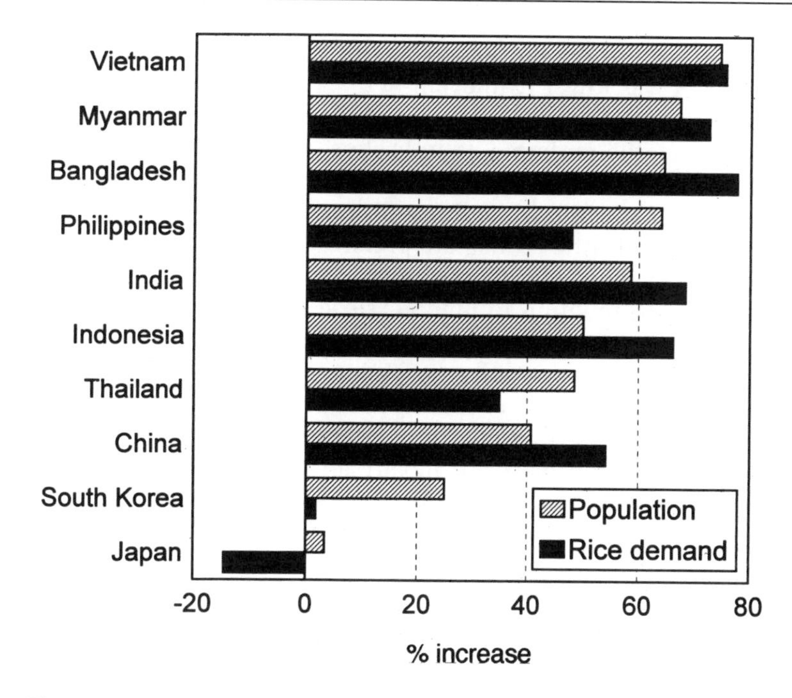

Fig. 6.2. Projection of population growth and demand for rice, 1990-2025. (Source: Hossain, 1995.)

is expected to increase by 18 percent during the 1990s, and by 53 percent in the next 30 years (Fig. 6.2). In many countries, population is projected to increase 60–80 percent by the year 2025. Most of the additional population will be located in urban areas and the marketed surplus of rice has to increase to meet the urban demand. Recent projections show that, at prevailing price levels, the demand for rice will increase by 70 percent during the 1990–2025 period (Rosegrant *et al.*, 1995), most of it due to feeding a larger population. This means that Asian rice production must increase to more than 800 million tons over the next 30 years from the present level of about 480 million tons, a demand growth rate of of 1.8 percent per year.

Changes in quality preferences

Another notable pattern of rice consumption is that, with growing incomes, people express preferences for higher-quality rice once the calorie needs have been met. High-income consumers spend more on rice by paying higher prices for varieties with preferred eating quality which they substitute for the lower-quality variety consumed when the income level was lower (Unnevehr *et al.*, 1992). Tastes and preferences for rice varieties, however, vary enormously across countries in Asia. In East Asia, the preferred grain is short and roundish, with low amylose content which becomes sticky after cooking, as it is convenient to eat with chopsticks. Medium- to long-grain rices, with intermediate amylose and little aroma, are preferred by most consumers in Southeast Asia. In South Asia, consumers generally prefer parboiled long-grain rice with medium to high amylose and strong aroma, Basmati being the most popular of such rices. Most of the modern varieties are considered low quality by Asian consumers. Traditional varieties are generally of higher quality and fetch premium prices in the market. Thailand still grows low-yielding traditional varieties extensively, as the export market is mostly for high-quality rice. When the income level was low, South Korea used to grow the modern 'tongil' variety, but replaced it by relatively low-yielding traditional Japonica rices as consumers expressed preference for them by offering higher prices. In China, rising incomes and easing of market controls have contributed to increased demand for better-quality rice especially among the urban consumers in the coastal provinces (Rozelle *et al.*, 1995). In response to these market signals farmers are eager to grow even low-yielding high-quality rices, because the higher prices more than compensate for their lower yields.

As rice scientists have had limited success in developing high-yielding cultivators with better eating quality, the price difference between the standard- and high-quality varieties has been growing in Asian markets. In Bangladesh, for example, the price difference between medium- and high-quality rice was only about 15 percent in the mid-1970s. The gap had widened to about 36 percent by the early 1990s.

Emerging Supply Side Trends

While the demand for rice continues to grow steadily, both in quantity and quality terms, there are early warning signs that a deceleration in the growth of rice supply has begun to set in. The annual growth in global rice production was only 1.7 percent per year during the 1985–1993 period, compared with 3.2 percent during 1975–1985, and 3.0 percent the previous decade (Fig. 6.3). The recent 1985–1993 growth in rice production has failed to outpace population growth in several countries in Asia (Table 6.2).

There are several reasons for the recent deceleration of rice productivity: (i) the decline in irrigation investments, both for

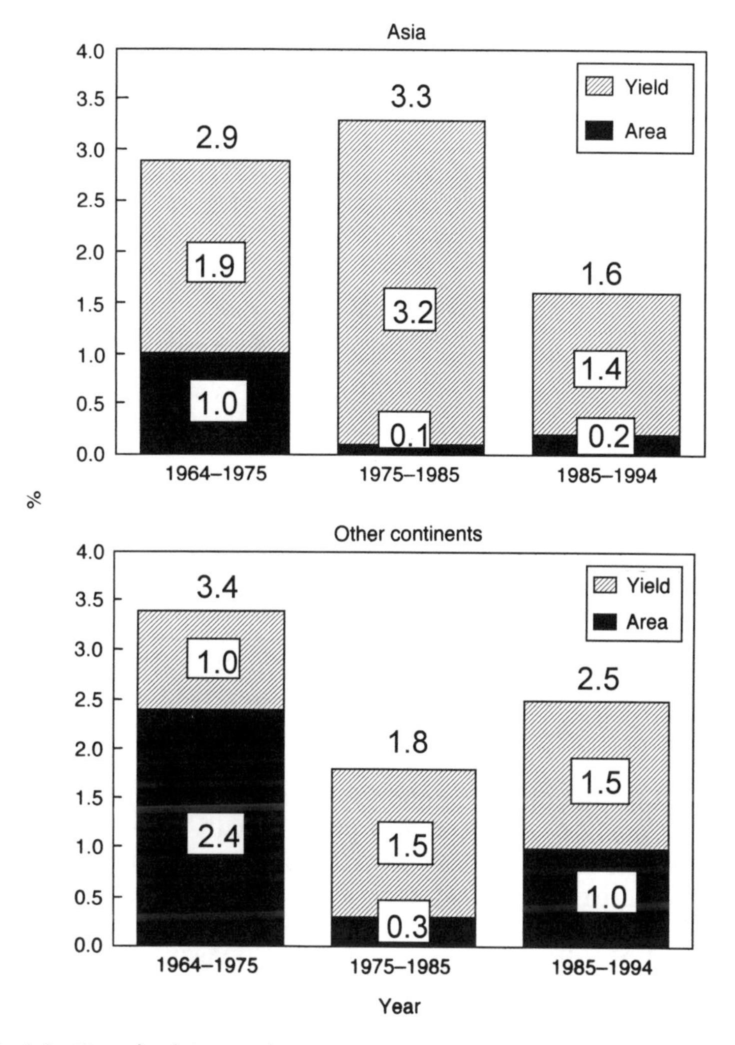

Fig. 6.3. Growth of rice production in Asia and other continents, 1964-1994. (Source: Hossain, 1996.)

expanding irrigated area and for maintaining existing infrastructure (Chapter 8); (ii) the diminished gap between the technological potential and farmer yields in the irrigated lowlands (Chapter 2), and the lack of productivity-enhancing technologies for the rainfed environments; (iii) intensification-induced productivity decline on the irrigated lowlands (Chapter 4); and (iv) reduced profitability of rice farming due to increased competition for inputs, especially land, labor and water (Chapter 9). Many countries of South and Southeast

Table 6.2. Recent trends in growth of population and rice production, major rice-growing countries in Asia.

	Rice harvested area, 1992 (million ha)	Population growth 1980–1991	Growth in rice production (% yr^{-1})	
			1975–1985	1985–1993
China	32.4	1.5	2.9	0.9
India	41.4	2.1	2.2	2.7
Indonesia	10.9	1.8	5.7	2.7
Bangladesh	10.2	2.2	2.3	2.9
Vietnam	6.5	2.1	3.6	4.3
Thailand	9.0	1.9	2.8	0.0
Myanmar	5.1	1.9	4.5	0.9
Japan	2.1	0.5	–1.6	–3.2
Philippines	3.2	2.4	3.5	1.3
Korea, Rep.	1.1	1.2	1.8	–1.6

Sources of basic data: Population data: World Bank (1993). Rice Production: FAO (1992).
Source: Hossain (1994).

Asia will find it difficult to maintain self-sufficiency in rice production over the next 10–20 years. The problem will be eased only if there is a further technological breakthrough in shifting the yield potential for the irrigated ecosystem, development of suitable high-yielding varieties for the rainfed environments, and a substantial increase in rice prices that makes further investment for water resource development profitable.

Technological progress running out of steam

The increase in rice yield over the last three decades has mainly come from the gradual adoption of modern rice varieties on existing irrigated lands, and from public and private investments in irrigation, flood control and drainage, which converted the rainfed into the irrigated ecosystem, and facilitated the adoption of modern varieties and improved farming practices. In the irrigated ecosystem, rice yield increased from 3.0 to 5.8 t ha^{-1} during the 1964–1990 period, but in the rainfed ecosystem it increased only marginally from 1.4 to 1.8 t ha^{-1} (Fig. 6.4) due to a lack of suitable improved varieties.

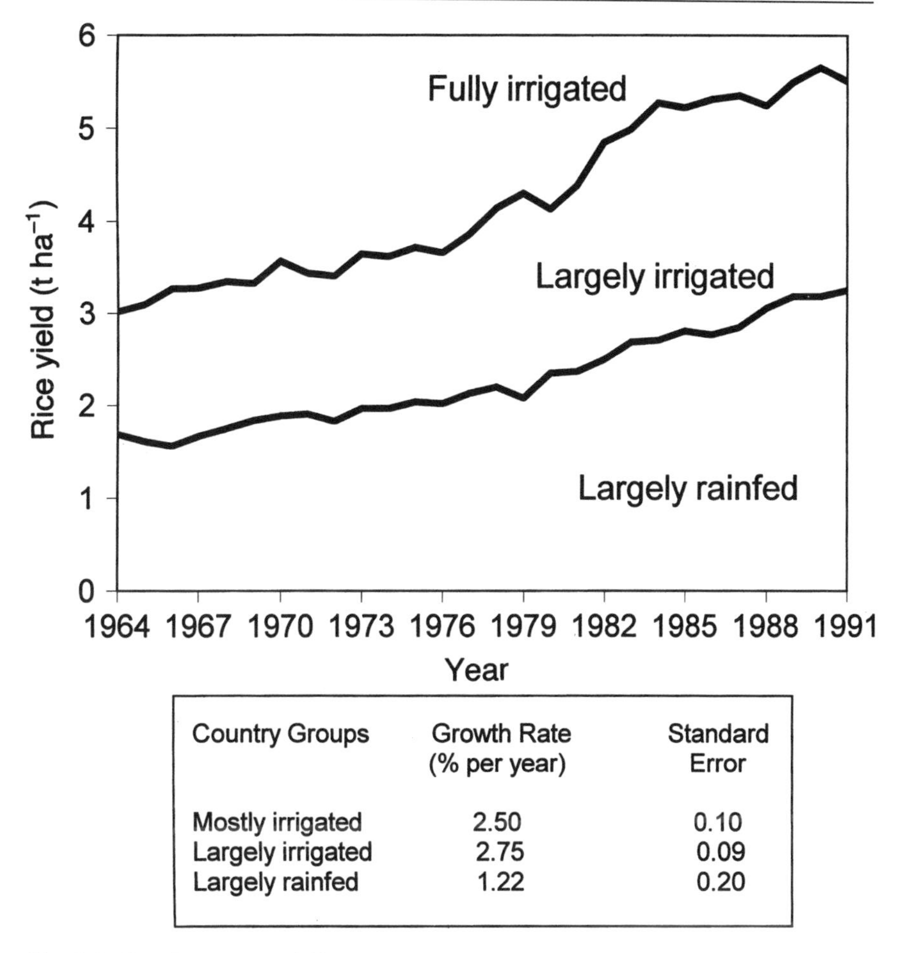

Country Groups	Growth Rate (% per year)	Standard Error
Mostly irrigated	2.50	0.10
Largely irrigated	2.75	0.09
Largely rainfed	1.22	0.20

Fig. 6.4. Trends in rice yield by ecosystem, 1964-1991. (Source: Hossain, 1995.)

In the irrigated rice ecosystem, which accounts for almost three-quarters of total rice supply, most farmers have already planted high-yielding modern varieties and the best farmers' yields are approaching the potential that scientists are able to attain with today's knowledge in that particular environment (Chapter 3). The yield potential of modern rices hardly increased after the introduction of IR-8 in 1966. The later modern varieties incorporated new traits such as resistance to insects and diseases, improved grain quality and shorter crop maturity period (Khush, 1995), but did not shift the yield frontier. Once the potential yield under the optimum management condition is reached on farmers' fields, it is difficult to increase the yield any further. In Japan and South Korea, rice yield has remained stagnant at around 6.0–6.5 t ha^{-1} after reaching that level in the late 1960s and mid-1970s respectively. The maximum achievable yield at the farmer level is lower than 6.0 tons in the humid tropics because of increased pest pressure, and frequent cloudy days with below-optimal sunshine in the humid tropics of Asia (Chapter 2). In regions with good irrigation infrastructure, rice yield is about to reach the 6.0 t ha^{-1} level. Future opportunities for shifting the yield frontier through the use of hybrid rice and a new high-yielding plant type are discussed in Chapter 10.

With intensive monoculture of rice in the irrigated systems, as discussed in Chapter 4, long-term decline in rice productivity is being observed (Flinn and De Datta, 1984; Cassman and Pingali, 1995). Quantitative evidence does not exist on the extent to which the observations, in selected locations, on farm-level productivity decline can be generalized across the irrigated lowlands, and the effect on aggregate productivity trends. It is safe to say, though, that if the current level of rice intensification continues without a shift in the technological frontier then one should expect the current trends in the slow-down of aggregate productivity to continue and become even sharper.

Would productivity growth in the rainfed lowlands be able to compensate for the slow-down in productivity in the irrigated environments? Productivity increases in the rainfed lowlands could come either through improved water control and supplementary irrigation or through varietal improvement, especially tolerance to drought and submergence. The scope for further expansion of irrigation infrastructure in the rainfed lowlands is becoming limited (see Chapter 9 for a detailed discussion). The cost of irrigation has increased substantially as easy options for irrigation development have already been exploited. Also, environmental concerns regarding adverse effects of irrigation and flood-control projects on waterlogging, salinity, fish production and groundwater quality have been growing. Already, there has been a drastic decline in investments for the development and maintenance of large-scale irrigation projects in many Asian countries (Rosegrant and Svendsen, 1992).

Where the groundwater table can be profitably exploited using pumps, one could expect investments in supplementary irrigation and a switch from low-yielding traditional varieties to high-yielding modern rice varieties. The rapid growth in the use of electric and diesel pumps in the Indo-Gangetic plains of India and Bangladesh is an example of this phenomenon. The recent growth in rice productivity in Eastern India has come from the increasingly widespread use of pumps. However, once the investments in pumps are made, it is not clear that rice-cropping intensities will increase. The wet-season crop will continue to be rice, given the high costs of drainage for non-rice crop production (Chapter 9); the dry-season crop, however, need not be rice. In fact, given the high water requirements for rice production where pumps are used for supplementary irrigation, rice would generally not be the crop of choice in the dry season, except when rice prices are very high.

Opportunities for varietal improvement for the rainfed lowlands are limited. This is because rice scientists have yet to succeed in developing appropriate high-yielding varieties that can withstand prolonged drought, temporary submergence and other climatic stresses common in the fragile rainfed environments (Zeigler and Puckridge, 1995). The heterogeneity of the rainfed environments makes the returns to breeding investments low, since varieties developed would be relevant for a very narrow agroecological domain. In any case, the probability of success in breeding for drought and submergence tolerance is very low, even with the use of modern biotechnology tools (Chapter 10). In the absence of opportunities for expanding supplementary irrigation, the prospects for rapid productivity growth in the rainfed environments are limited.

Economic growth and declining competitiveness of rice farming

With declining rice prices and increasing input costs the profitability of rice production systems is declining throughout Asia. While the trends are similar across Asia, factor prices are rising faster in the rapidly growing economies of East and Southeast Asia (Chapter 9). Factors of production that are most seriously affected are land, labor and water. As economies grow, existing ricelands are facing severe competition for urban and industrial uses, and for growing vegetables, fruit and fodder crops. Similar competition for water resources can also be anticipated with population growth, urbanization and industrialization. Although most Asian societies are densely populated, labor scarcity in rural areas is beginning to emerge, especially for peak season operations, as rural labor is bid away into the urban industrial and service sectors. A detailed discussion on factor market trends and outlook is provided in Chapter 9. It is important to note here that future productivity growth of rice would be affected if the unit cost of rice production rose due to increasing competition for factors of production.

The profitability of rice production systems can only be sustained if unit costs of production continue to remain below the output price. Unit production costs can be reduced through a shift in the yield frontier and/or through an increase in input use efficiencies. The prospects for shifting the rice yield frontier are discussed in Chapter 3, while the opportunities for improvements in input use efficiencies are discussed in Chapters 9 and 11. It is important to remember that farm-level incentives for improving input efficiencies and reducing unit production costs depend not only on the availability of technology but also on the policy environment that allows for the undistorted transmission of price signals.

Self-sufficiency versus self-reliance for food security

In most developing countries of Asia, rice is grown on small family farms. A typical Asian farmer plants rice primarily to meet family needs, and hence the marketed ratio for rice is small. Variable natural conditions cause shortages and surpluses to occur from year to year, and these in turn produce wide fluctuations in marketable surplus. This makes prices in both domestic and international market highly unstable. At the national level, an important political objective in most rice-growing countries is to achieve self-sufficiency in rice production, and to maintain stable prices for rice consumers (Timmer, 1989b). Rice is seen by many Asian governments as a political commodity, since it is the single most important element in the diet of the poor, and an important source of income for farmers, who are often a politically powerful group. As a result, governments intervene actively in their countries' rice markets. The interventions take many forms – subsidies and taxes on inputs and output, government control on international trade, and direct participation in marketing through procurement and distribution of grains.

In poorer countries, achieving self-sufficiency in foodgrain production is a major objective because of the lack of foreign exchange to finance major international purchases. Governments also consider it risky to depend on international markets for supply of staple food, because experience has taught them that prices tend to be high in years with a large deficit, and low when there is a surplus. Thus both farmers and governments lose from participation in markets. However, a commitment to self-sufficiency is not confined to the governments of poorer countries (Anderson and Hayami, 1986). High-income countries with no financial constraint to importing rice from low-cost sources have also tried to maintain self-sufficiency by providing support to rice farmers. If rice farming were abandoned, farming infrastructure such as irrigation facilities would not be maintained, and at times of crisis it would be difficult for the economy to revert back to rice farming. At high income levels consumers complain less about paying high prices to support relatively low-income farmers, because the cost of rice is a small part of their food bills and a tiny fraction of their incomes.

A country need not have self-sufficiency in rice production to achieve or sustain food security. In Asia, Singapore and Hong Kong produce very little foodgrains but have better records of food security than most countries of South and Southeast Asia. Malaysia meets almost 40 percent of its rice needs through imports. What is important is achieving food self-reliance. It requires at the national level a favorable export growth that permits import from food surplus countries, and at the household level productive employment and adequate incomes to acquire the needed food from the market (Hossain, 1995).

While countries with small populations, and/or small domestic rice demand can rely on the world market for a large part of their rice supplies, most large Asian countries would find it unwise to follow such a strategy. India, for instance, consumes 44 million tons of rice per year – this is 20 percent of the total amount of rice sold in the world market. If India tried to meet an unexpected shortfall of 3 percent of its supplies through imports, it could lead to a sharp increase in rice prices. If India, China and Indonesia entered the world market at the same time in search of additional supplies, there might not be enough for all three countries at affordable prices. Large Asian countries would therefore minimize their reliance on the world market for rice supplies and would continue to seek growth in domestic supplies through productivity improvements.

The Rice Market Prospects

Given the political importance of adequate and reliable domestic rice supplies, most governments favored investments and policies for increasing domestic production; consequently international trade in rice has been very limited. About 4 percent of the world's rice production is currently traded internationally, in contrast to nearly 20 percent for wheat and 11 percent for coarse grains. A large number of countries import rice, but on a very small scale. Of the major rice-consuming countries in Asia, few are major importers (except Malaysia and Sri Lanka), and it is only during times of natural calamities that they go to the world market. The rice-growing sector of Asia accounted for only 16 percent of the import market during 1992, a decline from about 68 percent in 1961 (Table 6.3). This reflects the achievement of self-sufficiency by large rice-importing countries such as India, Indonesia, Bangladesh and the Philippines.

The major rice importers today are Sub-Saharan Africa, the European Economic Community, Iran and Saudi Arabia. Thailand and the

Table 6.3. World rice market: changes in the pattern of trade, 1961-1992.

Regions	Share of exports (%) 1961	Share of exports (%) 1992	Share of imports (%) 1961	Share of imports (%) 1992
Asia	71.4	68.1	72.9	38.8
East Asia	8.7	6.9	10.3	4.0
Southeast Asia	55.5	46.8	32.1	9.5
South Asia	6.8	13.1	25.8	2.8
Middle East	0.4	1.3	4.7	22.5
Africa	4.6	1.3	8.0	24.8
Latin America	4.8	5.0	5.6	13.2
High-income countries	19.2	25.6	13.5	23.2
Volume of trade (million tons of milled rice)	6.8	15.8	6.7	15.2

Source of basic data: IRRI (1995).

US are the most important rice exporters, accounting for about 55 percent of the export market. China and Myanmar were important exporting countries in the 1960s, but have now become minor exporters. Thailand is gradually losing its comparative advantage in rice production and exports due to the rapid increase in farm wages. Only a decade ago, Thailand used to tax rice exports. It has already started providing export subsidy to raise prices in the domestic markets to sustain farmers' incentives in rice cultivation. The place of Thailand as a major rice exporter can easily be taken over by Myanmar, where there is surplus production capacity (Hossain and Oo, 1995). Vietnamese farmers responded favorably to the economic liberalization introduced in recent years (Pingali and Xuan, 1992), and Vietnam became the third most important exporter of rice in the world market due to rapid growth in rice production since the mid-1980s. But Vietnam has almost exploited its potential for increasing rice yields and may reduce future exports to accommodate the growing internal demand (Pingali *et al.*, 1996).

It is anticipated that over the next two decades Asian countries may substantially increase their rice imports in order to make up for growing domestic shortfalls. Domestic shortfalls in rice supplies could occur due to: (i) continuing and unabated population growth in the low-income developing Asian countries; (ii) rapid economic growth in the middle- and higher-income countries that makes the maintenance of self-sufficiency unprofitable; and (iii) trade liberalization and trade agreements that could lead to the Asian food sector becoming more open to international competition. Given the anticipated domestic shortfalls in rice production in Asia over the next two decades, we ought to ask the question where the additional rice supplies are going to come from. The opportunities for substantial additions to supplies from the traditional Asian rice exporters are limited, as discussed above. As the Asian import market grows, we ought to anticipate increased supplies from non-traditional sources such as Latin America and Africa.

The rice market is differentiated by quality preferences: the lower-quality, long-grain Indica rices and the higher-quality, medium-grain Japonica rices. Population-induced growth in demand from most developing Asian countries is expected to concentrate on Indica rices, hence these rices will continue to have the predominant share of the market. As incomes grow, however, the demand for Japonica rices is expected to grow, especially in the East Asian countries including China. The discussion below is therefore arranged according to these two primary segments of the rice market.

Meeting domestic shortfalls in Indica rice through international trade

Attention is currently focused on obtaining the additional future rice supplies from two sources: middle- and high-income Asian countries where per capita rice consumption is declining; and the expansion of production in Africa and Latin America.

As per capita rice consumption declines with economic growth, the middle- and high-income countries of Asia should have some surplus rice available for export to the low-income food-deficit countries. In Japan, peak rice harvest reached 18.8 million tons in 1967 but it started declining from that level and reached about 12 million tons in 1992. In Taiwan, the peak reached 3.6 million tons in 1976 while the present production is only about 2.0 million tons. These countries could have maintained their production levels through exports of surplus rice to other countries.

The competitiveness of rice farming was maintained initially through: (i) improved farm management practices that increase efficiency in the use of non-land inputs and thereby reduce the cost of production; and (ii) increased use of capital to replace labor through mechanization of farming operations so that labor productivity could be continuously raised when no further increase in land productivity is possible. But these technological changes were not adequate for raising incomes of the farmers at par with those of urban workers. The government had to come forward with protection of the domestic rice market so that the price mechanism could be used to transfer income

Table 6.4. Rice yield and unit costs of production, selected countries.

Country	Season/type	Rice yield (t ha^{-1})	Cost of production (US$)	
			Per hectare of land	Per ton of output
Bangladesh[c]	Wet season	3.37	327	97
	Dry season	4.56	513	113
Vietnam[b]	Autumn	3.80	353	93
	Spring	5.35	333	62
China[c]	Early season, Indica	5.34	416	78
	Middle season, Indica	6.49	399	62
	Japonica	6.58	513	78
Indonesia[a]	Irrigated	5.76	474	82
	Rainfed	3.57	389	109
Thailand[a]	Irrigated	3.78	369	98
	Rainfed	1.84	223	121
Colombia[a]	Irrigated	5.61	1144	204
	Rainfed	4.71	914	194
Costa Rica[a]	Irrigated	4.33	1020	236
	Rainfed	3.71	1117	301
Burkina Faso[a]	Irrigated	4.73	1707	361
	Rainfed	2.50	720	288
Zambia[a]	Irrigated	5.00	5515	1103
	Rainfed	2.50	808	323
Korea, Rep.[a]	Irrigated	6.50	4348	669
Italy	Irrigated	5.87	3188	543
Japan[a]	Irrigated	6.51	12935	1987
USA[d]	Long grain	5.94	1339	225
	Medium grain	8.57	1889	220

Sources: [a] Yap, C.L. FAO 1991 for 1987–1990 period.
[b] FAO, Rice Policy in Vietnam for 1993.
[c] Obtained through personal communication. China for 1991 and Bangladesh for 1993.
[d] USDA-ERS for 1991.

from the relatively well-off rice consumers to the relatively poor rice producers. In Taiwan, the government developed infrastructure facilities in rural areas to promote the growth of rural non-farm activities that made possible involvement of rural households simultaneously in farm and non-farm activities. As part-time rice farming increased, the household could compensate for the slow growth in farm incomes from the fast-growing incomes from non-farm sources. As the cost of rice cultivation continued to increase due to the rising opportunity cost of labor and land, the government had to raise rice prices and farm subsidies to maintain the balance between rural and urban household incomes. The protection of the domestic rice industry has led to the high-cost domestic production of rice. The cost of producing rice in Japan is about 17 times higher than in Southeast and South Asian countries and about 10 times higher than in the US (Table 6.4). Thus, having an exportable surplus of rice from the middle- and high-income countries of Asia is highly improbable. Rather, if these countries are exposed to international competition, as discussed in the next subsection, the area under rice cultivation may decline further and they will need to import rice to meet the domestic demand.

The opportunities for further expansion in Indica rice production from the United States and Australia are limited by the high costs of rice infrastructure, especially irrigation development (Childs, 1990). There is some potential for the expansion of rice area in humid tropics in Africa and Latin America. It is estimated that there are 20 million hectares of inland valleys in West Africa alone of which only 15 percent are currently cultivated. In tropical South America, an additional 20 million hectares of land may come under crop production by the year 2010 (FAO, 1993). Together, this accounts for 29 percent of the global rice harvested area. However, the current cost of rice cultivation is many times higher in Africa and Latin America than in low-income Asian countries (Table 6.4) because of labor scarcity and high wage rates. Also because of poor infrastructure facilities, the cost of transport and trade margin is much higher in Africa than in Asia (Ahmed and Rustagi, 1987). Thus, even if African and Latin American farmers could produce some surplus without significant technical change especially in labor-saving technologies, they might find it difficult to market that rice to feed the low-income Asian families.

Within Asia itself, Thailand, Myanmar and Cambodia have considerable excess capacity to meet potential shortages in other countries. The cost of production will remain competitive for a long time at least in Myanmar and Cambodia. If rice prices go up, farmers will be encouraged to increase rice production by investing in irrigation and chemical fertilizers, and adopting higher-yielding varieties. This might generate an additional 10 million tons of exportable surplus, less than 4.0 percent of the total rice consumption in Asia. Within the next 20 years, a doubling of the international rice market from its current level of approximately 15 million tons to 30 million tons could be supported by an expansion in production primarily from Asia, with additional supplies coming from non-Asian sources, as and when technical change occurs to reduce production costs.

For the low-income food-deficit countries – especially those with large rice-eating populations such as India, Indonesia and Bangladesh – achieving food security through international trade may not be possible due to the high volumes involved and foreign exchange constraints. Also, since rice production is a major rural economic activity, and land and labor cannot be easily diverted to other economic activities during the monsoon season, low-income households may find it difficult to access the imported food. If economic conditions of small farmers and landless laborers fail to improve due to stagnant productivity of rice, the increase in rice prices will only aggravate the poverty situation in the food-deficit countries. For the low-income countries of Asia, increasing productivity growth on existing ricelands will continue to be a top priority activity. The technological and policy options for increasing rice productivity in the above countries are discussed in Chapters 10 to 12 of this book. The bottom line is that rice imports cannot be seen as a substitute for increasing rice productivity for most developing countries in Asia.

Trade liberalization and the rice market

The Uruguay Round of trade negotiations has increased pressures to liberalize agricultural trade and to open up rice markets in the middle- and high-income countries of Asia (Pingali, 1995). As part of the recently concluded GATT agreements, Japan and South Korea have agreed to eliminate import bans on rice and establish import quotas of 427,000 tons in 1995 that will rise to more than 850,000 tons by the year 2000 and to nearly 1 million tons by 2005. The United States and European Union will reduce tariffs for rice by 36 percent and the US will establish a ceiling for the quantity of subsidized exports and reduce budgetary outlays for export subsidies of rice. Malaysia will continue to have zero import duty on rice. Details of the GATT agreement on agriculture and its implications for the rice market and rice research are discussed in Chapter 7. In the short to medium term, the effect of the GATT agreement will be felt more on the trade in the high-quality Japonica rice rather than in the lower-quality Indica rice.

A number of studies have attempted to estimate the possible effects of the trade liberalization on the world rice market (Anderson and Tyers, 1990; Childs, 1990; Brando and Martin, 1993). The findings show that market prices of rice will increase, more for the medium-grain Japonica rice than for the long-grain Indica rice, and more in the short run than in the long run. The volume of Japonica rice trade in the international market is expected to increase with larger imports coming from Japan, South Korea and Taiwan. Larger exports will come mostly from China, Australia and the US.

In the short run, the price of Japonica rice is expected to rise substantially because: (i) only about 12 percent of the world trade is in Japonica rice; and (ii) there is limited potential for expansion of Japonica rice production in traditional rice-exporting countries. In the US, the medium-grain rice is produced in about 300,000 hectares (25 percent of total rice area), mostly in California (Robinson, 1993). US producers would have strong incentives to increase production due to the increase in prices. However, the expansion capacity in the US is severely limited because of: (i) restrictions on the supply of low-cost irrigation water in California; (ii) the need to rotate area to control diseases; and (iii) higher cost of production and lower quality of Japonica rice in states other than in California. Australia is the largest exporter of quality Japonica rice, but it also faces severe land and water constraints and may not be able to expand production and exports substantially.

Thailand, Cambodia and Myanmar have low rice yields and have considerable excess capacity for rice exports. But they cannot produce high-quality Japonica rice due to unsuitable agroclimatic environment. These countries will benefit marginally, mainly from redistribution of world trade between Japonica and Indica rice. Australia and the United States will most likely divert all their Japonica rice to Japan and Korea, and the displaced importers will turn to exporters of Southeast Asia for high-quality Indica rice.

The country that is likely to gain most from the trade liberalization is China. Northeastern China – Manchuria and the Yellow River Basin – produces relatively good quality Japonica rice at a cost (US$ 150 per ton in 1990) much lower than that in California. China has had a significant freight advantage over the United States in servicing the East Asian market. In recent years, China has had a remarkable expansion in the cultivation of Japonica rice, mainly through an expansion of area under cultivation. In the three northern provinces of Jilin, Liaoning and Heilong, which are in the same latitude as Japan and Korea, ricefields expanded from 0.85 to 1.78 million hectares, and rice production increased from 4.2 to 10.9 million tons during the 1980–1992 period.

But China may not be able to maintain its competitive strength in the long run. In China, land rent is not considered as a component of the cost of production and wage rate is around one US dollar per day, only a small fraction of that in Japan and Korea. As the Chinese economy is also growing fast, the wage rate and the opportunity cost of land are going to increase. With the easing of government control on the rice market, the internal demand for Japonica rice has been increasing in recent years, and the gap in prices between Japonica and Indica rice (about 50 percent) has been growing wider. It is likely that China will soon have an upward pressure on rice prices to maintain the balance between farm and non-farm incomes.

As in the case of Indica rice, meeting the demand for Japonica rice in East Asia would require long-term growth in domestic supplies, even with trade liberalization. In Japan, Korea and Taiwan, productivity growth would have to come with a reorganization of agricultural production. It is likely that the size of rice farms in Japan and Korea will increase as urban residents give up the ownership of rural holdings. This will lead to more efficient utilization of large-scale agricultural machinery and a reduction in the number of part-time farmers now tied up in supervision and management of many small farms. The consolidation into larger holdings and the adoption of more efficient crop-management technologies may reduce the cost of maintenance of family labor, and help maintain farm household income when rice prices and profits decline. There is also limited scope of diverting the flooded ricefields to other uses which will put political pressure on keeping some trade barriers. Considering the above, the fear of extinction of rice farming in prosperous Asian countries following the move towards liberalization of agricultural trade may be misplaced. The agreements regarding the liberalization of agricultural trade may thus lead to only a marginal expansion in the world rice market.

Conclusions

The Asian rice market will continue to remain segmented. Because of the overwhelming importance of rice as a staple food and its political and cultural significance, Asian nations will continue to address the issues of rice producers' and consumers' welfare through inward-looking policies. The short-term disequilibrium created by the limited opening of the East Asian rice market will be adjusted in the long run by the increase in production cost and prices of Japonica rice in countries with lower costs, as well as by the accumulation in landholding and cost reduction in high-cost Asian countries. The price increase may *not* be transmitted to the market for Indica rice. The low-income, rice-consuming countries will continue to struggle to win the race between foodgrain production and population growth.

International support is needed to step up agricultural research, develop rural infrastructure, and address the food security problem in those parts of Asia which are characterized by extensive poverty and overexploitation of natural resources.

GATT and Rice: Impact on the Rice Market and Implications for Research Priorities

7

Introduction

This chapter focuses on the agricultural sector, specifically on rice, and identifies the short-term and longer-term consequences of the implementation of the General Agreement on Tariffs and Trade (GATT). The implications of trade liberalization on the rice market and rice research priorities are identified. This chapter argues that the short-term impact of GATT on rice production and trade could be modest. Over the long term, however, agricultural transformation and the commercialization of rice systems could lead to a substantial change in the world rice market (Chapter 8). In the middle- and high-income countries of Asia, changes in the organization of agricultural production, consolidation of small holdings, and vertical integration of the rice industry would help sustain rice profitability and supplies (Chapters 8 and 9). However, given the large demand for rice in Asia, imports of at least a part of its rice requirements may be necessary from other continents. Sustaining productivity growth in the irrigated lowlands and the intensification of the rainfed lowlands could reduce the magnitude of imports required.

This chapter provides the following:

1. A description of the agreement on agriculture signed as part of the Uruguay Round of GATT negotiations.
2. An assessment of the short- to medium-term impact of GATT on rice trade and rice prices.
3. A speculation of the implication of trade liberalization and economic reforms on the reorganization of agricultural production and the rice market.

4. Implications for research resource allocation and priorities for rice research.

GATT: Agreement on Agriculture

> The General Agreement on Tariffs and Trade – GATT – is a binding contract between 105 governments which together account for 90 percent of world merchandise trade. The objective of the contract is to provide a secure and predictable international trading environment for the business community and a continuing process of trade liberalization in which investment, job creation and trade can thrive. In this way, the multilateral trading system contributes to economic growth and development throughout the world.
>
> (GATT, 1992)

The final agreement on multilateral trade, resulting from the Uruguay Round of GATT negotiations, was signed by the contracting countries in December 1994 and went into effect on 1 January 1995. The above legally binding agreement provides a comprehensive set of rules for the conduct of trade in agricultural commodities and for the conduct of domestic agricultural policy to the extent that it impinges on international trade. An important factor determining the effectiveness of the Agreement is that it not only establishes general rules to be observed in international trade, but that all participating countries undertake specific commitments expressed in their schedules. This section discusses the major elements of the Agreement on Agriculture and highlights the agreed-upon exceptions to it. Tangermann (1994) and Josling *et al.* (1994) provide a comprehensive discussion on the Agreement on Agriculture. FAO (1994) provides an assessment of GATT for developing country agriculture.

Commitments for agricultural reforms were made in three broad areas: market access, export competition, and domestic support. The actual mechanisms for encouraging trade flows are through the reduction in tariffs, subsidized exports, and domestic price supports, and through minimum access commitments. Developing countries were treated differentially in terms of lower levels of commitments and longer periods for compliance.

The implementation of GATT starts in 1995, and the reduction commitments of the developed countries should be completed by the year 2000, whereas the commitments of the developing countries should be completed by 2004. The next round of GATT negotiations should start in 1999 and normally last for four years.

Market access

All participating countries have agreed to bind all tariffs, to convert all existing non-tariff barriers into bound tariffs, and not to introduce new non-tariff measures. For developed countries, bound tariffs have to be reduced by 36 percent over the six-year implementation period (1995 to 2000), on a simple (unweighted) average basis, with a minimum reduction rate of 15 percent for each tariff line. In the case of developing countries, reduction commitments are only two-thirds of those for industrialized countries and the implementation period is ten years (2004). In the case of least-developed countries, there are no reduction commitments although they also are expected to bind their policies at the base-period level (1986–1988). A few countries, notably Japan and Korea, had sought and obtained special treatment with respect to rice imports. The specific features of the 'rice clause' are discussed later in the chapter.

In order to hasten the entry into traditionally closed markets, minimum access provisions are being implemented. Where there are no significant imports, minimum access equal to 3 percent of domestic consumption in 1986–1988 will be established for 1995, rising to 5 percent of base-year consumption at the end of the implementation period. In cases where current access opportunities are more than the minimum, they will be maintained during the implementation period. Importation of minimum access quantities is not guaranteed, although reduced tariff rates are provided as an incentive to fill these quotas.

Export subsidies

Subsidized exports have been brought under the purview of international regulation as part of the GATT treaty on agriculture. Countries have individually accepted legally binding commitments regarding maximum export subsidies. Exporting countries have accepted commitments leading to a reduction in expenditure on export subsidies of 36 percent, as well as reduction in the quantity of subsidized exports by 21 percent during the six-year implementation period. Participating countries have also agreed not to provide export subsidies in the future for commodities that are currently not subsidized. Developing countries do not generally subsidize exports and hence are largely unaffected by the above rule.

Domestic subsidies

Domestic subsidies are measures implemented by a country to reduce the costs of production or increase the net revenues received by producers in the domestic market. Domestic subsidies come under GATT purview to the extent that they have trade-distorting effects. This is the first time that GATT has direct-intervention capacity with respect to domestic agricultural policy. Agricultural policies are divided into two groups: (i) those with trade-distortion effects; and (ii) permitted policies – Green Box policies – those with minimum distortion effects. The first set of policies, those that lead to production beyond the economic optimum, are to be quantified and are known as the aggregate measure of support (AMS). Developed countries will reduce their AMS by 20 percent over a period of six years, starting in 1995. Developing countries are to reduce their AMS by 13 percent over a period of ten years, also starting in 1995.

Green Box policies, those that encourage investments in agriculture and subsidize production inputs critical to the development of agriculture in developing member countries, are exempted from reductions. Green Box policies include investments in research, pest and disease control, training, extension and advisory services, inspection services, marketing and promotion and infrastructure services.

Sanitary and phytosanitary measures

Participating countries agreed to greater transparency in rules and screening procedures for sanitary and phytosanitary measures. The idea was to make it easier to distinguish between genuine health and safety concerns and disguised protection. Countries continue to have the right to set their own health and safety standards but these are to be based on 'sound scientific evidence', and international standards are to be followed to the extent possible. International standards are to be based on the guidelines provided by organizations such as the Codex Alimentarius Commission, the International Office of Epizootics and other similar agencies.

Impact of GATT on Rice Trade and Rice Prices: Short and Medium Term

This section provides a preliminary assessment of the impact of GATT rules on rice trade. The period under consideration here is 2000 to 2004, which is the short- to medium-term response period. Additional supplies are assumed to be coming from existing capacity, without substantial investments in increasing capacity, such as adding large areas under irrigated rice production. Changes in food preferences follow current trends over this period; no dramatic deviations from the trend are anticipated. Information on country-specific negotiated positions are summarized to the extent possible.

Market access – special treatment for rice

In the short- to medium-term a modest expansion in rice trade can be anticipated due to the opening up of traditionally closed rice markets, such as Japan and Korea. However, due to a negotiated special treatment for rice the extent of market opening is not as dramatic as it could have been. During the final stages of GATT negotiations, political considerations led to the adoption of a special clause for certain commodities. Because of the importance of rice-importing countries in this negotiation, it has been dubbed the 'rice clause'.

The special treatment given for rice is applicable to developing countries where rice is the predominant staple and to developed countries that import less than 3 percent of their consumption. Japan, South Korea and the Philippines availed the 'rice clause' and Indonesia negotiated a separate agreement on rice imports. The above countries are exempted from tariff reductions in exchange for minimum access quotas. For developed countries the quotas amount to 4 percent rising to 8 percent of domestic consumption over a six-year period. In the case of developing countries, the corresponding quota is 1–2 percent in the first five years, rising to 2–4 percent in the next five years.

Table 7.1 summarizes the negotiated rice imports of Japan, South Korea, the Philippines and Indonesia. Since the Philippines and Indonesia are currently importing amounts equal to or greater than the negotiated levels, their settlement does not lead to additional import

Table 7.1. Negotiated imports, selected Asian countries, 1995–2004 (000 tons).

	1995	2000	2004
Japan	379	758	800[a]
South Korea	50	100	200
Philippines	59	120	239
Indonesia	70	70	70

[a] To be renegotiated by the year 2000.
Source: Pingali (1995).

requirements. The Japanese and Korean settlements, however, put upward pressure on the rice market. Starting with imports of a little over 400,000 tons in 1995, the two countries are expected to import over a million tons by the year 2004. This is still a conservative estimate since the rice exception for Japan will be renegotiated by the year 2000. Japanese and Korean imports are primarily Japonica rice and will put an upward pressure on a market that is severely constrained on the supply side due to unique agroclimatic requirements.

The share of Japonicas in world trade is approximately 12 percent, with the United States, Australia and China as the main suppliers. Additional supply would have to come through displacing existing importers, reducing domestic consumption and/or through the expansion of area under Japonicas. In the short to medium term, Japan and Korea can outbid current importers and domestic users of Japonica rice from the United States and Australia, leading to an increase in its price (USDA, 1994). The US and Australian capacity to substantially expand areas under Japonicas is severely limited by climatic and water constraints. China can most readily increase Japonica exports and will gain from the higher prices and increased export volumes (USDA, 1994). Given the high domestic demand for long-grain rice in China, major area re-allocation to Japonicas will require high levels of long-grain imports.

Long-grain rice trade, which will also increase due to displaced importers of Japonicas, will be increasingly re-directed to Thailand, Vietnam and possibly Myanmar. The last is the only one with significant excess capacity for medium-term expansion once producer incentives have been re-established (Pingali and Siamwalla, 1993).

Reduced tariffs and export subsidies

The European Union and the US have both agreed to reduce their tariffs on rice by 36 percent by the year 2000. In the case of the US, tariff reduction is not expected to lead to an increase in exports (USDA, 1994). although the effect of tariff reductions on the imports of specialty rices, such as Basmati, needs further examination. For the European Union, tariff reductions could lead to an increase in the imports of high-quality rice. Table 7.2 indicates imports as much as 300,000 tons by the year 2004. Rice consumption in the European Union during the 1986–1988 period was approximately 1.7 million tons; anticipated exports equal 18 percent of baseline consumption.

The world rice market will not be affected significantly by the reduction in export subsides. Subsidies for rice are essentially provided by the US and the European Union. In both instances, subsidy

Table 7.2. Export subsidies for rice of selected countries, 1995–2000.

	Annual quantity[a] (000 t)		Annual outlay (000 US$)	
	1995	2000	1995	2000
US	272 (11.45)	39 (1.64)	15,706	2369
EU	177 (16.46)	145 (13.24)	58	40
Australia	none		none	
Thailand	none		none	
Vietnam	none		none	

[a] Quantity exported with subsidy.
Figures in parentheses are percent of subsidized exports in total exports for 1986–1988.
Source: Pingali (1995).

reductions have been negotiated. However, subsidized exports form a very small portion of total exports of the countries concerned.

Domestic policy – subsidies and producer support

In the case of developing countries, domestic subsidies take the form of fertilizer subsidies, provision of certified seeds and other inputs at below-market price levels. Price support mechanisms are also considered part of producer support. Where the sum total of support provided is less than 10 percent of the total value of production – the *de minis* level – reductions are not required. Most developing countries claim subsidies below the 10 percent level and hence the impact on production from this clause can be expected to be minimal. Developed countries have negotiated settlements on domestic support that will not lead to reductions in production (Tangermann, 1994).

Rice prices and supplies

The USDA projects that rice prices will rise by 11 percent relative to trend estimates by the year 2000 and by 14 percent by the year 2005 (USDA, 1994). The price increases are related to the shift in the demand for Japonica rice caused by increased purchases from Japan and Korea in the face of a relatively inelastic supply. In the short to medium term, the Japonica-rice price-rise leads also to a rise in the price of long-grain rice due to increased competition for international supplies. Major increases in production are not anticipated in the short to medium term. Long-term prospects for prices and supplies depend on: (i) the emergence of new rice suppliers, such as Myanmar, Latin America, etc.; (ii) economic incentives for sustaining current production levels in developing Asian countries; (iii) changes in food preferences relative to population-induced growth in rice demand; and (iv) further liberalization of the rice markets in East Asia. These issues are discussed further in the next section.

Implications for Trade Liberalization and Economic Reforms on the Rice Sector

This section takes a longer-term (beyond 2005) and a more speculative view of the transformation of the rice sector. In order to understand long-term changes in the rice sector it is important to recognize that: (i) trade liberalization is a continuous process; it does not end with the current GATT agreement; (ii) GATT has direct impacts on the relative profitability of rice versus non-rice agricultural enterprises; and (iii) employment generation and income growth in the non-agricultural sector due to GATT have significant impacts on the nature and organization of agricultural production. In essence, the long-term consequences of GATT can only be understood through a holistic understanding of the interlinkages between the various sectors of the economy. The point being made in this section is that increased competitiveness for production resources, both in the agricultural and non-agricultural sectors, could lead to movement away from rice self-sufficiency to self-reliance with imports at the margin.

Income growth in the non-agricultural sector

Over the long term, GATT implementation could lead to significant growth in the non-agricultural sector based on the principles of comparative advantage. Worldwide income growth is projected to increase by as much as 5 trillion US dollars over ten years (USDA, 1994). While the gains are certainly not expected to be uniformly spread across the developed and developing countries, one can still anticipate significant reorientation of production and income gains in the developing countries. The Philippines, for example, expects the following GATT-related annual benefits from the agribusiness sector alone: (i) a 3.4-billion peso increase in agricultural trade earnings; (ii) a 60-billion peso increase in gross agricultural value added; and (iii) the creation of 500,000 additional jobs (Department of Agriculture, 1994). Gains in the agribusiness sector could be expected to be relatively smaller than the anticipated gains from increased trade opportunities in the industrial sector.

There are several implications for both the demand and supply of rice in the face of rapid growth in the non-agricultural sector. On the demand side, over the long-term, one should expect a shift towards a more diversified diet that includes vegetables, meat and dairy products (Chapter 2). The downward shift in the demand for rice induced by income growth is tempered by continued rapid population growth (Chapter 6). Rice supplies could be expected to be negatively affected by competing demands, from the non-rice and the non-agricultural sectors, for land, labor and other factors of production (Chapter 9). Several countries that are now self-sufficient in rice may find that it is more profitable to import at least part of their rice requirements in exchange for diverting production resources to more remunerative activities. The reorientation of agricultural production away from self-sufficiency concerns is triggered more by the price responsiveness of individual farmers than by elaborate planning exercises performed by the state.

Re-alignment of land use in the high-potential areas

Diversification out of rice monoculture systems is most likely to occur in the irrigated lowland environments (Pingali, 1992). Crop diversification can be both in terms of permanent movement out of rice systems or in terms of seasonal diversification (Chapter 8). Where export markets are well established, permanent switches from irrigated rice systems to horticulture and aquaculture have been observed, as in the recent transition in the Central Plains of Thailand. Trade liberalization resulting from GATT could, over the long-term, create an environment that would be conducive to such permanent change in enterprises, although the area under such systems would be relatively small.

Domestic income growth that results from GATT could also lead to increased diversification trends in the irrigated lowlands. The demand for vegetables and fodder crops could lead to dry-season diversification, especially in peri-urban areas. Rice would continue to be the crop of choice in the wet season due to the high drainage costs of the alternatives (Chapter 8).

Re-alignment of land use in unfavorable areas

In environments that are unfavorable to rice production, the response to trade liberalization could be expected to be different for the uplands as opposed to the rainfed lowlands. In the uplands, improved transport infrastructure and market access could lead to a shift away from subsistence rice production. The movement away from upland rice is well under way in much of Asia today and one can expect the current trends to accelerate with GATT. Soil conditions permitting, upland areas will tend to specialize in high-value commercial production systems.

In the case of the rainfed lowlands, one ought to expect rice to predominate because the drainage requirements for growing a non-rice crop in the wet season are too high and uneconomical. There are several changes in the organization of rainfed rice production that are to be expected with GATT. Given increasing non-farm employment opportunities and the consequent withdrawal of labor from the agricultural sector, rainfed systems will be reorganized in order to make them competitive relative to other opportunities. The movement from subsistence to market-oriented rainfed production could follow a general pattern:

1. The abandonment of highly drought-prone environments, especially in areas where opportunities for groundwater exploitation are limited.
2. The shift from small subsistence farms to mechanized cultivation of large farms.
3. Where dry-season water supplies are available, increased areas under vegetables, fodder legumes and other high value crops. With GATT the rainfed lowlands will have a comparative advantage in rice production. However, given the productivity differences, the bulk of market supplies will still have to come from the irrigated environments.

Prospects for the rice market and regional re-alignments in rice production (within Asia and outside Asia)

Over the long-term the reorganization of production resources in the traditionally rice-growing environments of Asia, especially the diversion of high-potential lands to non-rice enterprises, could lead to a net increase in the import demand for rice in several Asian countries particularly where the protection of the market has led to high-cost domestic production. Over the long term, changing comparative advantages could lead the major rice-consuming countries of Asia to import at least 5 percent of their consumption requirements. If India, Bangladesh and China were to import 5 percent of their baseline consumption requirements (1986–1988 levels), it could lead to an increased pressure on the world rice market of approximately 8 million tons (Table 7.3). If the other rice-importing countries of Asia are included, rice-import requirements could rise by an additional 2 million tons. Potential import demands, in the long-term (beyond 2005), could lead to an expansion in the world rice market from the current level of 14 million tons to 24 million tons, an increase of 70 percent. This additional demand is exclusive of the increased import demands due to increased population growth and degrading current productive resources, such as the degradation of irrigation infrastructure in countries that are currently self-sufficient in rice.

Can the current rice exporters provide the additional 10 million tons required beyond the year 2005? If not, what are the prospects for additional supplies coming from countries that are currently minor exporters but with capacity for expansion (Chapter 6)? The major current rice suppliers are Thailand, the USA and Vietnam. Expansion of exports from the above suppliers is severely constrained by agroclimatic conditions, increasing costs of expanding irrigated areas, both economic and environmental, and increased competition for production resources devoted to rice. In Asia, rice export supplies could potentially come from Myanmar and Cambodia. Pingali and Siamwalla (1993) have argued that Myanmar in the short to medium term could export approximately 2 million tons with the existing rice infrastructure, if the policy reforms bring about improved production incentives for farmers. Over the long term, exports from Myanmar could double to 4 million tons of rice per year. Cambodia used to export half a million tons of rice in the mid-1960s; this historic share of the export market could potentially be reclaimed. With added investments in irrigation and other infrastructure, long-term prospects for exports from Cambodia could go up to a million tons.

Table 7.3. Potential trade access to South Asia and China.

Country	1986–1988		Potential access beyond 2005 (million tons)
	Consumption (million tons)	Imports (million tons)	
Bangladesh	14.67	0.329	0.734
India	51.59	0.242	2.579
China[a]	110.39	0.412	5.519

[a] China is not a signatory to GATT.
Source: Pingali (1995).

One could assume that approximately 4 to 5 million tons of the additional exports could come from within Asia. Even under such an optimistic scenario, approximately half the additional rice supplies would have to come from outside Asia. The only two regions with large unexploited potential are Latin America and Africa. Africa has the potential to expand rice area by 20 million hectares in inland river valleys (FAO, 1995). If there is assured demand from Asia to the tune of 4 to 5 million tons per year that is not expected to be met by current suppliers, investments could be made to develop infrastructure and technologies, and generate those supplies. Latin America is likely to exploit its potential earlier than Africa, since it has broader road and transport infrastructure, and also a strong research infrastructure already in place. A more detailed analysis is required on the costs of exploiting unused capacity in Africa and Latin America. The comparative advantage of the above regions in commercial rice production also ought to be assessed.

Equity and environmental implications

Equity implications of GATT ought to be evaluated in the broader context of income distribution changes that come about due to trade liberalization. Anticipated staple food price increases resulting from GATT implementation also raise equity concerns especially in terms of their impact on the rural and urban poor. It is not necessarily true that the benefits of GATT accrue only to large farmers and those in the high-potential environments. GATT, over the long term, could potentially lead to a reduction in income differential between the high-potential and low-potential environments by drawing surplus labor out of agriculture and into industrial-service sectors. The resulting increase in real agricultural wages benefits landless labor households remaining in agriculture. Given the evidence on equalization of wages across production environments (David and Otsuka, 1994) there is every reason to believe rising wages benefit the high-potential and low-potential environments. The possible rise in real food prices could have serious consequences on poor consumers. The supply responsiveness of non-traditional rice suppliers, such as Latin America and Africa, could dampen the magnitude of the price rise.

The environmental impacts of GATT could be both positive and negative. The phytosanitary agreement discussed above provides a basis for standardizing and promoting safer pesticide use and reducing the incidence of residues on food products, especially those exported. The use of persistent pesticides could decline with greater international trade in food. In Asia, the move from monoculture rice systems in the irrigated lowlands to a more diversified production system could have long-term productivity benefits by improving soil fertility status (Cassman and Pingali, 1995; Pingali, 1994). Increased chemical use for high-value crops and increased herbicide use as a substitute for handweeding continue to be matters of concern both from an environmental and health point of view (Pingali and Roger, 1995; Pingali *et al.*, 1995). Also an issue that needs careful attention is the impact of intensive cultivation of high-value crops in the uplands, especially in terms of soil erosion consequences. In the case of the uplands, an assessment is required on the role of improved property rights as a means of encouraging erosion control investments (Pingali, 1990).

GATT-related expansion of rice cultivation in non-Asian countries could also lead to environmental concerns, specifically in terms of water. Intensive cultivation of rice for exports would require expansion in irrigation and drainage infrastructure in Latin America and Africa. The problems with intensive irrigation-water use that were encountered in Asia are relevant to these continents as they intensify rice production. Specific attention ought to be given to water-induced paddy land degradation, such as salinization, soil toxicity buildup, waterlogging, etc. Given the experience of Asia, it could be possible to pay greater attention to these environmental factors in the design and development of irrigation infrastructure in Africa and Latin America.

Implications for Rice Research Priorities

Research resource allocation for irrigated versus rainfed systems

The allocation of research resources between the favorable and the fragile environments is generally based on congruence analysis, reflecting the importance of the above systems in terms of area and production. Strict congruence analysis would indicate a research resource allocation of 70 percent for the irrigated lowlands and 30 percent for the unfavorable environments. IRRI has modified the above allocations by explicitly incorporating equity concerns in the analysis, as the majority of the rural poor live in the unfavorable environments. The poverty modifier has resulted in a resource allocation of approximately 50–50 between the two broad rice environments. Given the long-term impact of GATT on the increasing competitiveness of the irrigated environments for other crops and other non-rice enterprises, there is a need to modify the above congruence analysis. The upland environments, as discussed in the previous section, also face increasing competition due to growing commercialization. At the same time, one would have to consider the productivity differences between the favorable and the unfavorable environments, and the probability of research success, the latter being small for the unfavorable environments.

While a revised priority setting exercise has yet to be conducted, one could speculate that the outcome would not be significantly different from a 50–50 split between the favorable and unfavorable rice environments. However, within the unfavorable environments themselves, one ought to expect a significant reduction in emphasis on rice in the upland environments and on the highly drought-prone rainfed environments.

Intensification versus diversification

With growing commercialization trends, the emphasis of the irrigated environments would shift from rice monoculture systems to diversified farming as discussed earlier. From a research point of view, understanding rice as part of a system in which several crop and non-crop activities occur becomes crucial. The profitability of component rice production technologies ought to be assessed within the context of a diversified farming system. The relevant measure of system performance in such a diversified system is no longer yield per hectare, but total household income and total factor productivity.

In the case of the rainfed lowlands, intensification continues to be the most important research objective, both in terms of adding crops as well as increasing yield per hectare per crop. Breeding and crop-management activities designed to reduce water stress are high-priority activities in the rainfed lowland system. The concentration on the uplands ought to shift from attempts to increase the productivity of the rice crop to sustainable management of a diversified production system.

Yield enhancement versus quality improvements

Even with the diversification trends, irrigated lowlands will continue to be the main sources of rice supply for the growing urban populations. Given the increasing diversion of some of the irrigated rice lands to non-rice activities and to non-agricultural uses, such as urban and industrial uses (Chapter 9), it is imperative to continue the high research emphasis on shifting the rice yield frontier. The new plant type can be expected to play an important role sustaining the yield productivity growth for rice. Even as the emphasis remains on shifting out the yield frontier, increased attention ought to be paid to enhancing rice grain quality. As incomes grow in Asia, the demand for higher-quality rice will increase and the research system ought to be able to respond with high-yielding, high-quality rices.

Knowledge-intensive technologies – increased opportunity cost of labor

Increasing opportunity costs of labor due to enhanced off-farm employment opportunities have significant implications for priorities in crop-management research. Recent work in crop-management research has concentrated on increasing input use efficiency through the use of knowledge-intensive technologies, such as integrated pest management, judicious use of irrigation water and improved fertilizer management (Chapter 11). All of the above technologies require the farmer to spend time in management, supervision and use of the technology. Farmer time can be expected to become an increasingly expensive input in farm production, and hence the profitability of knowledge-intensive technologies for enhancing input efficiency is brought into question. Research ought to concentrate on increasing input efficiencies while minimizing farmer time requirement.

Private sector versus public-funded rice research

With the implementation of GATT and the agreement on intellectual property rights there is a growing perception that the private sector could carry out much of the needed agricultural research. While it is true that the involvement of the private sector in research will increase, it is unreasonable to expect that it will substantially replace public-funded research. The private sector will tend to concentrate on research activities from which it can fully capture the returns to its investment. Investments in biotechnology, especially gene constructs and transgenic plant materials, and hybrid rices, are areas where the private sector can recoup its investments. In more traditional breeding activities and in studies on understanding process, the knowledge tends to be freely available once it is generated. Because of the 'public good' nature of most agricultural research, it will continue to be essential for the public sector to invest in it. This is particularly true for research on the less favorable environments which tends to be characterized by long gestation periods and low probability of success.

8

Agricultural Commercialization and Farmer Product Choices: The Case of Diversification Out of Rice[1]

Introduction

Economic growth, urbanization and the withdrawal of labor from the agricultural sector lead to the increasing commercialization of agricultural systems. Agricultural commercialization means more than the marketing of agricultural output, it means that product choice and input use decisions are based on the principles of profit maximization. Commercial reorientation of agricultural production occurs for the primary staple cereals as well as for the so-called high-value cash crops. Commercialization of agricultural systems leads to greater market orientation of farm production; progressive substitution out of non-traded inputs in favor of purchased inputs; and the gradual decline of integrated farming systems and their replacement by specialized enterprises for crop, livestock, poultry and aquaculture products.

As economies grow, there is a gradual but definite movement out of subsistence food crop production, generally in a monoculture system, to a diversified market-oriented production system. The process of diversification out of staple food production is triggered by rapid technological change in agricultural production, improved rural infrastructure, and diversification in food demand patterns. A slow-down in income-induced demand for cereals is accompanied by a shift of diets to higher-valued foods such as meat, fruit and vegetables. These dietary transitions are induced by declining income elasticities of demand for staples as per capita incomes rise, and by the rapid migration of population to urban areas.

Initially, diversification implies the addition of other crops and other enterprises at the farm-household level. As the level of commercial orientation increases, however, one observes mixed farming systems giving way to specialized production units that are designed to respond rapidly to market price and quality signals. Diversification at the agriculture-sector level is consistent with specialization at the farm level.

The general trend, in East and Southeast Asia, towards the withdrawal of labor from the agricultural sector, and the resulting increase in real agricultural wages, is leading to the commercialization and diversification of rice systems. While the speed of the above structural transformation differs substantially across countries they are all moving in the same direction. Timmer (1989a) provides a comprehensive discussion on the process of structural change and commercialization of agriculture. For a recent review on agricultural commercialization, see Pingali and Rosegrant (1995). Empirical evidence on commercialization trends is provided by: Dyck *et al.* (1993) for East Asia; Huang and Rozelle (1994) for China; Koppel and Zurick (1988) for Southeast Asia; and Naylor (1994) for Indonesia.

This chapter: (i) describes the process of commercialization of rice systems in Asia and its implications for the choice of outputs produced by farmers; (ii) identifies the agronomic and economic opportunities and constraints to diversification out of rice monoculture systems; (iii) documents farm-level mechanisms for alleviating constraints to diversification; and (iv) provides a set of policy and research options for aiding farmers through the transformation.

From Subsistence to Commercial Production Systems

Asian rice systems can be characterized as subsistence, semi-commercial and commercial systems (Table 8.1). Increased commercialization shifts farm households away from traditional self-sufficiency goals and towards profit and income oriented decision making; farm output is accordingly more responsive to market needs. The returns to intensive subsistence production systems that require high levels of family labor generally decline relative to production for the market with predominant use of hired labor. The proportion of farm income in total household income declines as family members find more lucrative non-agricultural employment opportunities. At the same time, the share of agriculture in farm household income declines.

The process of structural transformation is well under way across much of Asia, although the speed at which it is occurring varies by country (Table 8.2). East Asia, except China, is nearing completion of the transformation process. Southeast Asia is rapidly transforming its agricultural systems, with Malaysia, Indonesia and Thailand taking the lead. South Asia continues to lag behind, although the

Table 8.1. Characteristics of food production systems with increasing commercialization.

Level of market orientation	Farmers' objective	Sources of inputs	Product mix	Household income sources
Subsistence systems	Food self-sufficiency	Household generated (non-traded)	Wide range	Predominantly agricultural
Semi-commercial systems	Surplus generation	Mix of traded and non-traded inputs	Moderately specialized	Agricultural and non-agricultural
Commercial systems	Profit maximization	Predominantly traded inputs	Highly specialized	Predominantly non-agricultural

Table 8.2. Stage of commercialization by country.

Country	Subsistence	Semi-commercial	Commercial
East Asia			
Japan			×
Korea		×	×
Taiwan			×
China	×	×	
Southeast Asia			
Thailand	×	×	
Malaysia		×	×
Indonesia	×	×	
Philippines	×	×	
Vietnam	×	×	
Laos	×		
Cambodia	×		
Myanmar	×		
South Asia			
India	×	×	
Pakistan	×	×	
Bangladesh	×	×	
Nepal	×		

macroeconomic liberalization policies initiated by these countries – especially India – in the early 1990s could lead to a faster rate of transformation. Relatively larger populations and continuing rapid population growth rates will continue to dampen the speed of transformation in South Asia.

Product choice: specialization amongst diversity

Commercialization, while leading to an increase in the diversity of marketed products at the national level, also leads to increasing regional and farm-level specialization. The case of rice is illustrative of the market and agroclimatic determinants of diversification and specialization.

Asian ricelands are categorized in terms of toposequence and hydrology into irrigated lowlands, rainfed lowlands, deepwater environments, and the uplands. The irrigated lowlands by their nature are inherently more market-oriented because of their ability to generate a surplus and because of better transport infrastructure. Rice monoculture systems in the irrigated lowlands could be characterized as semi-commercial systems even in slow-growing economies. A large share of the output is marketed and, compared to the other rice environments, a relatively larger share of the inputs is purchased. Increasing commercialization trends lead to both a seasonal diversification out of rice monoculture systems – to include non-rice crops in rotation with rice – and specialized enterprises for horticulture, aquaculture, poultry and hog production (Table 8.3). Specialized enterprises for high-value horticultural products, hog production and aquaculture tend to be concentrated in the irrigated environments because of a relatively more reliable and cheaper supply of water. The opportunities and constraints to seasonal crop diversification are discussed in the next section of this chapter.

Commercialization trends could also lead to cattle and small ruminant production in the rainfed lowlands – taking advantage of seasonal grazing lands – and intensive aquaculture systems in the deepwater and tidal wetland environments (Table 8.3). In the deepwater environments, the traditional flooded rice crop tends to be bypassed for a dry-season irrigated rice crop using shallow pumps (Dey and Hossain, 1994). Wet-season rice will continue to be the dominant source of income for all three lowland environments (see next section).

The upland environments change dramatically in response to commercialization. Upland areas, with soils that are relatively less susceptible to erosion, tend to move with improved market infrastructure from subsistence cereal and root-crop production to a variety of commercial enterprises. These include horticulture, tree crops, dairy and cattle ranching. In Asia, commercial utilization of the uplands has generally resulted in the movement out of upland rice production.

The extent to which agroclimatic constraints can be overcome – for example through corrective investments, water control and drainage for ricelands – depends on the nature of demand elasticities.

Table 8.3. Changing product mix by rice environments.

Level of market orientation	Irrigated lowlands	Rainfed lowlands	Deepwater and tidal wetlands	Uplands
Subsistence systems		Single rice crop Subsistence livestock and poultry	Single rice crop Subsistence fish production	Shifting cultivation of subsistence livestock and poultry
Semi-commercial systems	Intensive rice systems, peri-urban dry season vegetables, dry-season fodder crops	Single/double rice crop Cattle/small ruminant production	Dry-season irrigated rice	Horticulture Maize, rootcrops Vegetables
Commercial systems	Year-round vegetables Horticulture Aquaculture Intensive poultry and hog production	Intensive poultry production	Intensive aquaculture	Intensive dairy farming Cattle ranching Tree crops

Investments in land modification and the expansion of area under a particular crop are profitable where farmers face an elastic demand for that crop. Urbanization and trade liberalization lead to more elastic demand for high-value non-cereal food products, such as vegetables, fruit, meat, poultry and dairy products.

The future of integrated farming systems

Across all rice environments, integrated farming systems, such as rice–livestock, rice–aquaculture, and rice–fruit-tree systems, generally become infeasible at a commercial scale because of product-specific requirements in technical and managerial skills and infrastructural investments. For example, commercial crop production and high-quality livestock production would not generally occur on the same farm. While the physical size of the farm may not be a constraint to multi-product enterprises, the time of the farmer-entrepreneur becomes the ultimate binding constraint to the generation of multiple outputs.

Table 8.4. Substitution of traded for non-traded inputs.

Level of market orientation	Power	Soil fertility	Fodder	Human nutrition
Subsistence systems	Animal/ human	Farmyard manure	Crop residues	Predominantly home produced
Semi-commercial systems	Motor/ animal/ human	FYM and chemical fertilizers	Crop residues	Home produced and purchased
Commercial systems	Motor	Chemical fertilizers feed	Purchased	Predominantly purchased

The diminishing advantage of integrated farming systems is also the result of increased substitution of non-traded for traded inputs (Chapter 9, Table 8.4). Power, soil fertility maintenance, fodder for farm animals and household nutrition are the primary activities for which non-traded inputs are used in subsistence societies. Under low-wage conditions, integrated crop–livestock production systems are the most economical means of providing the above inputs for the farm production system. As wages and the opportunity cost of family labor rise, the use of purchased inputs becomes substantially cheaper than home-produced inputs. Chapter 9 discusses the substitution out of animal power to mechanical power, and the substitution out of farmyard manure to chemical fertilizers with the growth in commercialization. We address below two other commonly cited reasons for the existence of integrated farming systems – fodder production, and milk and meat for home consumption. In both cases, we argue that the trends are towards increased specialization.

Emergence of fodder markets

The early attraction of combining crop and livestock activities within the same enterprise was the availability of cheap fodder for livestock, essentially crop residues from the farm. Stall feeding of cattle and small ruminants with crop residues was seen as a means of preventing the overgrazing of marginal pasture lands, especially common access lands. This argument does not hold in more advanced economies due to the emergence of commercial fodder production and lower transport costs. Livestock producers no longer need to grow their own fodder. Stall feeding can be sustained on an economical basis with purchased fodder of higher quality. The type of diet provided for commercially produced livestock also changes, becoming based on grain, root-crop or oilseed, rather than straw-based, in response to the demand for higher-quality livestock products. With changes in fodder composition and quality, the perceived economies of home-grown fodder no longer exist.

Han (1992) documents the growth in the feed industry in Asia with increased commercialization of the livestock sector. By the early 1990s, a total of 101 million tons of compound feed was produced in the Asia–Pacific region. China and Japan are the largest feed producers in the region, at 39 million tons and 29 million tons respectively. Feed production is mainly for swine, poultry and cattle. Commercial feed production is expected to grow across Asia as the demand for livestock products increases. In South Korea, for example, commercial feed production increased from 100,000 tons in 1963 to 11.5 million tons in 1991, an increase of 115 times. Farm-produced fodder will not be able to compete with commercially produced feeds as wages increase and transport costs become lower.

Milk and meat for home consumption

As economies grow, increasing opportunity cost of family labor, and the increasing availability of commercially produced milk and meat, makes subsistence production of livestock products uneconomical. Rural societies are no longer remote self-contained units that need to produce all of their food requirements. Food consumption studies across Asia have shown the increased reliance of farm households on purchased food, and this trend will become stronger as incomes grow (Bouis, 1994). Improved transport and market infrastructure makes subsistence food production non-viable in all but the remotest locations.

Northeast Thailand provides a striking example of changing food consumption patterns in rural areas. In the 1960s, meat consumption was limited to special occasions such as festivals, perhaps once or twice a year. This pattern has changed, in favor of greater purchased meat consumption, with improved transport infrastructure and increased family income due to seasonal migration for urban employment (Valayasevi and Winichagoon, 1992).

As economies grow, the returns to intensive subsistence production systems that require high levels of family labor are generally lower than exclusive reliance on purchased inputs. The benefits of integrated production systems also decline with the rise in the opportunity cost of family labor, and specialized crop, livestock or aquaculture enterprises emerge[2]. Few Asian economies have an exclusively specialized production system today, but enclaves do exist and the trends are in favor of increased commercialization and specialization[3].

Diversification of Rice Lands: Opportunities and Constraints

The last section took a broad sweep at the possible evolution of Asian rice systems. This section is narrower in focus. It attempts to assess the flexibility of ricelands and rice farmers to respond to the commercialization trends through seasonal or permanent diversification out of rice monoculture systems. The potential for diversification out of rice production depends on both physical and economic factors. The feasibility and cost of substituting other crops vary across the four rice ecosystems: irrigated and rainfed lowlands, floodprone wetlands, and uplands. Each of these systems also presents different rainy- and dry-season profiles and requires different levels of physical and human capital investment to switch from rice to non-rice crops and back.

Do farmers have the flexibility to respond to changing relative prices and relative profitability in their crop choice decision making? Flexibility can be described in terms of the level of investments (both physical and human capital) required in switching from rice to non-rice crops and back. For instance, non-rice crops are grown year-round in Indonesia in a Sorjan (ditch and dike) system which involves high levels of investments in drainage control. Flexibility is low because moving out of monoculture rice to upland crop production on elevated dikes or moving back into monoculture rice production involves high physical investments. Upland areas, however, can switch between rice and non-rice crops with minimum additional investments. Table 8.5 presents the flexibility of crop choice by ecosystem and season.

Wet-season crop flexibility is extremely low in all but the upland environments, because the investment requirements for drainage are high and not easily reversible in the lowlands. Switching between rice, maize and other crops is possible in the uplands because the fields are not bunded and do not require to be puddled before crop establishment.

Table 8.5. Flexibility of crop choices by ecosystem and season.

Ecosystem	Wet season	Dry season
Irrigated lowlands	Low	Moderate to high[a]
Rainfed lowlands	Low	Low to moderate[b]
Deepwater and tidal wetland	Low	Low to moderate
Uplands	High	Moderate[b]

[a] This period includes the post-rice period (late wet season) or the pre-rice period (dry–wet transition).
[b] Conditional on rainfall level and distribution.

During the dry season, crop choice is constrained by water availability and drainage. The irrigated lowlands have the most reliable water supply. These areas, depending on the severity of the drainage constraint, have the highest flexibility in dry-season crop choice. Switching from dry-season rice to non-rice crop production will involve a certain amount of investment in temporary drainage structures and in learning about non-rice technology, cultivation practices, and irrigation water management. Onion farmers in the Upper Talavera River Irrigation System (UTRIS), Philippines, for example, construct multipurpose ditches and levees in the rice paddies for facilitating the drainage of excess water (Tabbal *et al.*, 1990). Other examples of temporary drainage structures can be found for the Philippines in Moya (1992), Alagcan and Bhuiyan (1992) and Maglinao and Valdeavilla (1992); for Indonesia in Pusposutardjo *et al.* (1993); and for Bangladesh in Ghani *et al.* (1993). The amount of land modification required is related to soil texture; heavy soils require elaborate drainage structures while light sandy soils may not require any drainage structures at all. The returns to these investments are highest for the irrigated lowlands with moderate- to well-drained soils and, hence, these areas will tend to diversify more than the other ecosystems as the relative profitability of non-rice crops improves.

Irrigated lowland soils can be classified into: well-drained soils; moderately drained soils; and poorly drained soils. Flexibility of crop choice for each of these soils by season is presented in Table 8.6. For the wet season, only the well-drained soils have possibilities for non-rice crop production; investments in a bed and furrow system or a Sorjan system are required for successfully growing non-rice crops. On the other hand, for the dry season the flexibility of crop choice in irrigated ricelands is high for all but the poorly drained soils. Only heavy textural waterlogging-prone irrigated rice soils have little option but to specialize in rice production. For this last category the amount of drainage investments that have to be made prior to growing non-rice crops is often prohibitive. Irrigated areas in South and Southeast Asia

Table 8.6. Fully irrigated lowlands: flexibility of crop choice.

	Well-drained soils	Medium drainage	Poorly drained soils
Wet season	Moderate[a]	Low	Low
Dry season	High	High	Low to moderate[b]

[a] Conditional on rainfall levels and effective water control.
[b] Depending on the level of drainage investments.

that have a long history of dry-season diversification have all limited their non-rice crop production to well-drained soils while intensive rice production has continued concurrently on poorly drained soils. See Masicat *et al.* (1993), Tabbal *et al.* (1993) and Maglinao *et al.* (1993) for case studies of irrigation systems from the Philippines; Wardana *et al.* (1993) for Indonesia; and Ghani *et al.* (1993) for Bangladesh.

The length of the period of irrigation water availability is also an important determinant of dry-season diversification. The large partially irrigated areas which cannot support a dry-season rice crop have a natural advantage in diversifying into upland crops during the dry season. But crop choice may again be limited on heavy textured, poorly drained soils, in which water control to avoid waterlogging or drought is difficult.

The progression to crop and income diversification has taken place smoothly in countries where product markets operate relatively freely. In Suphan Buri, Thailand, for instance, the adoption of non-rice enterprises was closely associated with recent rice price trends. Panel data for 146 households in Suphan Buri, Thailand, indicate that between 1985 and 1988, 79 percent of the households first adopted non-rice enterprises (Table 8.7). Rice prices in Thailand were on a declining trend between 1980 and 1986, reaching their lowest level during 1985/86. The non-rice enterprises adopted included: non-rice crops such as vegetables and fruit orchards; non-crop farm enterprises, such as shrimp farming and livestock production; or non-farm activities, such as rural industries or urban employment (Table 8.7). By 1987, 91 of the 146 households had adopted diversified farming systems. It is interesting to note that a third of the households switched back to exclusive rice production in 1988 when rice prices went back up following the drought of 1987.

Diversification of Suphan Buri ricelands took two forms, dry-season diversification and year-round diversification. Dry-season diversification was into vegetables and other seasonal crops, such as maize and sweet potatoes; 39 percent of the households adopted a dry-season non-rice crop. Land investment requirements for establishing these crops is minimal and when the rice prices improved in 1988 these lands quickly returned to rice production. Year-round diversification was into sugarcane, shrimp and fish farming, and fruit orchards; 14 percent, 3 percent and 4 percent of the households respectively, adopted these enterprises. Investment requirements for year-round diversification out of rice are very high and would only be made if expectations of relative long-term profitability are in favor of the particular non-rice enterprise. For fish and shrimp production, for instance, the initial investment costs are about 110,000 Baht per hectare (approximately US$ 4400).

Table 8.7. Number of households adopting non-rice enterprise classified by type of enterprise and the first year of adoption.

Type of enterprise	Year of first adoption: before 1980	1980	1981	1982	1983	1984	1985	1986	1987	1988	Total households
Cattle	2	0	1	0	2	3	2	6	5	13	34
Poultry	4	0	0	3	2	1	2	4	1	2	19
Prawn and fish	0	0	0	0	0	0	0	2	3	0	5
Vegetables	0	0	0	2	1	0	5	10	13	5	36
Fruit trees	0	0	0	2	1	0	1	1	0	1	6
Seasonal crops (short periods)	0	0	0	0	1	1	3	6	8	2	21
Sugarcane E	0	0	0	0	0	0	1	2	7	3	13
Sugarcane C	0	0	0	0	0	0	0	0	5	2	7
Off-farm work (avg.)	0	0	0	1	0	1	0	0	0	0	2
Total	6	0	1	8	7	6	14	31	42	28	143
Price of paddy (Baht ton^{-1})	—	—	na	2470	2415	2273	2230	2398	3122	3726	

Source: Sriarunrungreauang (1989).

Input and labor requirements are also higher for non-rice enterprises. This includes both the dry-season and the year-round enterprises. Table 8.8 provides data on the relative input requirements and the profitability of rice and non-rice enterprises. Sriarunrungreauang (1989) using the panel data for Thailand finds that, if the rice price dropped by 20 percent, dry-season non-rice crops would be relatively more profitable than rice, but year-round diversification would not be a profitable alternative to rice in the irrigated lowlands.

The opportunities for dry-season diversification in the rainfed lowlands and the deepwater areas are limited by water availability for post-rice crop production. In the humid and subhumid zones, rainfall level and distribution are such that a post-rice or a pre-rice crop in the rainfed lowlands is possible. Post-rice cropping of legumes (e.g. mungbean), cereals (maize) or vegetable crops may be possible on late-season rains and residual moisture. This practice has become much more feasible on that portion of rainfed ricelands which now produce earlier maturing rice cultivars, which are harvested before the onset of the dry season. In the Cagayan Valley of the Northern Philippines, the replacement of traditional rainfed rice varieties of six-month duration with early maturing modern varieties has led to double cropping of rice in the lower elevations and the introduction of a pre-rice crop of

Table 8.8. Relative input requirements and profitability of rice and non-rice enterprises, Thailand, 1988 (Baht year^{-1}). (There may be small discrepancies due to rounding of figures.)

	Enterprises			
	Rice–rice	Rice–vegetable	Sugarcane	Prawn
INPUTS				
Fertilizer (B ha^{-1})	2,915	27,174	995	627
Pesticide (B ha^{-1})	964	19,224	280	—
Other costs (B ha^{-1})	294	17,037	8,389	18,589
Feeds (B ha^{-1})	—			42,049
Sub-total	4,173	63,435	9,664	61,265
Labor (md ha^{-1})				
Family	42	595	17	90
Hired	41	445	60	2
Total	83	1,040	77	92
Labor costs (B ha^{-1})	5,739	71,916	5,325	6,362
TOTAL COSTS (B ha^{-1})	9,912	135,351	14,989	67,627
GROSS RETURNS (B ha^{-1})	28,427	160,517	32,399	104,485
NET RETURNS (B ha^{-1})	18,515	25,166	17,410	36,858

Source: Sriarunrungreauang (1989).

mungbeans on the upper elevations (Garrity *et al.*, 1988). Pre-rice crops in the lower elevations are only possible on ridges to prevent waterlogging (Pernito and Garrity, 1988). The strategy of increasing cropping intensities in the rainfed lowlands will only be successful if modern rice varieties adapted to these problem hydrologies (i.e. drought-prone, flood-prone, and drought- and flood-prone conditions) are available.

In rainfed environments where there is a sharp and prolonged dry season (especially the semi-arid zones) post-rice crops are not possible without supplementary irrigation. In the rainfed lowlands of South Asia, Northeast Thailand, Cambodia and Laos, dry-season crops on residual moisture would not be possible even if traditional rice varieties were replaced by appropriate short-duration modern varieties. Markets permitting, there is potential for a short pre-rice crop followed by a short-duration rice crop; suitable candidates are mungbean, and green manure crops such as Sesbania and forage legumes.

Where supplementary irrigation is available, as with pumps, opportunities exist for a dry-season rice or non-rice crop. In Nueva Ecija, Philippines, where there is a six-month dry season, the introduction of deep tubewells has led to the adoption of maize followed by mungbeans in the dry season after a rainfed wet-season rice crop (Gines *et al.*, 1989). It ought to be emphasized that diversification occurred only on the upper paddies with light textured and easily drained soils ('turod'). On the other hand, the lower paddies with heavy textured soils that are prone to waterlogging ('lungog') were used for cultivating a dry-season rice crop. While two rice crops are also possible on the 'turod' soils with the dry-season crop being irrigated by pumps, the private and social returns to a diversified cropping system dominate the rice–rice cropping system. This is so primarily because the costs of irrigation for rice are high and a significantly smaller area can be irrigated efficiently (Gines *et al.*, 1989). Engelhardt (1984) reports for the semi-arid tropics of India the emergence of a diversified cropping system with the introduction of deepwell pumps. Rainfed rice in the wet season is followed by either groundnuts, sorghum or vegetables. In Bangladesh, approximately 60 percent of the dry-season cultivated area is irrigated by tubewells and pumps (Hakim and Parker, 1990). Much of this area is planted to a rainfed wet-season rice crop followed by an irrigated dry-season nonrice crop. Wheat, potatoes, gram and onions are popular alternatives to irrigated rice (Mondal *et al.*, 1990; Islam, 1992).

Dry-season diversification in the upland areas similarly depends on the level and distribution of rainfall. In areas with a sufficient growing period, a post-rice crop can be grown. Maize, sweet potatoes, and vegetables are common sequential crops. In Northern Mindanao, Philippines, for instance, where the average annual rainfall of 2,350 mm is evenly distributed over an eight-month period, double cropping of maize is practiced on a quarter of the upland area (Mandac *et al.*, 1987). In Northern Laos, where the average annual rainfall is 1,400 mm, diversification from one upland rice crop to two non-rice crops is not feasible due to risk of drought stress for the second crop (Fujisaka, 1990). For the lower rainfall upland areas in much of the subhumid and semi-arid zones, rice production is generally not profitable due to the risk of drought stress. Where irrigation is not available, wet-season sorghum, millet and pulses such as pigeon pea and chick pea are commonly grown (see Walker and Ryan, 1990, for a description of cropping patterns in India's semi-arid tropics).

Diversification out of rice production in response to changes in the relative profitability between rice and non-rice crops would be most feasible in the dry season. The rice ecosystems in which it will be most profitable and feasible are the irrigated lowlands, because of greater reliability of water supply and higher return to diversification investments.

Dominant crop and non-crop options by ecosystem

Table 8.9 shows the dominant income-generating activities for each season and environment. Empirical evidence on the sources of income by rice environment is provided in Tables 8.10 and 8.11 for the Philippines and Thailand, respectively. During the wet season, rice will continue to be the dominant source of income in all but upland environments. This is not to imply that rice is not an important source of income for the uplands, but rather to stress the fact that the uplands have always been very diversified because they do not face the drainage constraint.

In the irrigated lowlands, dry-season rice will continue to be the major source of income. Areas with good market access and those near urban centers will increasingly diversify to non-rice crops and vegetable production. The dominant dry-season activity for the rainfed lowlands will essentially be non-crop activities, off-farm employment, livestock production and cottage industries. There is scope for post-rice crops on residual moisture, or pre-rice crops during the early wet season. However, the share of total income from this activity would be relatively lower than from the other activities. Dry-season cropping activities in the rainfed areas are limited because of technical problems related to timely and effective crop establishment, limited moisture (or excess moisture in some cases), and generally modest yields and high-yield instability. Off-farm activities are often more dependable income sources, suggesting that dry-season cropping intensities will remain low even if technical problems in crop production are solved. For the flood-prone environments, fish production in the wet season followed by dry-season irrigated rice production could be the dominant activity, especially in areas where tubewell irrigation is feasible.

Table 8.9. Dominant crop and non-crop option for sustaining incomes by environment.

	Wet season	Dry season
Irrigated lowlands	Rice	Rice/non-rice crops
Rainfed lowlands	Rice	Off-farm employment
Deepwater and tidal wetlands	Rice	Off-farm employment
Uplands	Diversified production systems	

Table 8.10. Sources of household income, rice farms classified by environment, Philippines, 1988.

Source of income	Rainfed farms: Percent of total income	Rainfed farms: Number of households	Upland farms: Percent of total income	Upland farms: Number of households	Irrigated farms: Percent of total income	Irrigated farms: Number of households
Rice income	48.1	48	3.0	39	56.7	129
Non-rice income	6.3	41	31.0	39	0.4	1
Non-crop income	9.0	39	16.0	50	6.5	60
Off-farm income	1.0	20	10.0	28	10.2	35
Non-farm income	35.6	53	40.0	15	26.2	95
Total value of income	22,748 (₱) 1,078 US$	49*	15,777 (₱) 748 US$	54*	33,975 (₱) 1,610 US$	132*

* Total number of sample for each category, some households have two or more sources of income.
Source: Social Sciences Division, IRRI.

Table 8.11. Number of villages, number of rural households, sources and levels of net household income by province, Thailand, 1980–1981.

Region and province	Number of villages	Number of households	Sources of net household income (Baht): Farm	Non-farm: Other sources	Non-farm: Wage	Other	Total
Northeast							
Khon Kaen (rainfed)	8	141	13,275 (47.4)	3,385 (12.1)	6,627 (23.7)	4,713 (16.8)	28,000 (100.0)
Roi Et (rainfed)	5	75	4,889 (22.4)	6,047 (27.7)	5,514 (25.2)	5,404 (24.7)	21,854 (100.0)
North							
Chiang Mai (upland)	9	163	6,046 (18.8)	10,629 (33.0)	11,417 (35.5)	4,095 (12.7)	32,187 (100.0)
Center							
Suphan Buri (irrigated)	3	42	29,232 (70.8)	–409 (–1.0)	9,027 (21.8)	3,461 (8.4)	41,311 (100.0)
All provinces	25	421	10,643 (35.5)	6,284 (21.0)	8,544 (28.5)	4,481 (15.0)	29,952 (100.0)

Source: Onchan and Chalamwong as cited in Pingali (1992).

The above discussion leads to the conclusion that irrigated environments, while having an absolute advantage (relative to the other environments) in a rice–rice cropping pattern, may also have a comparative advantage in a rice–non-rice cropping pattern. The extent of comparative advantage for the irrigated lowlands in dry-season diversification depends on the physical constraints and the market opportunities for non-rice crop production. On the other hand, during the wet season, the upland environments have both an absolute and a comparative advantage in non-rice crop production.

Diversification Constraints

The profitability of diversification is constrained by market size, land quality, irrigation infrastructure and labor supply. Where output demand is relatively elastic, the returns to investments in land, technology, and time spent learning about new crops are relatively higher.

Size of market and price risk

Diversification from a rice monoculture system to a system that includes non-rice crops could lead to increased variability in farm-household incomes. Variability in incomes come from yield or price fluctuations. Timmer (1992) has argued that the two sources of income variability are seldom negatively correlated to a sufficient degree to produce income stability and neutrality of farmer decision making. Timmer cites the case of Indonesia, where price risks from growing rice are significantly smaller than for other crops. Risk aversion is a significant impediment to what would seem to be rational diversification on the basis of average profitabilities of alternative crops. Behavior in the face of risk aversion is affected by farmer attitudes and the nature of technology, but the fundamental problem is the failure of local credit and risk institutions to provide any potential for farmers to transfer their risks to other parties. Avoiding risk thus becomes an internal household strategy, and many households concentrate on growing a familiar crop with known technology and guaranteed prices rather than risk their livelihood on new crops with sharply fluctuating prices. For a more general discussion on diversification and price risks see Pingali and Rosegrant (1995).

In the Upper Talavera River Irrigation System (UTRIS) of the Philippines, seasonal tenancy arrangements and forward-selling contracts are means by which farmers diffuse the price risks associated with non-rice crop production. Small onion farmers divide their farms into two, cultivate one part and give the other to a seasonal tenant who pays a fixed rent of 3,000 Philippine pesos per hectare. This way the landowner gets a certain income from a part of his land and gambles on the remainder (Pingali, 1992). Farmers also agree to sell their onion harvest to traders at the start of the season. The price is fixed and the traders provide credit to the farmer. No interest is charged for the credit, as long as the output is exclusively sold to the trader at the agreed upon price. The traders benefit from the substantial price difference between the harvest and post-harvest months, while the farmers benefit from the avoidance of price risks (Pingali *et al.*, 1989).

Land suitability and land rights

The ability to profitably convert ricelands for non-rice crops is constrained by drainage requirements for the lowlands and erosion control investments in the uplands. It is important to understand that not all lands can be converted out of rice production. Even for lands that can be converted, substantial investments in land improvements need to be made to sustain long-term productivity and profitability of non-rice crop production. Investments in land improvements are likely to be made only where secure rights to land exist.

Within an irrigated micro-environment, lands most preferable for rice production are heavy clay soils and lands that have the best access to irrigation water (lands in the head section and paddies close to the irrigation canal). Rice yields almost always decline from the head to the tail of the irrigation system (see Chapter 11). As the relative returns to dry-season non-rice crops rise, one observes an increase in preference for lands normally considered marginal to rice production. Within the irrigated lowlands, the following could be considered marginal to dry-season rice production: upper paddies that are difficult to irrigate; well-drained soils; sloping lands; and stony gravelly land. All these lands would be more suitable for dry-season non-rice crop production due to good drainage characteristics (Table 8.12). Investment requirements for drainage are lower on these lands as compared to low-lying paddies, heavy clay soils and land with better water access.

Wardana *et al.* (1990) document for Cikeusik Irrigation System in West Java, Indonesia, differences in yields and net returns for rice and non-rice crops (Table 8.13). They find the relative profitability of non-rice crops to increase on lands further away from the head of the system, to a point where water scarcity could be a problem. Pingali *et al.* (1989) document similar cropping pattern choices for the Upper Talavera River Irrigation System (UTRIS) in the Philippines. Two crops of rice are grown on the upper portions of the system, while

Table 8.12. Market infrastructure versus physical constraints as determinants of the profitability of diversification. Irrigated lowlands: dry season.

	Well-drained soils	Poorly drained soils
Good market access	High	Moderate to low[a]
Poor market access	Moderate to low[b]	Very low

[a] Conditional on the level of investment requirements for drainage control.
[b] Conditional on input supply conditions.

Table 8.13. Costs and returns per hectare onion by section. Cikeusik Irrigation System, Cirebon, West Java, Indonesia, 1988 DS.

Item	Head (*n*=26)	Middle (*n*=29)	Tail (*n*=24)	All farms (*n*=79)
Mean yield per hectare (t ha^{-1})	9.7	10.5	8.4	9.5
Mean price of onion ($ kg^{-1})	0.16	0.17	0.16	0.16
Total value of production ($ ha^{-1})	1676	1822	1332	1590
Costs of production ($ ha^{-1})				
Seeds	494	421	301	396
Fertilizer	137	134	86	116
Insecticide	177	231	143	181
Labor				
Hired labor	556	468	423	477
Family labor	414	215	239	284
Other costs	150	76	168	134
Total paid out costs of production ($ ha^{-1})	1514	1330	1121	1304
Total variable costs of production ($ ha^{-1})	1928	1545	1360	1588
Returns above paid-out costs ($ ha^{-1})	162	492	211	286
Gross margin ($ ha^{-1})	(252)	277	(28)	2

[a] US$1 = Rp 1800.
Source: Wardana *et al.* (1990), Table 8.12.

onions, chillies and vegetables are common in the midsection. *Dry-season crop choices at the tail of the system are conditioned by reliability of water supply. Where farmers have access to pumps, non-rice crops are grown.*

The following generalization is possible: In the irrigated lowlands, when the dry-season returns to non-rice crop production dominate the returns to rice production, the demand for and the price of land with the least constraints to diversification out of rice will be the highest. Pingali *et al.* (1989) examined the changing land preferences in the Upper Talavera River Irrigation System (UTRIS) in the Philippines. Over the 1983–1988 period, UTRIS observed dramatic changes in the preferences for dry-season cultivation land and consequently changes in land values. UTRIS consists of areas of heavy clay soils and sandy loam soils that are more suitable for dry-season onion and vegetable production. In the last five years land preferences have switched from the heavy clay soils to the sandy loam soils.

Table 8.14 presents, for the uplands, the physical and market constraints to diversification. If market access is good, the profitability of diversified field crop production on soils not highly susceptible to erosion is high. For soils susceptible to erosion, profitability of field crop production is determined by the level of erosion-control investments required. Where high levels of erosion-control investments are required tree crops may be a more viable option than field crops, particularly after land degradation has been allowed to occur through field crop production. In upland areas with poor market access the returns to diversification out of subsistence rice production are limited in areas of either type of soil.

Table 8.14. Market infrastructure versus physical constraints as determinants of the profitability of diversification. Uplands: wet season.

	Serious soil chemical[a] constraint and/or erosion hazard	Without major soil constraint
Good market access	High-input diversified cropping systems or agroforestry systems	Diversified farming or cash cropping
Poor market access	Shifting cultivation	Subsistence cropping systems

[a] Includes highly acid soils with potential aluminum toxicity/P deficiency.

The relationship between the flexibility of crop choice and erosion control investments becomes pronounced on the sloping uplands, which are extremely susceptible to soil erosion. There are various options for erosion control to maintain permanent cropping on these lands, ranging from grassy strips to stone wall terraces. Farmer's choice of erosion-control strategy depends on population pressure on the land, on market access, and on the appropriate erosion-control techniques available. Fujisaka and Garrity (1988) and Pingali (1990) argue that farmer interest in erosion-control measures is directly related to land values and market access and is conditional on suitable technologies being available to them (see Chapter 5 for further details).

Secure rights to land create the incentives farmers need to invest in land improvements that conserve and increase the long-term productivity growth which can be induced by the start of commercialization (Pingali and Rosegrant, 1995). Secure land rights increase the probability that farmers recoup the benefits from long-term investments, thereby increasing their willingness to make them. Land titles act as collateral to loans and thereby increase lender willingness to offer credit, leading to easier financing of purchased inputs and land improvements (Feder and Onchan, 1987; Hazell and Rosegrant, 1994).

Irrigation infrastructure as a constraint to diversification

Large-scale diversification of cropping systems necessarily involves diversified production in the irrigated lowlands, because of the importance of irrigation to overall agricultural production (see above). Many observers have argued that existing irrigation systems constrain diversification because of the rigid design of infrastructure and inflexible water delivery systems (Schuh and Barghouti, 1988). It is argued that this inflexibility prevents appropriate allocation of water to non-rice crops, constraining farmers to rice monoculture (see Rosegrant and Yadav (1993) for a detailed treatment on this set of issues). Based on these arguments, technology-based solutions to diversification within irrigation systems are advocated, mainly capital investment in improved conveyance, diversion and drainage systems. An alternative argument would be that the failure to diversify within irrigation systems is the result of incentive failures resulting from centralized allocation of un-priced irrigation water. Policies that establish markets in tradable water rights could establish incentives to economize on water and choose less water-intensive crops (in the dry season), by inducing water users to consider the full opportunity cost of water (Rosegrant *et al.*, 1995). Establishment of transferable water rights can provide maximum flexibility in responding to changes in crop prices and water values as demand patterns and comparative advantage change and diversification of cropping proceeds (Rosegrant and Binswanger, 1994).

Labor constraint

Does diversified cropping increase labor requirements? Yes; relative to rice, the per hectare labor requirements for onions, vegetables and other high value crops are substantially higher. Meeting labor requirements for providing temporary drainage structures is an essential activity immediately following the rice harvest. Planting, weeding, harvesting and post-harvest operations are also extremely labor-intensive for these crops. Recent research by the International Irrigation Management Institute (IIMI) in the Philippines estimated the mean labor demand for rice, mungbean, onion and garlic as 85.7, 68.7, 468.5 and 241.0 mandays per hectare, respectively (Wijayaratna, 1991). Labor requirements for non-rice crops are higher at the head of the system relative to the lower portions. This is presumably because of the greater need for drainage investments in the former (Wardana *et al.*, 1990). Comparisons of labor requirements, for dry-season rice versus non-rice crops can be found for the Philippines, Indonesia and Bangladesh in Miranda and Maglinao (1993). Given the higher crop and drainage labor requirements, non-rice crops on irrigated lands are grown on extremely small plots, in general about a quarter of the paddy area.

Does diversified cropping aggravate labor peaks between the harvest of the rice crop and the planting of the non-rice crop? As discussed above, additional labor is required for constructing temporary drainage structures; and additional labor or mechanical power is required for land preparation. The land preparation activity for non-rice crops following rice would require breaking the paddy hardpan (the compact soil surface caused by puddling paddy soils). If this hardpan is not broken, there would be problems with root penetration and hence the establishment of a non-rice crop (Zandstra, 1992). The power requirements for this soil modification are higher on heavy clay soils than on the lighter soils. Mechanization to an extent can alleviate this labor peak. However, the machine power required for upland crops is substantially greater than that required for puddling rice paddies. The incompatibility in machines for tillage of rice versus upland crops can be overcome by contract-hire operations, which, however, would be profitable only when large areas are grown to non-rice crops.

In addition to crop labor requirements, the supervision time required of the farmer is significantly higher: this may be the dominant labor constraint to high-value non-rice crop production given the highly inelastic nature of management labor available in the farm household, compared with hired labor augmented by seasonal migrants. In the Philippine UTRIS project, onion farmers overcame the supervision constraint by dividing their farms in half, cultivating one part and renting the other to a seasonal tenant farmer. Farmers with heavy clay soils often leased sandy loam soils for the dry season. Landowners provide half the purchased inputs and get 50 percent of the output.

Conclusions

Commercialization of agricultural systems is a universal phenomenon that is triggered by economic growth. While the rate at which the above transformation occurs varies by continent and by country within continents, the direction of change is the same across the world. Structural adjustment and trade liberalization policies that are currently being implemented in much of the developing world can be expected to further enhance the speed at which the commercialization process occurs.

Commercialization trends require a paradigm shift in agricultural policy formulation and research priority setting. The paradigm of staple food self-sufficiency that has been the cornerstone of agricultural policy in most developing countries becomes increasingly obsolete with economic growth. The relevant development paradigm for the twenty-first century is one of food self-reliance, where countries import a part of their food requirements in exchange for diverting resources out of subsistence production. Future emphasis of agricultural policy ought to be on maximizing farm household incomes rather than generating food surpluses.

Governments have a difficult task to perform: on one hand, continued food security needs to be assured for populations that are growing in absolute terms; on the other hand, research and infrastructural investments need to be made for diversification out of the primary staples. The tendency of governments to react to short-term 'crisis situations' may be counter-productive in terms of meeting long-term goals of food security and income growth. 'Ultimately the process of rural diversification must be consistent with the longer-run patterns of structural transformation' (Timmer, 1988).

It is important to remember that even with increased commercialization and diversification trends, rice will continue to be the most important staple food in Asia, in relative and absolute terms. Given growing populations and income-induced demand for increased rice consumption, there continues to be a strong need to seek higher productivity levels for rice.

Agricultural commercialization should not be expected to be a frictionless process, and significant equity and environmental consequences should be anticipated at least in the short to medium term, particularly when inappropriate policies are followed. The absorption of rural poor in the industrial and service sectors has significant costs in terms of learning new skills and family dislocations. Commercial systems could also face higher environmental and health costs, especially in terms of higher chemical input use. Higher opportunity costs of labor increase farmer reliance on herbicides for weed control for rice and other staple food crops that are currently managed through handweeding. Insecticide and fungicide use for high-value crops, such as vegetables and fruit, is substantially higher than for staples, and improper use can increase the incidence of pesticide-related illnesses. Also, where property rights are not clearly established, high-value crop production in the upland environments could lead to higher risks of soil erosion and land degradation.

Appropriate government policies can alleviate many of the possible adverse transitional consequences arising from the process of commercialization and diversification. Long-term strategies to facilitate a smooth transition to commercialization include investment in rural markets, transportation and communications infrastructure to facilitate integration of the rural economy; investment in crop-improvement research to increase productivity, and crop management and extension to increase farmer flexibility and reduce possible environmental problems from high input use; and establishment of secure land rights to land and water to reduce risks to farmers and to provide the incentives for investment in sustaining long-term productivity.

Notes

1. This chapter builds on concepts and material, presented in Pingali (1992), Pingali (1994), and Pingali and Rosegrant (1995).
2. 'We are now at the beginning of a second agricultural revolution which is inaugurating a new farming system based upon the specialized production of milk, meat, eggs, vegetables and fruit' (Astor and Rowntree, 1946, writing about post-war agriculture in Europe).
3. In the provision of high-volume, high-quality livestock products, small-farm operations are necessarily at a disadvantage. They will not be in a position to gain from the scale economies inherent in specialized large-scale operations for processing, transport and marketing of livestock products. Integrated farming systems will not have a comparative advantage in a commercial food-supply system geared towards urban consumers (Pingali, 1993).

9

Strategic Look at Factor Markets and the Organization of Agricultural Production beyond 2025

Introduction

The growing economic prosperity in Asia is a crucial factor that determines the availability of labor, water and land for rice cultivation. The competing demands for these inputs in various economic activities affect their relative scarcities and prices, and change relative profitability depending on the intensity of input use in various economic activities. The future across Asia is one of increased factor scarcities. The factors particularly at risk are labor, land and water. The forces driving this change are: (i) increased off-farm wage rates and opportunity costs of family farming; (ii) increasing withdrawal of highly productive irrigated lands from rice production and/or from agricultural uses; and (iii) reduced availability of irrigation water due to declining investments in irrigation systems, the loss in capacity due to the degradation of existing infrastructure and increased diversion of irrigation water for urban and industrial uses. Growing opportunity costs for the above factors of production contribute directly to declining future profitability of rice systems.

Escalating wage rates and rapid urbanization make the opportunity cost of subsistence-oriented family farming very high. One should see a gradual shift towards the increased commercialization of Asian rice farming systems. Intensive rice monoculture systems may give way to a diversified production system that is responsive to market signals. Further discussion on the opportunities and constraints to diversification was presented in Chapter 8.

This chapter identifies:

1. The probable changes in factor markets (including water) towards the year 2025, due to economic prosperity and increased competition from the non-agricultural sectors.
2. The necessary changes in rice farming systems to accommodate increased modernization and commercialization of the agricultural sector.

Economic Prosperity and Emerging Factor Scarcities

The remarkable progress made possible by the Green Revolution in rice cultivation went hand in hand with enormous economic progress in many parts of Asia. Economic growth in Asia has been many times faster than that for other regions in the developing world. Since 1970, the average annual rate of growth has been a robust 6 percent per year in Asia, compared to 3.2 percent in South America, and 2.4 percent in Africa (Fig. 9.1). A rate of population growth of over 2.5 percent per year has eaten up the meager growth in Africa, so most African nations have seen a deterioration or stagnancy in the economic conditions of

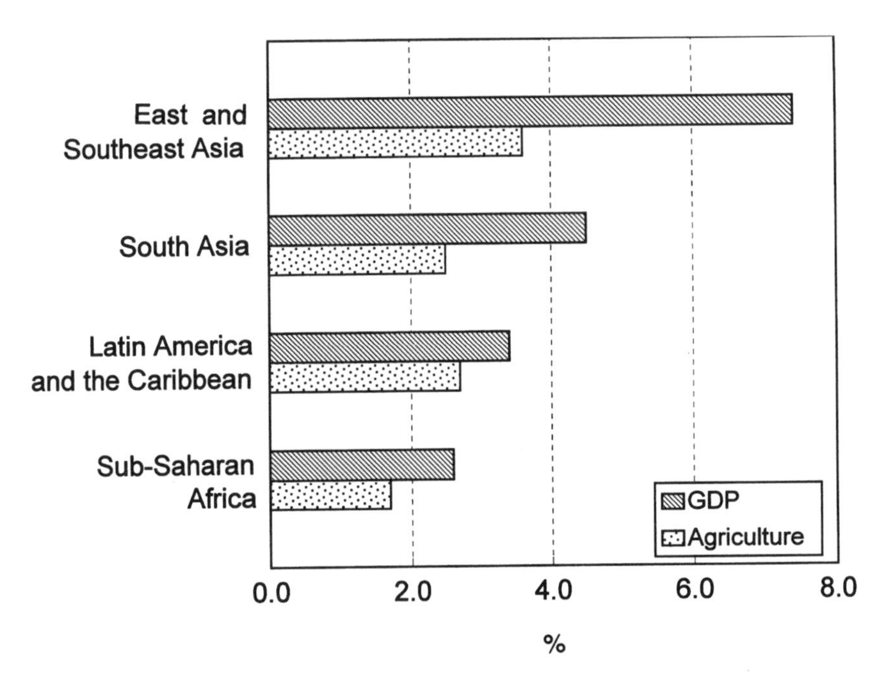

Fig. 9.1. Growth of economy and agriculture in developing countries, 1970–1993. (Source: Hossain, 1995.)

the population. In contrast, economic growth in many Asian nations has also contributed to notable progress in population control, which has been reflected in faster growth in per capita incomes. With an annual rate of growth of about 5 percent per year, the per capita income of Asian people has doubled every 14 years since the early 1960s.

The rate of economic progress has, however, been uneven across countries in Asia. The growth has been much faster in East Asia compared to Southeast Asia, which in turn has grown faster than South Asia. In Southeast Asia, economies in the Philippines, Myanmar, Cambodia and Laos grew much slower than those of Thailand, Malaysia and Indonesia. There was also a positive relationship between overall economic growth and the development of the agricultural sector (Fig. 9.2). But it is the faster growth of the non-agricultural sectors that contributed to the vast economic prosperity in Asia, particularly in East Asia and more recently in Southeast Asia.

At the beginning of the second half of this century, Asian countries, except for Japan, had almost similar economic standing. How-

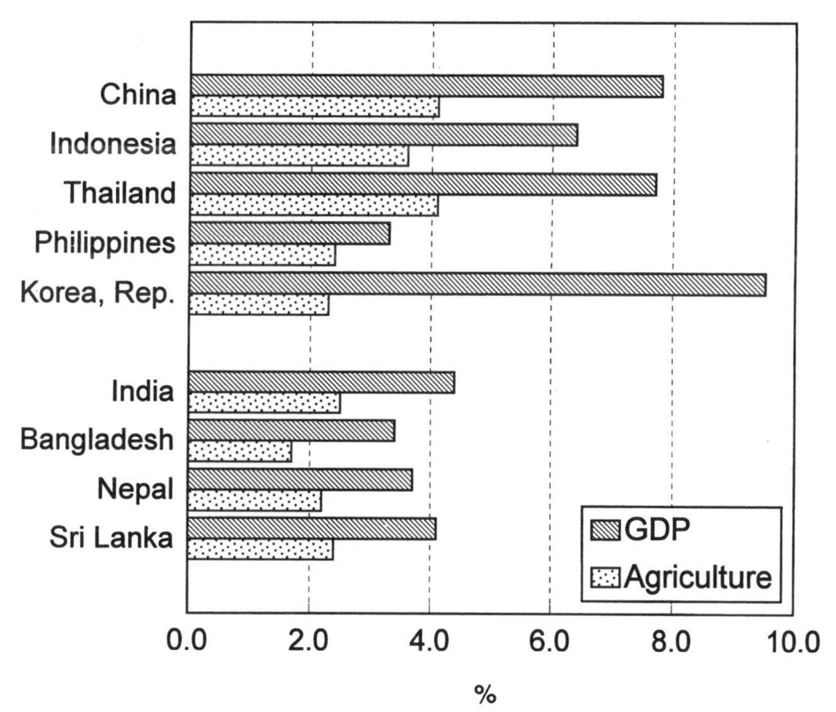

Fig. 9.2. Growth of economy and agriculture in selected Asian countries, 1970–1993. (Source: Hossain, 1995.)

ever, uneven economic growth has created substantial economic disparity. For example, in the early 1960s, the Philippines had higher levels of income than South Korea. Now, the Koreans have eight times higher incomes than the Filipinos. At that time, Indians had the same level of income as the Indonesians. Now the income gap between the two countries is 2.5 times. As the economy grew, the importance of agriculture declined, vibrant manufacturing and service sector activities pulled labor force and population from rural to urban sectors, and higher income levels and larger participation of women in economic activities reduced the demand for children and lowered population growth. The economic disparity among Asian nations, the importance of agriculture, and the structure and growth of population can be seen from Table 9.1.

Economic growth and urbanization

Economic growth and development in Asia is leading to a rapid urbanization and a demographic shift from a predominance of rural populations to one in which urban populations will be in the majority. The

Table 9.1. Economic disparity and the growth and structure of population in major rice-growing countries, 1993–1994.

Country	Production of rice unhusked 1994 (million tons)	Per capita income 1993 (US$)	Agriculture's share of the economy, 1993 (%)	Population growth 1990–1995 (% per year)	Urban population 1993 (%)
China	178.3	490	19	1.2	29
India	118.4	300	31	1.9	26
Indonesia	46.2	740	19	1.6	33
Bangladesh	27.5	220	30	2.0	17
Vietnam	22.5	170	29	2.2	20
Myanmar	19.1	59	63	2.2	26
Thailand	18.4	2,110	10	1.4	19
Japan	15.0	31,490	2	0.3	77
Philippines	10.2	850	22	2.4	52
Korea, Rep.	7.1	7,660	7	0.9	78

Sources: World Bank (1995), FAO (1995).

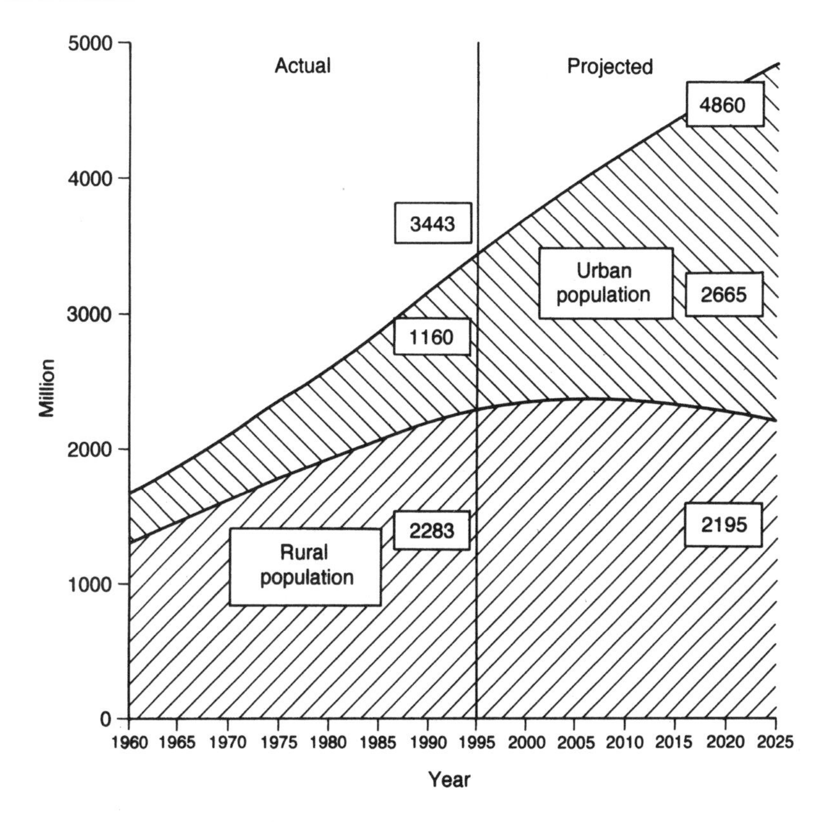

Fig. 9.3. Growth of population and urbanization in Asia, 1960-1995, and projection up to 2025. (Source: Hossain, 1995.)

World Bank projects urban populations to be twice as large as rural populations by the year 2030. Developing-country cities will grow by 160 percent during the 1990–2030 period while rural populations will grow only by 10 percent during the same period. For Asia as a whole, urban population growth is expected to be at the rate of 3 percent per year (IRRI, 1995). By 2025, 53 percent of the people in Asia will live in urban areas compared to 30 percent in 1990 (Fig. 9.3), approximately 2.6 billion people (Feder and Keck, 1996).

> Asia has and will continue to have the largest urban population in the world, even though urban growth rates in Asia are lower than those in Sub-Saharan Africa, North Africa and the Middle East. Close to 1 billion of the 1.4 billion urban residents in developing countries in 1990 lived in Asian cities. China alone comprised one-third of Asia's urban population.
>
> (Naylor, 1994)

Unrelenting growth of urban populations in Asia leads to a quantum increase in purchased food requirements, while 'the ability to provision the cities with food from domestic production often declines' (Naylor, 1994).

The movement of populations from a rural base to urban areas makes the provisioning of food supplies the single most important challenge for policymakers in Asia. While urban populations demand a more diversified diet, the magnitude of rice required for the urban market would be several times greater than it is today. Providing adequate rice supplies at affordable prices to urban consumers often becomes a political nightmare for most governments. Countries with smaller populations but high incomes manage the problem of urban food supplies and food prices through imports, like Malaysia for example; however, large-scale imports are not an option for countries with large populations, such as China, India, Indonesia and Bangladesh. These countries would have to rely on increasing domestic productivity growth to meet the growing urban demand for food. Provisioning the cities cannot be done in a traditional subsistence production mode. The entire strategy ought to be towards surplus generation, and the favorable rice-growing environments will be under intense pressure to provide for the urban masses. The required reorganization of agricultural production systems in order to sustain the profitability of rice farming is discussed in the next section of this chapter.

Urbanization also affects factor markets in terms of competition for factors of production and consequently their prices. The factors particularly affected are land, labor and water. Factor scarcities resulting from the structural transformation of economies is discussed in the next few subsections.

Labor supply and agricultural wages

Economic growth brings dramatic changes in the structure of employment, adoption of labor-saving technology, and an increase in labor productivity. The working-age population which is the source of labor supply is determined by the rate of population growth and female participation in economic activities. In most developing countries, the rate of growth in working-age population has been around 2 to 3 percent per year (World Bank, 1995). A faster rate of economic growth has led to severe competition for labor among economic sectors in Asia, since demand grew at a faster rate than supply. With opportunities for more remunerative employment rising elsewhere, workers move out of low-productivity, low-wage activities.

The relationship between the sectoral distribution of the labor force and the level of economic growth analyzed by the International Labor Organization (ILO) is shown in Fig. 9.4. On the average, agriculture's share of employment falls from 90 percent of the total in low-income countries to roughly 5 percent in the high-income countries. The share of industry – which includes manufacturing, construction and mining – increases from 4 to 35 percent, and that of services from 6 to 60 percent. As the economy grows faster, the pull of labor in the manufacturing and services sectors away from agriculture becomes stronger and, after some time, the labor force employed in agriculture declines in absolute terms.

Although the agricultural sector tries to address the problem of labor shortage by adopting labor-saving technologies, productivity differences between the sectors continue to grow with economic prosperity. In South Korea, for example, labor productivity in manufacturing increased by 4.3 times during the 1966–1990 period, compared to only 1.2 times in the agricultural sector. The total agricultural labor force

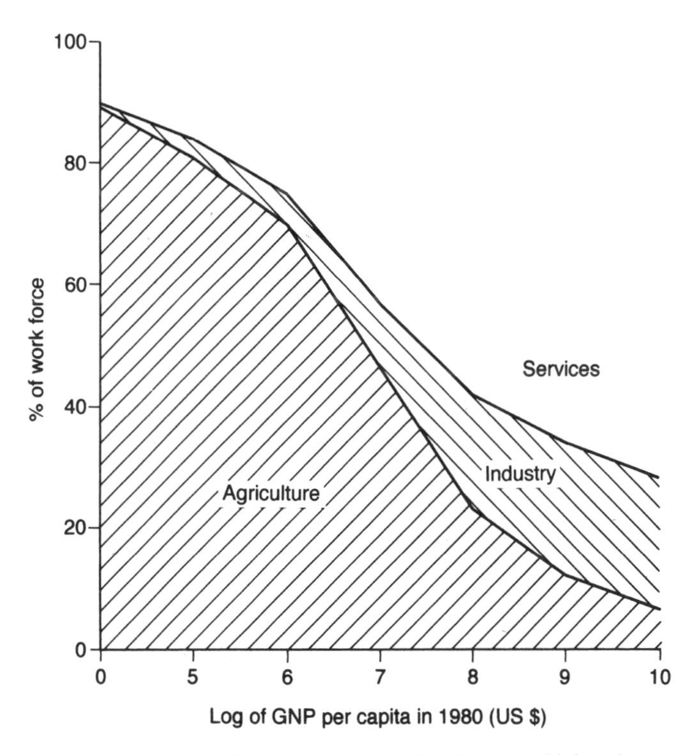

Fig. 9.4. Relationship between sectoral distribution of labor force and the level of economic growth. (Source: Hossain, 1995.)

increased from 4.5 to 6.1 million persons between 1966 and 1975, then started declining in absolute terms and reached 3.2 million by 1990.

The scarcity of labor is reflected in its price, the wage rate. The effect of the differential rate of economic growth on the trend of real wages in the manufacturing sector in various regions of the developing world is shown in Fig. 9.5. In Sub-Saharan Africa where economies have remained stagnant, real wages hardly increased during the 1970–1990 period. In contrast, in East and Southeast Asia which experienced more than 5 percent per year growth in per capita incomes, real wage rate increased by 170 percent over the 20-year period. In South Asia, where economic growth was moderate, real wage rate increased by only 50 percent.

In addition, rural–urban migration, increases in labor productivity, and escalating wage rates in non-farm sectors put an upward pressure on the rural wage rate, affecting the costs and profitability of rice production. The growth in agricultural wage rates over the 1961–1991 period for selected Asian countries is shown in Table 9.2. In the early 1960s, the difference in wage rate across countries was only marginal. In slow-growing countries such as Bangladesh, India and the Philippines, agricultural wage rates had hardly increased, but these were escalating in Japan and South Korea. In 1991, the cost of agricultural labor was more than 20 times higher in Korea, and 50 times higher in Japan, compared to that of Bangladesh.

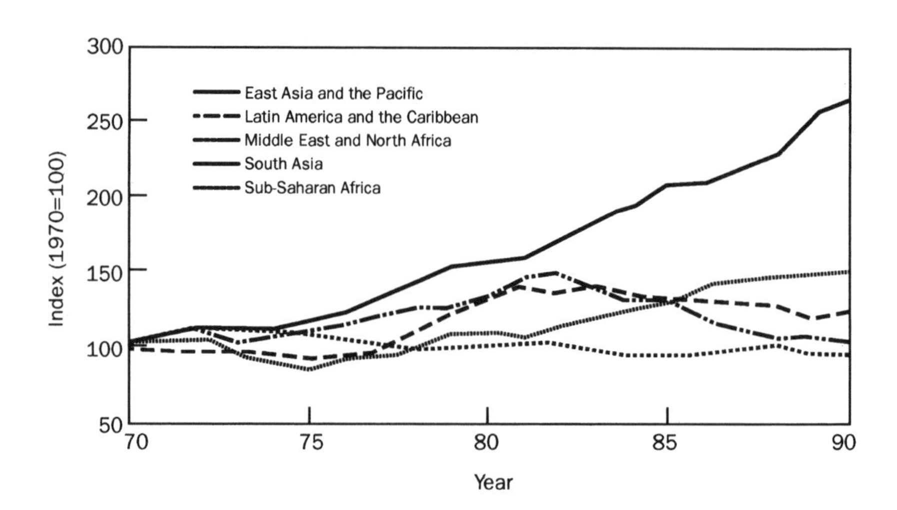

Fig. 9.5. Real wages in the manufacturing sector in various regions of the developing world. (Source: World Bank, 1995.)

Table 9.2. Long-term trend in wage rates, selected Asian countries (US$ day^{-1}).

Country	1961	1971	1981	1991
Bangladesh	0.46	0.44	0.86	1.39
Philippines	1.39	0.59	1.51	2.16
Korea, Rep.	0.82	1.86	10.84	32.59
Japan	1.22	8.19	24.16	51.93

Source: IRRI (1995).

Competing demands for ricelands

An important implication of growing urbanization is that some of the fertile agricultural land has to be diverted to meet the demand for housing, factories and roads. Also with urbanization and the associated change in food habits, markets for vegetables, fruits and livestock products will grow stronger. Changes in relative profitability will induce farmers to divert ricelands to grow more profitable non-rice crops such as vegetables, fruits and fodder. About 84 million hectares of additional land will need to be converted to urban uses in Asia (excluding Japan) between 1990 and 2025, or about 62 percent of the unused cultivable land. Obviously not all expansion in land for urban uses will take place on cultivable land, but, given the historical tendency of urban concentrations to develop close to agricultural production areas, most urban expansion will compete directly with agricultural uses (Feder and Keck, 1996).

The area under rice has already started declining even in low- and middle-income countries such as China, Philippines, Indonesia (Java) and Bangladesh. In China, harvested rice area declined from 37 million hectares in 1978 to 32 million hectares in 1993. In the Philippines, harvested rice area declined from 3.7 to 3.2 million hectares within the same period.

Most of today's megacities of Asia were purposefully located in the fertile plains or close to them. A large number of the provincial 'towns' that are now rapidly expanding are also located in the vicinity of well-developed and highly productive irrigated lowland areas. Irrigated lowland areas tend to be highly desirable for conversion to urban uses because of established water, power, transport and communication infrastructure. In the irrigated rice bowl provinces that surround Metro Manila, for instance, 17,305 hectares of irrigated lands have been converted for urban uses within the last five years (Moya *et al.*, 1994). Rice production forgone due to this conversion is in the order of 180,000 tons per year. Similar magnitudes of agricultural land conversion are recorded for other Southeast Asian countries.

In Java, Indonesia, around 170,000 hectares of irrigated lands were converted to non-rice uses between 1980 and 1990, approximately 50 percent of which was for non-agricultural purposes. The rate of conversion has increased in the later part of the 1980s (between 1987 and 1992) with an average annual conversion of ricelands at about 23,000 hectares, 56 percent of which was for non-agricultural purposes (Pasandaran and Sayaka, 1996). Ricelands in the provinces surrounding Bangkok are being rapidly converted for residential and industrial uses. Since 1988, land conversion rates in Nonthaburi, Pathum Thani and Samut Prakarn were 7.2, 5.9 and 2.6 percent respectively, increasing from the historical rate of only 1 percent per annum (Poapongsakom, 1996).

Once the speculative pressure on agricultural land starts due to anticipated conversion for urban uses, it becomes very difficult to sustain agricultural production. Land values rise dramatically and it becomes highly profitable for farmers and long-term tenants to give up their rights to the land in exchange for lumpsum payments. In Thailand, for instance, the land transactions process has tremendously increased the farmers' wealth. After selling the land, Thai farmers usually have enough money to buy a larger plot of land, to build a new house, to buy a car or motorbike, and still have some savings in the bank account (Poapongsakom, 1996). (See Box 9.1 for case study evidence from the Philippines on the process of land conversion.)

As urban incomes grow, the demand for rice increases particularly among the lower-income groups, but in the higher-income groups there is a shift to a more diversified diet. Monoculture ricelands become increasingly diversified with growing urban demand for high-value fruit and vegetables. Diversion of ricelands can be seasonal, as in dry-season non-rice crop production, or it can be permanent, as in year-round vegetable and fruit production, once investments in drainage and water control have been made (see Chapter 8 for more details on the economics of crop diversification). In Thailand, for instance, the area devoted to paddy production slightly decreased from 11.4 million hectares in 1976 to barely 11 million hectares in 1991. In the same period, the area planted to tree crops doubled from 1.65 to 3.21 million hectares, and the area devoted to vegetables and flowers increased threefold from 57,163 hectares in 1976 to 137,309 hectares in 1991 (Poapongsakom, 1996).

There is nothing wrong, if the price is right, with the diversification of ricelands for non-rice production systems or for the conversion of ricelands to non-agricultural uses. It just means that the opportunity

Box 9.1. The conversion of agricultural land for urban uses – who gains and who loses?

For the entire Philippines, 40,644 hectares of agricultural land (0.5 percent of total cropland) were legally converted to non-agricultural uses between 1988 and 1993. If illegal and unreported conversions are also considered, the total converted area could increase by 50 percent. These lands were converted mainly to industrial and residential estates, which commanded the highest price.

A detailed assessment across Laguna, Cavite, Bulacan and Pampanga indicated that only 30 percent (about 5,300 hectares) of the converted lands were formerly ricelands. The drop in rice area is about 8 percent of the total rice area in the same provinces, even if it is assumed that 50 percent of the converted lands were formerly ricelands. For the nation as a whole the best estimate for riceland conversion is 15,000 to 20,000 hectares per year, only 0.6 percent of total riceland. If a national figure of 20,000 hectares of riceland converted to non-rice activities and an average yield of 4 t ha^{-1} for the wet season and 5 t ha^{-1} for the dry season is assumed, then the annual decline in rice production is about 180,000 tons. This is about 2 percent of total rice production per year in the Philippines. While the area converted may be proportionately small, the loss in production is significant because the conversions have occurred mainly in high-yield-potential, irrigated areas that are the main suppliers of rice for urban consumers.

Rice areas converted to residential sites command a much higher displacement compensation than areas converted to industrial sites. A mean displacement compensation of US$ 38,840 per hectare of area converted to residential sites was paid to farmers who owned or tilled the land. In contrast, farmers whose farms were converted to industrial sites received a compensation of US$ 8,480. The total compensation package usually consists of cash payments plus good-sized residential lots, but in some cases the compensation is in cash. In areas converted to industrial sites, the incoming companies are committed to hire the displaced farmers and their families, partly explaining the difference in the amount of compensation that the farmers receive.

The majority (54 percent) of the farmer-respondents felt they are better off today than when they were still farming. About 31 percent rated their economic status as the same as before and 15 percent felt they are worse off than before the land conversion. This last group are those with no present source of livelihood, and whose major portion of compensation was spent in subsistence or consumption expenses. The farmer-respondents spent part of their compensation to either: construct or renovate a house; purchase some necessary furnishings; invest in income-generating activities, such as small village stores, rice and fruit marketing; and transport service operation. The farmers commonly shared part of the cash compensation with their children and deposited the remaining portion in the bank. Only a few farmers spent their cash compensation to clear debts, acquire new land, or to fund their children's education.

In the end, does the conversion of agricultural land resources to urban uses lead to a net social cost or a net social benefit? While there may be no absolute answer to this question, a number of socioeconomic factors should clarify the whole issue of costs and benefits of land conversion: (i) the agricultural potentials of converted land; (ii) existence of alternative sources of food in the region or elsewhere, considering the possibilities of inter-regional and international trade; (iii) the resulting increase in labor income; and (iv) the distributional consequences of land conversion. At a broad positive aspect, land conversion creates vast employment opportunities, provides systematically designed residential spaces, increases land productivity, as well as promotes market and countryside development. However, with an annual decline in total rice production of about 2 percent per year, land conversion may substantially increase the level of rice imports in order to support the increasing Philippine population.

Source: Moya *et al.* (1994).

costs of rice production become higher, together with the costs of a ton of rice produced.

Competing demands for water and reduced supply

Water is an indispensable input of rice cultivation. Agriculture accounts for nearly 90 percent of the annual withdrawal of the renewable water resources. Water resource development has been the key to increasing rice production in virtually all Asian countries where land is a scarce factor of production. Among all activities involving exploitation of natural resources, irrigation is by far the most important. Asia accounts for 55 percent of an estimated 253 million hectares of irrigated land in the world; China and India alone have over 100 million hectares of irrigated land (Fredericksen *et al.*, 1993).

Population growth, urbanization and industrialization trends in Asia are leading to increased water demands and competition with agriculture for water resources. At the same time, the supply of irrigation water is declining due to: reduced spending on irrigation infrastructure; degradation of existing infrastructure; and the overexploitation of groundwater resources. Irrigation water has been traditionally provided to farmers at a price considerably lower than its social cost. As water scarcity increases, however, the non-monetary costs associated with ensuring adequate water supplies at the farm level will increase. These costs include increased farmer time in water-related conflicts, farmer participation in water allocation decisions and management, and farm-level supervision of water deliveries.

There are no good estimates of the competing demands for water and any particular country's ability to meet these demands (see Gleick, 1993, for the current state of knowledge). The following propositions could be used to anticipate future trends: (i) domestic and household demand for water has a positive income elasticity; and (ii) as the share of agriculture in gross domestic product (GDP) falls, the opportunity cost of water for agricultural use rises and the share of national water resources allocated to agriculture falls. Given a finite supply of renewable water resources, the above propositions imply growing competition for water in Asia. The problem will be more severe in the more arid zones and in locations depending on underground aquifers with a slow rate of recharge.

Postel (1993) states that Beijing's water needs will increase by 50 percent between 1990 and 2000. Water tables beneath the city have been dropping by 1–2 meters per year, with already a third of its wells gone dry. Shifting water from farm to city may be the only way to balance the region's water budget. Farmers in the vicinity of Beijing could

Table 9.3. Water availability and water-use indicators, major rice-growing countries in Asia, 1990.

Countries	Renewable water resources (m^3 yr^{-1} $person^{-1}$)	Withdrawal per capita (m^3)	Withdrawal as percent of availability	Share of agriculture in total withdrawal (percent)
China	2,427	462	19.1	87
India	2,464	612	24.8	93
Indonesia	14,020	465	3.3	94
Bangladesh	11,740	211	1.8	96
Thailand	3,274	599	18.3	90
Vietnam	5,638	81	1.4	78
Myanmar	25,960	103	4.0	90
Philippines	5,180	693	13.4	61
South Korea	1,452	298	20.5	75
Pakistan	3,962	1,250	31.5	98
Nepal	8,686	155	1.8	95
Malaysia	26,300	765	2.9	47
Sri Lanka	2,498	503	20.1	96
Cambodia	10,680	69	0.6	94

Source: Fredericksen *et al.* (1993).

Box 9.2. The Angat Irrigation Project: a case study in the Philippines.

The Angat Irrigation Project is a multipurpose project covering three major components: irrigation, power supply and domestic water supply for Metro Manila. The Angat reservoir has a capacity to store 1,028 million cubic meters of water and a service area of about 28,000 hectares of ricelands. The main canal supplies water for rice irrigation, whereas an auxillary canal supplies domestic water supply to the urban center of Metro Manila.

Between 1980 and 1995, there has been a steady increase in the supply of water for urban use, at the expense of irrigation (Fig. 9.6). The average annual urban water supply increased from 300 million cubic meters to about 800 million cubic meters, or by more than 10 percent per year. Since the total water supply in the reservoir did not increase during this period, the higher deliveries to the urban sector caused reductions in irrigation water supplies. So while the urban sector enjoyed a steady growth in water supply, allocations for irrigation had to be adjusted to the variability of the total available water in the reservoir, and had suffered serious fluctuations parallel to those in the total supply.

The extent of area receiving irrigation water supply in the dry season, and the drought-affected area within the project's service area, were determined by the rate and amount of water released between January and April. As expected, the project suffered serious reductions in both service area and rice production especially in years when dry-season water supply was substantially reduced. In 1990, when the January-to-April irrigation water supply was less than one-third of the long-term average, about 13,500 hectares of ricelands could not be planted to the dry-season crop, and more than 2,200 hectares of ricelands had significant yield reduction. The average rice yield in that season was the lowest in the 16 years of record analyzed. During the past three years, 1200 to 2800 hectares of rice area were affected by drought in the dry season.

Source: Bhuiyan and Tabbal (1996).

lose 30–40 percent of their current water supply within the next 10 years. Such scenarios are becoming commonplace around other Asian cities. In order to meet the growing water requirements of Manila, water from the Angat Dam, the primary source of irrigation water to the rice bowl region of Central Luzon, is being increasingly diverted away from farm use in the last 15 years. (See Box 9.2 for detailed evidence on this report.)

In absolute terms, annual water withdrawals are by far the greatest in Asia (WRI, 1992). They vary from 600–1250 m^3 per capita for Pakistan, India, Malaysia, Philippines and Thailand, to 400–500 cubic meters per capita for China, Indonesia and Sri Lanka (Table 9.3). Agri-

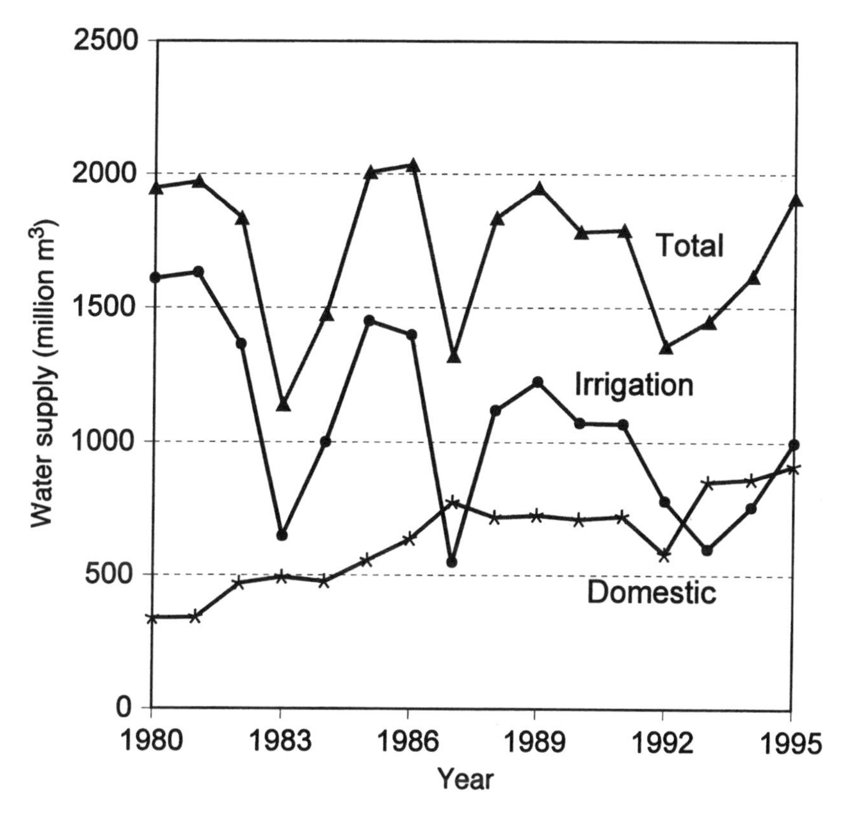

Fig. 9.6. Annual total and sectoral water supplies in the Angat Irrigation Project, Bulacan province, Philippines, 1980–1995. (Source: Bhuiyan and Tabbal, 1996.)

culture accounts for 86 percent of total annual water withdrawal for Asia, compared to 49 percent for Northern and Central America and 38 percent for Europe. The per capita availability of water resources declined by 40–60 percent in most Asian countries over the 1955–1990 period (Feder and Keck, 1996). By common convention, countries are defined as water-stressed when per capita availability is between 1000 and 1700 cubic meters. Projections based on constant availability of water and increasing population suggest that China, India, Sri Lanka, Pakistan and South Korea are expected to reach near stress levels by 2025 (Fig. 9.7). If calculations are based on water availability during the dry season, most Asian countries will be under stress within that time horizon.

Declining irrigation investment and degradation of irrigation infrastructure

Irrigated rice environments not only face increased competition for water but also face reduced water supplies due to system degradation

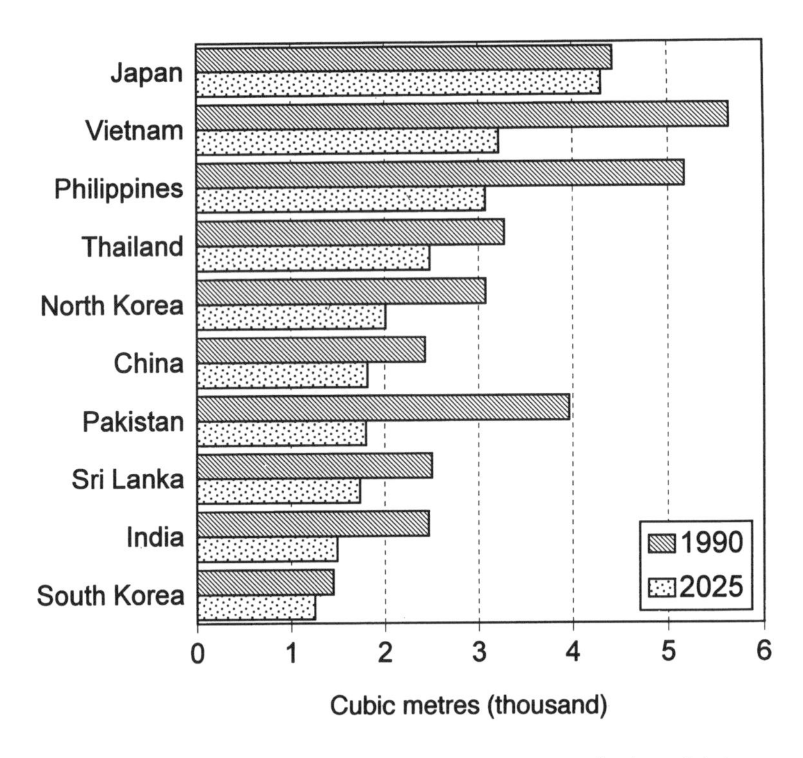

Fig. 9.7. Projected change in per capita water resources of selected Asian countries. (Source: Gleick, 1993.)

and reduced investments in infrastructure development. For Asia as a whole, there has been a sharp decline in the rate of growth in irrigated area in recent years (Rosegrant and Svendsen, 1992). In East Asia, the growth in irrigated area was 2 percent annually through the mid-1970s, but was virtually stagnant in the 1980s. In South Asia, the growth rate dropped from 2.8 percent during 1975–1980 to 0.1 percent during the 1985–1988 period. The recent sharp reduction in irrigation investment is likely to further slow the rate of growth in area irrigated. Aggregate lending and assistance for irrigation development in South and Southeast Asia for the 1970s and 1980s is shown in Fig. 9.8. For South Asia, the peak lending period was 1980–1982, followed by a 50 percent drop in annual lending by 1986/87. The decline was even more precipitous in Southeast Asia from a peak in annual average lending of US$ 630 million in 1977–1979 to US$ 202 million in 1986/87 (Rosegrant and Svendsen, 1992).

Some countries, notably India, have insulated their irrigation investment programs from cutbacks in international lending by increasing the use of domestic funds. However, most countries in

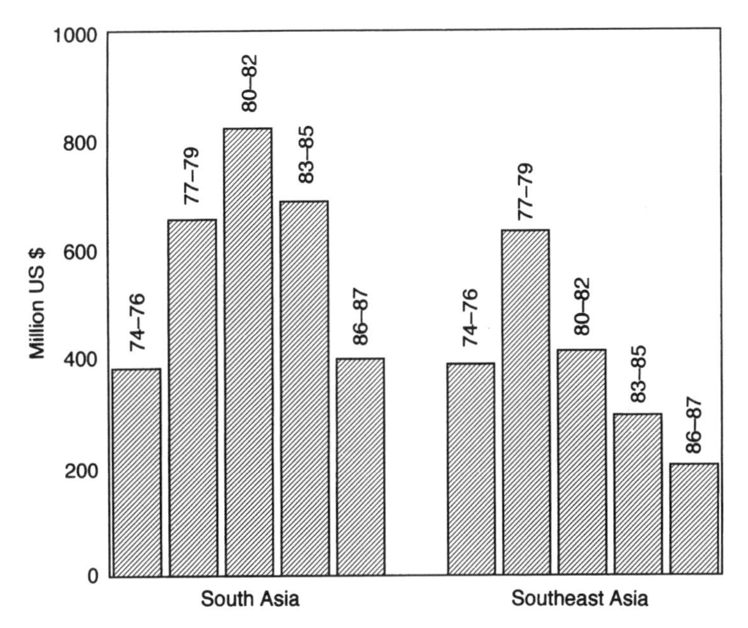

Fig. 9.8. Average annual lending for irrigation projects by donor agencies. (Source: Hossain, 1996.)

Asia, for example Indonesia, the Philippines and Thailand, have sharply reduced total spending on irrigation as international lending has decreased. The full effects of declining irrigation investment on rice production are just beginning to be felt due to the lags in irrigation construction. The decline in irrigated area is not easily reversible because of the large gestation period of most projects.

Among the factors contributing to the reductions in investment are the large public and foreign debt loads carried by most of the countries in the region, the declining share of unexploited irrigation development, and concerns about the environmental implications of irrigation projects (Rosegrant and Svendsen, 1993). However, the most important causes of declining investment appear to be the decline in world rice prices and the increasing real costs per hectare of new irrigation development (Rosegrant and Pasandaran, 1990; Rosegrant and Mongkolsmai, 1990; Svendsen and Ramirez, 1990). Compared to the late 1960s, the estimated investment costs per hectare of land for recent years is almost three times for Sri Lanka, double for India and Indonesia, and nearly 50 percent higher for the Philippines and Thailand (Fig. 9.9). Today, large-scale irrigation development in Asia can cost anywhere from US$ 3000 to US$ 6000 per hectare, which could once again double by 2025.

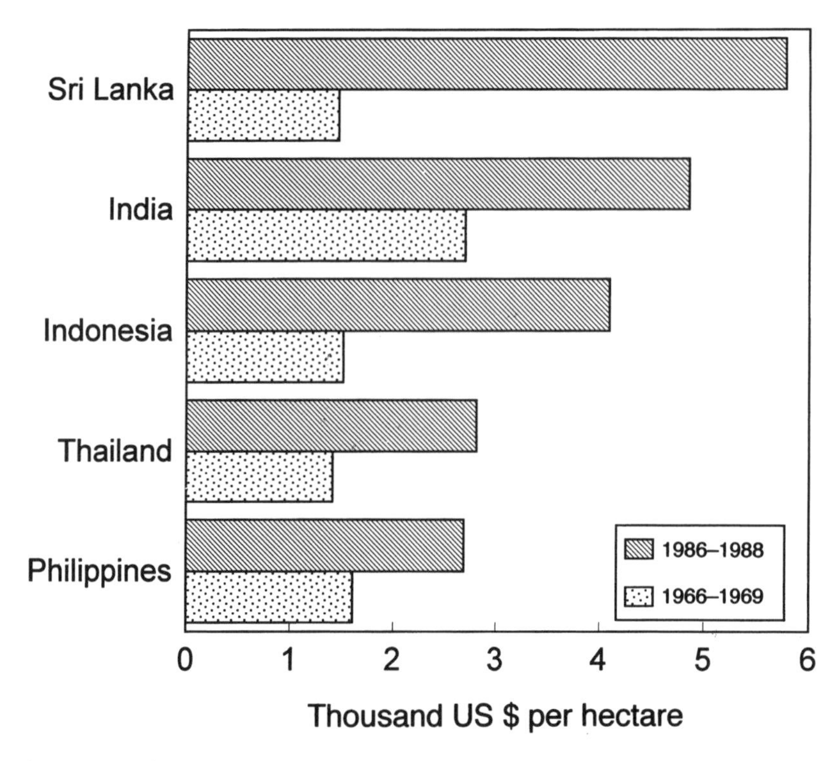

Fig. 9.9. Real capital cost for construction of new irrigation systems. (Source: Rosegrant and Svendsen, 1993.)

The problem of declining investments in expanding irrigation area is exacerbated by the poor maintenance of much of the existing irrigation infrastructure. Despite a relative shift in overall irrigation investment in the 1980s from new construction to rehabilitation and operations and maintenance, there is evidence of continued decline in the quality of existing irrigation infrastructure.

In China, more than 930,000 hectares of irrigated land have come out of production since 1980, an average loss of 116,000 hectares per year, due to siltation and salinity buildup (Postel, 1993). An analysis of 92 irrigation systems in Luzon Island of the Philippines shows that almost a third of them have declining trends in wet- and dry-season irrigated areas, and in wet- and dry-season rice yields (Masicat *et al.*, 1990). Between 1979 and 1989, the absolute wet- and dry-season irrigated area in Luzon declined by 20,466 hectares and 36,175 hectares, respectively. De Vera (1992) estimates, for an average irrigation system in the Philippines, that a 1 percent drop in expenditures for operations and maintenance could lead to a 0.26 percent drop in wet-season irrigated area and a 0.46 percent drop in dry-season irigated area (see Chapter 5 for more details on this study).

For Asia as a whole, further decline in irrigated area can be expected over the next decade. The decline will be not only in absolute area but also in the quality of irrigation. Asian farmers will face increasingly unreliable water supply and poor water quality due to upper watershed and industrial externalities. Concerns about water quality are discussed in greater detail in Chapter 5. The net effect is increasing costs of water at the farm level, even if water pricing does not change, and declining profitability of irrigated agriculture.

Implications for the Organization and Management of Rice Farming Systems

The first section of this chapter argued that increased competition with the non-agricultural sectors is leading to an increase in the prices of the primary factors of food production: labor, land and water. Long-term rising trends in the above factors of production not only affect the profitability of intensive rice production systems but along with growing urban demand for food lead to a fundamental transformation in the organization and management of rice production.

Agricultural development strategies, for the most part, assume that Asia's farmers are simply farmers producing agricultural goods and services in agrarian economies. (This paradigm is rapidly becoming obsolete.) The future of Asia's agriculture is wedded to comprehensive and fundamental transformations under way in what Asia's rural societies 'are' and what these societies are 'becoming'. Agricultural development and the future of agriculture rest in the context of these broader transformations (Koppel and Zurick, 1988).

The fundamental transformation expected is the definitive movement away from subsistence production systems to commercial operations. Increasing commercialization of agricultural systems implies: greater market orientation and specialization of farm production; and progressive substitution out of non-traded inputs in favor of purchased inputs. Chapter 8 provides a detailed discussion on the anticipated changes in the product market as a result of the movement from subsistence to commercial agricultural systems. This section deals with anticipated changes in the management of rice production – reduced reliance on family labor and non-traded inputs. Such changes in the management of rice production, which have already taken place in the advanced economies of Japan, Korea and Taiwan, will begin taking place in the rapidly growing economies of Southeast Asia and will eventually be seen throughout the high-intensity rice bowls of Asia. The driving force behind the transformation of rice production systems is the rapidly increasing opportunity cost of labor. While growing land and water scarcity can be compensated for with increasing scientific knowledge and farmer management, farmer time required for sustaining productivity and profitability of intensive rice systems will become increasingly scarce. The collapse of subsistence systems will come about because of competing demands for farmers' time.

Specific changes in farm organization and management that ought to be expected are: (i) larger operational holdings; (ii) increased incidence of part-time farming; (iii) changing demographics of family farm operators; and (iv) the widespread adoption of labor-saving technologies.

Operational holding size

As the opportunity cost of family labor rises, small family-farm operations for subsistence production become increasingly unprofitable. Landless tenant farmers will gradually find their way to the urban industrial sector. Small landowners will likewise find it more profitable to sell or lease their holdings rather than to cultivate them. The extent to which farm ownership size increases depends on land reform legislation. Strict rules against land accumulation designed for a different period may prevent changes in farm ownership, at least in the short run, as is the case in Japan and Taiwan. Over time, one should expect the operational holding size to increase as land consolidation occurs. In Japan, Taiwan and South Korea, it is increasingly important to distinguish between the ownership of riceland and the operation of a rice-farm enterprise (Dyck *et al.*, 1993). Rice farms are increasingly managed by professional contractors who share profits with the landowners.

Part-time farming and the role of family labor

Where small-scale farming survives under high-wage conditions, it will necessarily be only on a part-time basis; the majority of the household income will come from non-agricultural sources. In Korea, for instance, off-farm income accounted for 46 percent of total farm-household income in 1991, compared with 32 percent a decade earlier. In Japan it was around 80 percent, and in Taiwan it grew from 50 percent in 1960 to 87 percent in 1990 (Dyck *et al.*, 1993). Income and expenditure surveys conducted in South and Southeast Asia in the 1980s reveal that the proportion of off-farm income was as much as 40 percent in Indonesia, Thailand and the Philippines (Koppel and Zurick, 1988).

The future scenario for Asia is one of continued tightening of the rural labor markets and increasing real wage rates, especially for peak-season agricultural operations. The extent to which wages rise depends on population growth and the growth in off-farm employment opportunities. Even households that continue to cultivate will find that a large proportion of their total income comes from off-farm sources. Given the declining share of agriculture in total farm income, households will tend to economize on the amount of time spent in farm operations. Labor- and management-intensive rice production systems will not be profitable under rising real wage conditions.

Changing rural demographic profiles

Observation of farm households in the newly industrialized countries of Asia indicates significant changes in their age and gender profile. Farming operations are increasingly carried out by older family members and women. Young male family members generally migrate out of the family farm to seek economic opportunities in the urban/industrial sectors. The experience of South Korea is illustrative. In 1990, 34 percent of Korean farmers were between 20 and 50 years old, as compared to 60 percent in 1963. Younger members of the rural community have been leaving at the rate of 400,000 persons per year, while older farmers remain behind (Park, 1996).

In Thailand, increasing job opportunities in the non-agricultural sector have caused rural people to switch jobs, resulting in labor shortage in rural activities. Sustained economic growth not only affected the casual agricultural labor market, but has started showing pronounced effects on the labor-market structure. Firstly, wet-season agricultural employment has been declining since 1989, from when the number of female farmers also decreased by more than one million persons up to 1991. Dry-season agricultural employment has also declined substantially, from 60 percent in 1987 to 50 percent in 1991 (Poapongsakom, 1996).

In the Central and Northeastern regions, people who remain in the villages are the middle-aged and the elderly. An increasing number of households were also found to have only two adults – the husband and wife – working alone on the farm. Most of the younger generation, especially female and educated persons, have found better employment opportunities in the non-agricultural sector, and have migrated to work in the cities. On average, 80 percent of the 15–25-year-old age group in the Central Plain villages are working in the cities. If this trend continues, the proportion of agricultural employment will most probably decline to 40 percent by the year 2000 (Poapongsakom, 1996).

Migration in Thailand, being an adaptive strategy for agricultural families faced with inadequate household income and output, has led to two major issues. First is the concern for the loss of labor, especially of male labor, which may result in the deterioration of farming where only females and small children are left to mind the farm, and second is the concern for remittances which can be spent on farm improvement, to compensate for the loss of labor. (Refer to Shinawatra and Pitackwong, 1996, for a detailed discussion on these issues.) In general, migration of one or more members of a farm family will certainly change the pattern of work within the household. The burden of additional work resulting from not having a family member will depend on labor availability of the households.

In Japan, rapid industrialization in the 1950s and 1960s found women participating in farming more than men, who were recruited to work more in the non-agricultural sector. At the same time during the 25-year, 1950–1975 period, the working population in the farming sector dropped by half, and the composition of the agricultural labor force changed to include mainly the aged and female members in the farm family (Kada and Kada, 1985, as cited in Shinawatra and Pitackwong, 1996). However, despite the drop in the ratio of the farming population, there was no drop in agricultural output as improved technology has compensated for the loss of manpower (Shinawatra and Pitackwong, 1996).

The switch from non-traded to traded inputs

As economies grow, the returns to intensive subsistence production systems that require high levels of family labor are generally lower than exclusive reliance on purchased inputs. Increasing opportunity cost of family labor with the growth in off-farm employment opportunities leads to a substitution of non-traded for traded inputs (see Table 8.4, Chapter 8). Moreover, with the changing age and gender composition of the rural household, and with the anticipated rise in operational holding size, the ability of the household to supply adequate quantities of non-traded inputs declines. Power, soil fertility maintenance, and crop care are the primary activities for which non-traded inputs are used in subsistence societies.

Choice of power sources: human and animal versus mechanical

Agricultural operations can be grouped according to the relative intensity with which they require power, or energy, in relation to the control functions of the human mind or judgment (Pingali *et al.*, 1987). Operations such as land preparation, transport, milling, grinding and threshing are power intensive, while weeding, sifting, winnowing and fruit harvesting, for example, are control-intensive operations. Animal power was traditionally used for power-intensive operations while human power is still used for control-intensive operations. One should expect that as wage rates rise, animal-powered technologies will increasingly give way to motor-powered technologies. This trend is clearly seen across Asia, especially with the advent of rental markets for machinery. The substitution of animal power with machines for power-intensive operations is profitable even in countries with slower income growth. The persistence of animal draft power in these countries is not because motorized power per energy unit is more expensive but because of policies that impeded the growth of machines and the provision of rental markets. With increasing commercialization, mechanical and chemical technologies will also substitute for human labor for the more control-intensive operations, such as weeding and harvesting.

Chemical fertilizers versus farmyard manures

Intensification of land use is only possible with nutrient replenishment to the soil to sustain its productivity. In subsistence societies soil nutrient supply is replenished by farmyard manures. Output growth in intensive, commercially oriented food production systems is not possible in the absence of chemical fertilizer use. There are several reasons for this: (i) the physical quantities of farmyard manure required for sustaining soil fertility would make it uneconomical relative to chemical fertilizers because of the labor requirements for manure production – the feed requirements for maintaining the number of livestock required to meet manure requirements and the high cost of transporting it to the field; (ii) on efficiency grounds, high-bulk, low-value materials – manure, agricultural by-products and crop residues – do not repay labor-intensive management simply because they have such dilute concentrations of useful ingredients relative to chemical fertilizers (McIntire *et al.*, 1992); and (iii) with falling fertilizer prices relative to labor and with improved transport infrastructure, chemical fertilizers are the dominant choice for soil fertility maintenance.

Studies comparing chemical fertilizers and farmyard manures have shown that, per unit of nutrients, the yield response to farmyard manures and chemical fertilizers is similar. In other words, manure does not produce responses different from fertilizers at equal concentrations when applied with similar methods and conditions. Moreover, the contention that manures have a cumulative effect and therefore can replace future fertilizer applications has also been found to be without an empirical basis (see McIntire *et al.*, 1992, for an assessment of the literature). From a sustainability point of view there are other technological alternatives that are cheaper than farmyard manure use. For high-intensity production systems, further promotion of farmyard manure or any other organic-based (labor-intensive) system for replenishing soil fertility would not be successful.

The shift out of labor-intensive crop care and management

The traditional role played by hired labor in conducting intensive farm operations will not be profitable under escalating farm wage conditions. Substitution of hired labor with mechanical and chemical alternatives can be expected as wages rise. The last section discussed the opportunities for labor-saving technologies in intensive rice systems. In Southeast Asia, substantial labor savings for transplanting, weeding and harvesting can be anticipated with the increasing adoption of direct seeding with herbicide use, and with the anticipated introduction of small harvest machines. Operations where substitution out of labor is not possible could be contracted out to professional labor companies, saving the farmer supervision labor. Thailand and Indonesia should be expected to move along the path of significant labor substitution over the next five years. The Philippines, China and South Asia would be slower in this transition, although one would expect the rapidly industrializing Southeastern provinces of China may move ahead of the rest of the country, and similar patchy transformations may occur in some provinces of India and the Philippines.

From family labor to farm manager

As the opportunity cost of family labor rises, it will be used less as a source of power and more as a source of knowledge (technical expertise), management and supervision. This would be particularly true since the returns to knowledge-based inputs increase over time. Sustaining current yield and productivity gains in the crop sector depends on the farmers' ability to efficiently manage their resource base.

> Productivity gains accrue to farmers who: have the ability to learn about the new technologies; discriminate among technologies offered to them by the research system; adapt the technologies to their particular environmental conditions; and provide supervision input to ensure the appropriate application of the technology.
>
> (Pingali *et al.*, 1990, p. 19)

The problem facing family-managed intensive rice systems is that sustaining productivity requires greater knowledge and informed decision making on the part of the farmer. In other words, sustaining productivity requires a greater amount of farmer time, the most expensive item in future farm budgets. The only viable solution for sustaining the farm resource base, given high or rising wages, is to reduce the level of intensity.

Huang and Rozelle (1994) report for China that the area planted to two-season rice has fallen from 66 percent in 1980 to 58 percent in 1990. They attribute the fall in double-cropped area to the rising opportunity cost of labor during the early 1980s, in response to the market-oriented diversification policies. The switch to single-season rice has been sharpest in the coastal provinces of Jiangsu and Shanghai provinces, falling from 47 to 2 percent and from 97 to 19 percent, respectively, in the 1980s. Intensive crop production practices, such as the use of organic manure, also dropped to negligible levels in these provinces by 1990. Malaysia has deliberately chosen not to further intensify its rice production systems and import supplies as needed. Thailand may find itself trading off its status as the largest rice-exporting country if wage rates continue to rise dramatically.

10

Post-Green-Revolution Seed Technology for Intensive Rice Systems

Chapter 4 presented evidence that the yield potential of modern semi-dwarf varieties has remained relatively stagnant since the release of IR-8 in 1965. While substantial improvements have been made in later-generation varieties in terms of improved pest resistance, grain quality and reduced crop duration, yields have increased only marginally. Recent progress in plant-breeding research indicates that a significant shift in the yield frontier is possible both in the medium and in the longer term. In the medium term, yield increases of around 20 percent could be possible through the adoption of hybrid rice. The longer-term prognosis is for a 'new plant type' that could yield about 12.5–13 t ha^{-1} and, as a parent of the hybrids, could increase this yield to 15 t ha^{-1} of grain. This chapter describes the prospects for these yield-increasing technologies and their potential farm-level impact.

Yield improvements through the development of hybrid rices and the new 'super' high-yielding rices can be expected through the application of conventional breeding techniques; the role of biotechnology in the immediate future for yield enhancement is limited. The prospects for using recent advances in biotechnology are more in the area of improving the efficiency of breeding for pest resistance. Opportunities for using biotechnology tools for shifting the yield frontier would be available only in the very long term.

This chapter provides a discussion of:

1. The recent advances in hybrid rice technology and the prospects for its adoption in Asia.
2. Progress towards a new 'super' high-yielding plant type and its anticipated impact.

3. The prospects for using modern biotechnology tools, especially for reducing yield variability, through breeding for pest resistance and stress tolerance.

Prospects for Hybrid Rice in Tropical Asia

The first-generation offspring of a cross between genetically different parents of a plant species is called 'hybrid'. Hybrid rices commercially exploit the phenomenon of hybrid vigor or heterosis and exemplify one of the most important applications of genetics in agriculture. The phenomenon of hybrid vigor can be best described as the tendency for offspring of diverse parents to perform better than their parents. Hybrid rices yield more than inbreds because they have a larger total biomass and more grains per unit area (Virmani and Dedolph, 1994)[1].

China has managed to sustain its output growth since the mid-1970s by switching from conventional modern varieties to hybrid rice. In the irrigated ricelands of China, hybrid rice was rapidly adopted and resulted in a 15–20-percent higher yield. China's success with hybrid rice technology has generated interest in other Asian countries, especially India. However, hybrid rice varieties from China cannot be easily transferred to the rest of tropical Asia. Under tropical conditions these varieties tend to be more susceptible to insect and disease damage. The duplication of China's success in hybrid rice for the rest of Asia would require intensive research effort to produce hybrids suitable for tropical conditions.

In considering the prospects for hybrid rice in Asia, there are valuable lessons to be learnt from the Chinese experience. On the demand side, it is important to understand the conditions under which hybrid rice would be profitable relative to true-breeding modern varieties. On the supply side, it is important to understand the constraints associated with technology generation, seed production and distribution.

The successful development of hybrid corn in the 1930s in the United States provided an important impetus for breeders of other crops. Suggestions that heterosis be commercially exploited by developing hybrid rice were made from time to time. However, unlike corn and sorghum, rice is a self-pollinating plant with tiny florets, and it is thus impossible to produce hybrid seed in bulk by hand emasculation. The most effective way is through exploiting the phenomenon of cytoplasmic male sterility. The commercial production of hybrid rice seed involves a complicated three-line method: (i) locating a cytoplasmic male-sterile parent plant; (ii) crossing it with a maintainer line to produce offspring showing male sterility but with desirable genetic characteristics; and (iii) crossing the sterile plants with a 'restorer' line to produce F_1 seeds which when planted produce fertile plants (Yuan, 1985). Because of the difficulties of hybrid seed production, most researchers were discouraged from continuing their research. Breeders in China were the exceptions (Virmani and Edwards, 1983).

China's hybrid rice research, under the initiative and leadership of Professor Yuan Long-Ping, started in Hunan Province in 1964. A breakthrough was made when Yuan and his assistant found a male-sterile plant in a wild rice population in the same year. Then, the search for 'maintainer lines' and 'restorer lines' became a concerted nationwide program involving more than 20 research institutes in several provinces. The first maintainer variety was discovered by Yuan and another researcher in Jiangxi Province in 1972, and the first restorer variety was discovered by a breeder in Guangxi Province in 1973. A hybrid combination with marked heterosis was developed in 1974. Regional production tests were conducted simultaneously in hundreds of counties in 1975. In 1976, hybrid rice began to be commercially released to farmers (Zhu Rong, 1988).

By 1991, over 50 percent of China's rice hectarage was devoted to F_1 hybrid rice (Fig. 10.1). Rigorous household-level studies have found that the yield advantage of hybrid rice over the conventional semi-dwarf varieties is about 15 percent, without major differences in material costs and labor requirements (He *et al.*, 1984; 1987). He *et al.* compared farmers growing hybrid rice and farmers growing conventional modern rice varieties in Jiangsu province for the 1984 crop season. They found hybrid rice yields to be at least a ton higher than conventional variety yields which were around 6.5 t ha^{-1} (Table 10.1). Returns to labor were higher for hybrid rice, but returns to non-labor inputs and total costs were similar to that of conventional varieties.

The above results were substantiated by a more recent survey of 500 farm households in Hunan province for the 1988 crop season by Lin (1990). While the mean yields of hybrids were significantly higher than conventional varieties for middle- and late-season rice, the difference was not statistically significant for early-season rice (Table 10.2). The yield advantages of hybrid rice were found to be partly offset by added requirements for chemical inputs and more expenditure for seeds. Therefore, the advantage of growing hybrids compared to conventional varieties depends largely on the need for increasing yield to meet the rice needs of the growing population, and on prices of chemical inputs and seed. However, hybrid rice does not require more labor input than conventional rice.

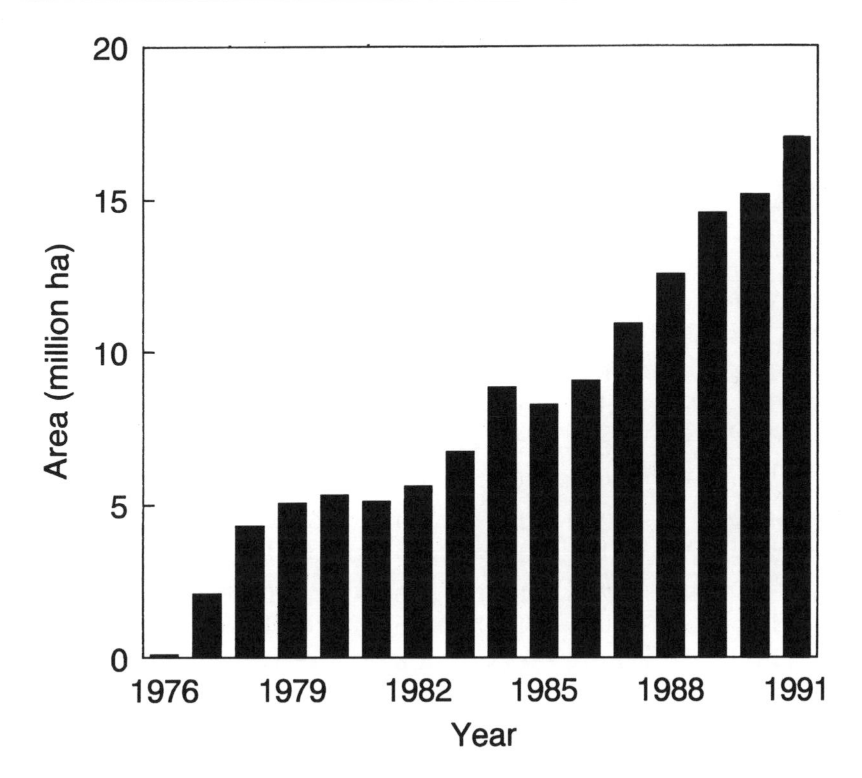

Fig. 10.1. Areas under hybrid rice in China, 1976-1991

Table 10.1. Comparison of the yields and resource use of hybrid rice (HR) and Japonica (conventional) rice (JR), Jiangsu, China.

			Difference	
	HR	JR	Quantity	Percent
Mean yield (t ha^{-1})	7.8	6.8	1	1.5
Seed (kg ha^{-1})	21	96	–75 *	–78
Pesticide (kg ha^{-1})	39	22	17	77
N from organic and inorganic fertilizer (kg)	260	285	–25 ns	–9
Total labor inputs (md ha^{-1})	273	297	–25 *	–8
Irrigation expenses (US$ ha^{-1})	26	26	0 ns	0

* Significant at the 5 percent level.
ns, Not significant.
Source: He *et al.* (1987).

Table 10.2. Comparative mean yields and input use of hybrid and conventional rice, Hunan, China, 1988.

	Early-season rice		Middle-season rice		Late-season rice	
	Conv. (*N*=392)	Hybrid (*N*=13)	Conv. (*N*=34)	Hybrid (*N*=116)	Conv. (*N*=213)	Hybrid (*N*=308)
Seed	173.13	46.27	102.99	29.85	92.54	29.85
(kg ha^{-1})	(3.7)	(2.2)***	(3.1)	(0.9)***	(3.3)	(1.1)***
Fertilizer cost	313.43	410.45	222.39	282.09	338.81	368.66
(yuan ha^{-1})	(9.4)	(7.9)*	(7.9)	(10.4)*	(13.2)	(10.7)*
Pesticide cost	76.12	108.96	61.19	92.54	85.07	108.96
(yuan ha^{-1})	(4.7)	(2.7)	(3.5)	(4.1)*	(3.9)	(5.4)***
Labor	228.36	253.73	332.84	302.99	207.46	214.93
(days ha^{-1})	(6.4)	(3.8)	(13.6)	(7.4)	(5.8)	(5.4)
Draft animals	22.39	28.36	44.78	55.22	16.42	16.42
(days ha^{-1})	(1.1)	(1.0)	(3.0)	(2.1)	(1.1)	(0.7)
Machine	7.46	19.40	20.90	0.90	8.96	8.96
(days ha^{-1})	(0.8)	(1.4)**	(6.9)	(0.31)*	(0.8)	(0.9)
Rice output	5264.18	5752.24	4029.85	6455.22	4829.85	5770.15
(kg ha^{-1})	(97.3)	(172.1)	(117.8)	(124.2)***	(90.2)	(85.3)***
Straw	3088.06	4374.63	3162.69	4416.42	3594.03	4419.40
(kg ha^{-1})	(109.9)	(57.4)**	(92.4)	(89.6)***	(95.7)	(98.1)

*, **, ***, Significantly different at the 10, 5, and 1 percent levels of confidence, respectively.
Figures in parentheses are standard deviations.
Source: Lin (1990).

Advances in hybrid rice for tropical Asia

Since 1979, when hybrid rice research was initiated at IRRI, 24 national and international agencies have made resource commitments for advancing hybrid rice technology. India has been making the greatest effort, with 15 institutions (three of them private) actively pursuing hybrid rice research. Numerous IRRI-bred hybrids are now being tested across Asia and several IRRI-bred cytosterile lines are being used as parents in crosses with local material (Virmani *et al.*, 1993).

Table 10.3 summarizes yields of the highest-yielding rice hybrids compared to the highest-yielding inbred rice varieties of corresponding growth duration in replicated yield trials conducted at IRRI during 1986 to 1993. The best hybrids outyielded the best inbred varieties by about 17 percent. Similar yield advantage was observed in trials of IRRI-bred hybrids in the Philippines, India, Vietnam and Malaysia (Table 10.4). There is no significant difference between the yields of

Table 10.3. Comparison of highest-yielding F_1 rice hybrids and inbreds in hybrid rice yield trials conducted at IRRI during 1986–1993.

Season	Trial	Hybrid	Yield (t ha^{-1})	Difference from the best check	% of check	Growth duration (d)
1986 DS	I	IR54754A/IR46R	7.4	1.2*	119	126
	II	IR54754A/ARC 11353R	7.9	2.3*	142	133
1986 WS	I	IR54752A/IR64	3.9	1.0*	134	126
	II	IR19728A/IR25167-9-2	3.6	0.6*	120	122
	III	IR46830A/IR50R	4.1	0.7*	120	110
1987 DS	I	IR46830A/IR29723-143-3-2-1R	6.4	1.8*	139	112
	II	IR54752A/IR2797-125-3-3-2R	7.8	1.8*	130	130
	III	IR54752A/ARC 11353R	6.8	1.0*	117	128
1988 DS	II	IR46830A/IR9761-19-1R	4.8	0.0	100	108
	IV	IR54752A/IR64R	5.3	0.9*	120	120
	V	IR54752A/IR15324-13-3-3-2R	5.9	0.8ns	116	122
	VI	IR54752A/IR13146-45-2-3	6.8	1.1*	119	116
1988 WS		IR46830A/IR9761-19-1R	3.2	1.0*	145	105
1989 DS	I	IR54752A/IR9761-19-1R	6.3	1.5*	131	111
	II	IR54752A/IR28228-119-2-3-1-1R	6.5	1.0*	119	124
1989 WS	I	IR54752A/IR42686-C2-118-6-2R	3.1	0.4ns	114	120
	II	IR54752A/IR54742-22-19-3R	3.5	0.8*	131	136
1990 DS	I	IR62829A/IR9761-19-1R	4.0	0.6ns	117	110
	II	IR62829A/IR31805-20-1-3	4.5	-0.6ns	88	116
	III	IR58025A/IR10198-66-2R	4.8	0.5	112	116
	IV	IR58025A/IR29723-143-3-2-1R	5.6	1.0*	121	128
1990 WS	I	IR58025A/IR54752-22-19-3R	3.0	0	100	135
	II	IR58025A/IR40750-82-2-2-3R	3.2	0.2	107	126
1991 DS	I	IR62829A/IR35366-62-1-2-2-3	4.7	0.7*	118	112
	II	IR58025A/IR54745-2-45-3-2-4R	5.4	1.2*	128	122
1991 WS	I	IR58025A/IR19058-107-1R	6.4	1.2*	123	113
	II	IR62829A/IR47310-94-4-3-1R	5.1	1.1*	128	120
1992 DS	I	IR62829A/IR54883-43-1-3	6.4	0.5ns	108	112
	II	IR62829A/IR46R	7.2	0.7ns	111	115
	III	IR58025A/IR46R	6.3	0.8*	114	117

Table 10.3. *Cont.*

Season	Trial	Hybrid	Yield ($t\ ha^{-1}$)	Difference from the best check	% of check	Growth duration (d)
1992 WS	I	IR64608A/IR42686-C2-118-6-2	3.7	-0.6ns	86	111
	II	IR58025A/IR48725-B-B-141-2	4.3	0.2ns	105	116
	III	IR58025A/IR37712-90-3-3-3-2	4.1	0.4ns	111	119
	IV	IR58025A/IR54056-64-2-2-2	4.4	0.7*	119	126
1993 DS	I	IR58025A/IR50404-57-2-2-3	6.4	0.7ns	112	110
	II	IR58025A/IR34686-179-1-2-1R	7.4	0.8*	112	128
Mean DS			6.1	0.9	118	
WS			4.0	0.6	115	
Grand mean			5.2	0.8	117	

* Significant at the 5 percent level using LSD test.
ns, Not significant.
Source: Virmani *et al.* (1994).

hybrids and inbreds at low nitrogen levels, but at high nitrogen levels the yield difference is striking (Virmani, 1996). The yield advantage of hybrids was greater in the more high-yielding environments as compared to the lower-yielding environments (Virmani, 1992). There is no difference in insect and disease resistance and grain quality between the elite hybrids and inbreds.

Commercial release of tropical hybrids has taken place in India,

Table 10.4. Yield performance of best experimental F_1 rice hybrids and best check varieties in international trials, 1980–1986.

Country	Trials (no.)	Yield of best hybrids ($t\ ha^{-1}$)		Percent of check	
		Range	Mean	Range	Mean
Indonesia	15	4.1–8.9	6.2	102–143	117
India	21	3.3–9.8	6.1	91–143	116
Malaysia	2	4.2–5.0	4.7	89–127	108
Philippines	8	4.8–7.4	5.4	92–133	114
Vietnam	4	5.3–6.6	6.0	91–122	108
Overall	50	3.3–9.8	5.9	89–143	115

Source: Virmani *et al.* (1993).

Vietnam and the Philippines, and other Asian countries may also follow by the year 2000. Current research emphasis for hybrid rice is on increasing the profitability of seed production (Virmani, 1993). Indian farmers planted hybrid rices in 10,000 hectares in 1995; by the year 2000, India plans to cover at least 2 million hectares. Vietnam has been planting 40–50,000 hectares to hybrid rices introduced from China, and plans to cover about 0.5 million hectares with this technology by the turn of the century.

Potential for hybrid-rice adoption in tropical Asia

The demand for yield improvements associated with hybrid rice technology would generally come from the irrigated lowlands rather than from the rainfed lowlands and the upland rice environments. In China, hybrid rice is grown exclusively in irrigated areas, where the switch from conventional high-yielding varieties to hybrid rice, from a farmer's point of view, involves a change in variety only. Meanwhile, a change from current practice to hybrid rice in the uplands and rainfed lowlands would involve changes in variety, input use and cultivation practices. In other words, in the latter cases, it involves an entire switch in the farming system. For example, it is unlikely that a rainfed lowland farmer cultivating a traditional rice crop with almost no purchased inputs would find high input hybrid rice production profitable[2].

In the irrigated lowlands, yield improvements associated with hybrid rice would be demanded only after yield gains associated with conventional high-yielding varieties have been exhausted. The proportion of irrigated riceland in a country is therefore an important determinant of the potential for hybrid rice. The potential for hybrid rice may be higher in countries with a high proportion of irrigated rice – such as Vietnam, India and Indonesia – relative to countries with a lower proportion of irrigated area – such as Bangladesh, Thailand and Myanmar.

The relative profitability of hybrid rice over conventional rice is determined by the ratio of rice price to hybrid seed price. Given the high labor requirements in seed production, one can anticipate that seed prices will be relatively lower in lower-wage countries. Since agricultural wages are determined by labor–land ratio, among other factors, the relative profitability of hybrid rice production is directly related to the labor–land ratio.

Potential demand for hybrid rice technology can be identified by cross-classifying countries in terms of the proportion of irrigated area and labor–land ratio (Table 10.5). Countries with high labor–land ratio and high proportion of irrigated area are likely to have the highest potential demand for hybrid rice technology. In tropical Asia, these

Table 10.5. Classification of countries by labor–land ratio and irrigated ricelands.

	Labor–land ratio	
Proportion of irrigated riceland	Low	High
High	Malaysia Pakistan	Philippines Indonesia India Sri Lanka Vietnam
Low	Myanmar Thailand	Bangladesh Nepal

Source: Lin and Pingali (1994).

are: India, Indonesia, Philippines, Sri Lanka and Vietnam. Countries with high proportion of irrigated area but low labor–land ratio, such as Malaysia and Pakistan, would not find hybrid rice production profitable since agricultural wages would be relatively higher in these countries. Countries with low proportion of irrigation infrastructure would have low potential demand for hybrid rice technology irrespective of wage rates. Bangladesh, Nepal, Myanmar and Thailand are examples. Under non-irrigated conditions, yield improvement and output growth would be achieved initially through other means. Hybrid rice technology may play a role only in the medium-term and in favorable rainfed lowland ecosystems.

In evaluating the potential supply of hybrid rice technology, we need to assess the absolute magnitude of the irrigated rice area and labor endowments by country. The returns to investment in hybrid rice research and the development of hybrid seed industry depend on the absolute size of the irrigated rice area in the country. Thirty-one percent of the total irrigated area in tropical Asia is in India and 13 percent is in Indonesia. These two countries would potentially be the most important suppliers of hybrid rice technology. Given the size of the market, seed producers in these countries would benefit from substantial economies of scale. Scale economies would accrue in: the provision of hybrid seed adapted to specific ecoregions; input supplies, especially gibberellic acid; and technical skills.

The Chinese hybrid rice experience has shown that yield potential of hybrid rice can only be achieved if each ecological region develops its own hybrids or screens hybrids developed in other regions to meet specific local conditions (see Box 10.1). In countries such as India and Indonesia it is economically feasible to set up a regional research and seed production infrastructure. Region-specific technical skills could

Box 10.1. Regional differences in hybrid rice adoption in China.

Chinese national statistics on the rapid rise in hybrid rice production tend to mask dramatic geographic differences. The adoption rates in Sichuan and Shaanxi provinces were more than 50 percent in 1987, while the adoption rate in Heilongjiang and eight other provinces was zero in the same year (Fig. 10.2). Lin (1990) attributes the regional differences to differences in agricultural research infrastructure, adaptive research investments and the share of rice in provincial agricultural output.

Agricultural research infrastructure tends to be concentrated most in the provinces with high population densities. Agricultural research in these provinces tends to concentrate on innovations that increase yield per unit of land, i.e., land-saving innovations. Provinces with high labor-to-land ratios also have high levels of irrigation infrastructure and have high levels of modern variety adoption. In provinces with high population densities and proportionately large rice area, the provincial academy's allocation of scientists and funds to research of both conventional and hybrid rice tends to be high. Therefore, the supply of yield-increasing rice technologies tends to be positively associated with labor–land ratios.

Like other types of improved seed, the F_1 hybrids proved to be sensitive to local ecological conditions. To obtain the yield potential, each region had to develop its own hybrids or screen hybrids developed in other regions to suit specific local conditions. Adaptive research on hybrid rice is thus conducted in the provincial and lower-level research institutes in each rice-producing province. With any given yield advantage, the marginal returns to the innovation of a new rice variety are a positive function of the size of the rice-producing area and the price of rice. Therefore, the larger the rice area or the higher the rice price in a region, the more resources research institutes in that region will allocate to rice research, including research for F_1 hybrids. Consequently, the larger the rice area or the higher the rice price in a region, the more F_1 hybrid varieties will be available in that region. Since the availability of adaptable F_1 hybrids is a positive function of the size of rice area and the price of rice in a region, the adoption rate of F_1 hybrids will also be a positive function of the size of rice area and the price of rice in that region.

In a centrally planned economy such as China's, prices are regulated by the state. Regional rice price variation, if it exists at all, is negligible. Therefore, rice price cannot become a factor in determining regional discrepancies in the availability and adoption rate of hybrids. However, there are significant regional variations in rice-production areas. The resources devoted to hybrid rice research, the availability of hybrid varieties, and the adoption of hybrid rice in a region will thus depend on the rice-producing area in that region.

Because the allocation of manpower and expenditure to both hybrid and conventional rice research in a provincial academy depends on the rice-producing area in that province, there exists a positive significant relation between the number of new varieties of hybrid rice and conventional rice developed by a provincial academy and the area devoted to rice in that province.

The proportion of rice area to total cultivated area in a province is also an important determinant of profitable hybrid seed production. The higher the proportion of rice area, the lower the marketing costs for seed companies, therefore the higher the availability of hybrids.

The profitability of hybrid rice production is higher in provinces with a larger rice area. This is due to the greater availability of hybrid rice varieties and lower seed costs. The returns to farmer time and other investments for adopting hybrid rice are higher.

Source: Lin and Pingali (1994).

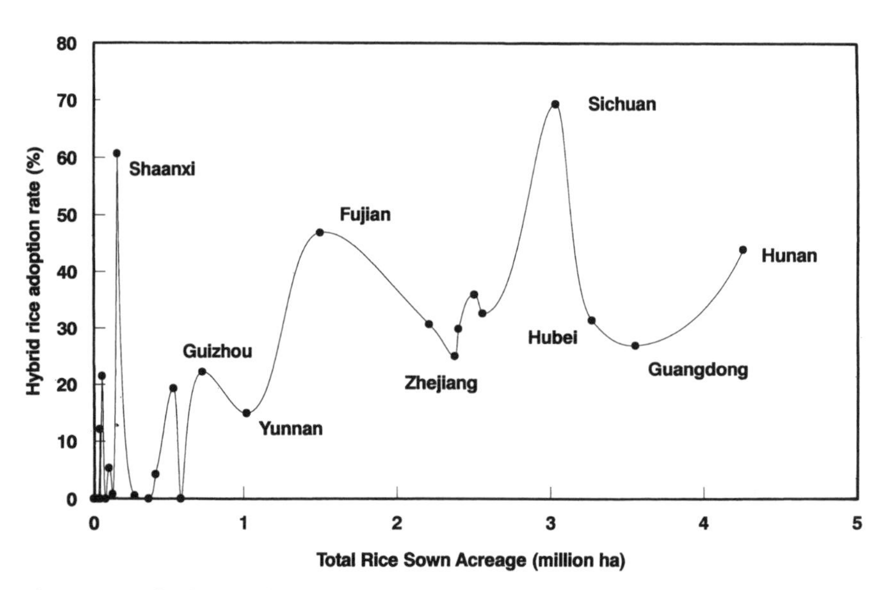

Fig. 10.2. Hybrid rice adoption rate and total rice area, China, 1987. (Source of basic data: Lin and Pingali, 1994.)

be enhanced under such a system. Region-specific institutions would not be cost effective in countries with smaller irrigated areas. In the Philippines, for example, a hybrid research and seed production system would not be economical for all the major ecoregions, with the possible exception of Central Luzon. In the case of input supplies, gibberellic acid is uniquely required for hybrid seed production; the unit costs of this input would decrease with an increase in the scale of production.

Countries with smaller rice areas, rather than set up their own hybrid rice research and seed infrastructure, could possibly benefit from the spillovers from neighboring efforts. Sri Lanka, for instance, could benefit from technological spillovers from hybrid rice research conducted in Southern India. Similarly, Malaysia could benefit from the efforts in West Java, Indonesia. The costs of developing region-specific hybrid rice technology for tropical Asia would be substantially lower if regional cooperation based on ecological similarities can be fostered.

Hybrid seed industry

The development and use of hybrid technology is closely associated with the seed industry. Historically, the seed industry in developed countries has expanded by developing and selling hybrids. Nearly 40 percent of the total global commercial seed business of about US$ 15 billion is accounted for by hybrid sales in various crops (Sehgal, 1992). In tropical Asian countries where prospects of hybrid rice are being explored, the seed industry is either in its embryonic stage of development (as in Indonesia, Philippines and Vietnam) or just beginning to enter the growth phase (e.g. India, Thailand).

In recent years many new companies have entered seed business in developing countries due to greater availability of hybrids in several crop species and easing of government policies pertaining to private sector entering the seed industry (Sehgal, 1992). One of the most notable examples of easing of government policies is India's 'New Seed Policy' which was announced in 1988. As a result of this policy, many new national and multinational companies entered the seed business in India. Similar deregulation in Thailand and the Philippines has led to an expansion in their hybrid seed industries during recent years (Sehgal, 1992). Sri Lanka is also considering strengthening its seed industry during the next five years by encouraging private investment. In several other countries there is a realization of the need to improve seed production infrastructure which would be helpful for the introduction of hybrid technologies.

Among the tropical rice-growing countries (e.g. India, Vietnam and Philippines) which have released rice hybrids for commercial cultivation, India has adequate seed industry infrastructure in the public as well as the private sector to undertake production, processing, certification and distribution of hybrid rice seeds. Although Vietnam does not yet have an adequate seed industry, it can mobilize resources in the public sector to handle the seed production processing and distribution of hybrid rice. The Philippines can mobilize private seed companies (Cargill Seeds, for example) and seed-growers' cooperatives for hybrid rice development. Indonesia also has a reasonably good seed industry infrastructure in both public and private sector which could be mobilized after suitable rice hybrids are identified.

Experience in other crops has indicated that the seed industry in Asia is capable of providing quality seed to farmers as they need it. The same will be true of hybrid rice (Sehgal, 1992). On an optimistic note, given the prominence of rice in Asia, rice hybrids could have the same catalytic effect on the development of the region's seed industry that hybrid corn had on the seed industry development in North America and hybrid sugar beet had on the seed industry development in Europe (Sehgal, 1992). Policymakers can continue to create incentives for the development and expansion of the seed industry as is currently happening in several developing countries. The private sector is to be taken as a partner, rather than an adversary, in the effort to develop seed industry in a country. After all, a good seed industry is useful not only for hybrid varieties but also for inbred varieties.

Progress Towards the New 'Super Rice'

While hybrid rice can meet the requirements for increasing productivity growth in the short to medium term, longer-term prospects for productivity growth would have to come from a substantial shift in the yield frontier for rice. Research is under way at IRRI to design a plant type that will lead to a quantum leap in yields. The prototype of a rice plant that could yield up to 30 percent more (yield potential of 13–15 t ha^{-1}) than the current modern high-yielding rice plant type was designed in 1989 (IRRI, 1993). Experiment-station yield trials began in 1994; farmer field testing could begin in 1998.

The modern semi-dwarf rice varieties have 20–25 tillers. Only 15–16 of these produce small panicles and the rest remain unproductive (sterile). The unproductive tillers compete with the productive tillers for nutrients. It may be possible to increase the efficiency of the rice plant by breeding for larger panicle size but reduced tiller number. The future rice plant is expected to have 4–5 productive tillers but with larger panicles (250 grains), as compared to 100–120 grains per panicle of current varieties, and thick sturdy stems in order to bear the weight of the larger panicles and greater grain weight (Khush, 1995). Figure 10.3 shows a diagrammatic comparison of the semi-dwarf modern varieties and the new plant type. This new plant type would be more amenable to dense planting and therefore would increase land productivity significantly. This is particularly true for the increasing

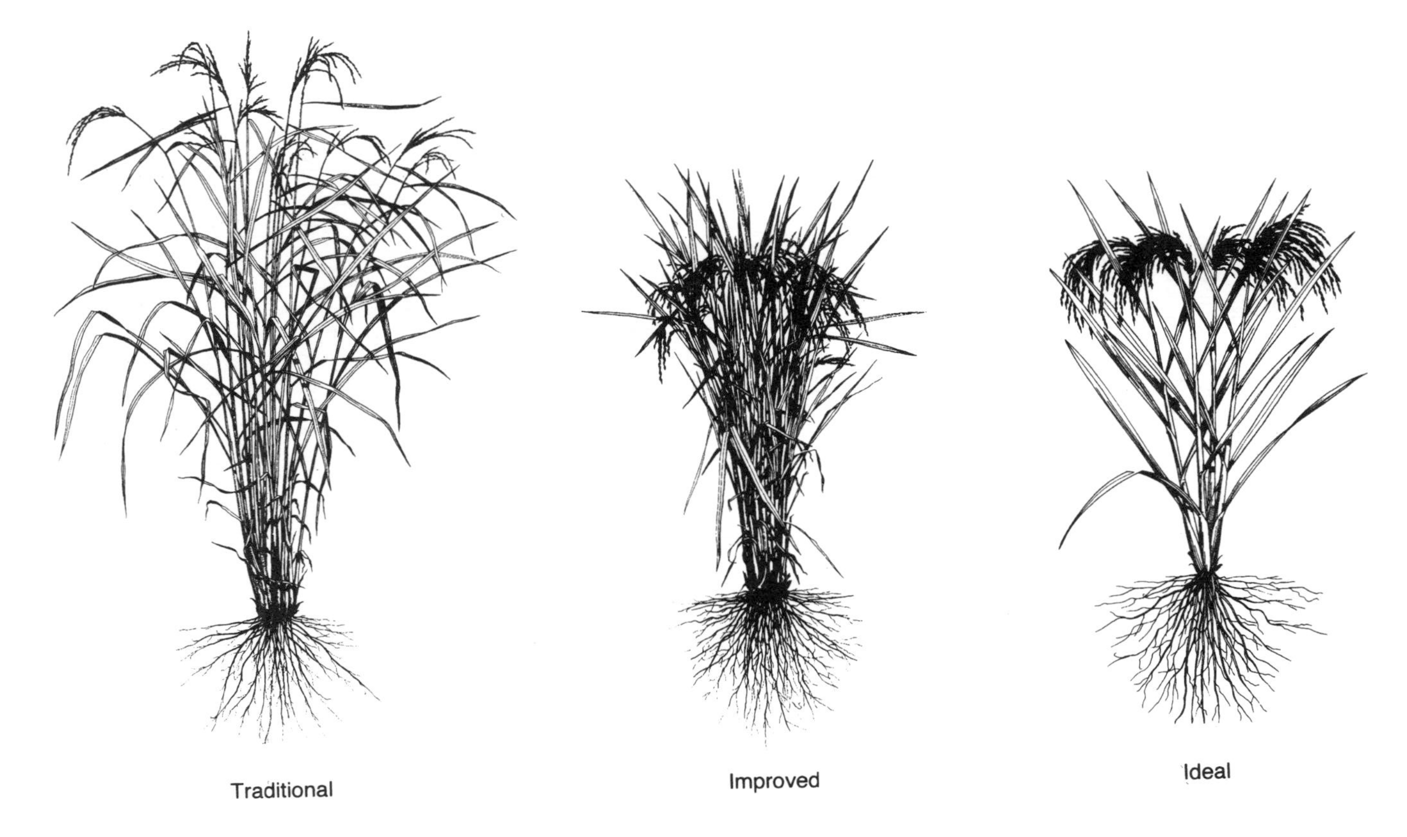

Fig. 10.3. Irrigated rice plant types. (Source: Khush, 1995.)

trend towards direct seeding as a substitute for transplanting in the irrigated ricelands of Asia.

Donor varieties for larger panicles, very sturdy stems and low tillering have been identified and are being used for developing new plant types with higher yield potential (IRRI, 1993). Donors come from tropical Japonica varieties from Indonesia, Malaysia, Philippines and other Southeast Asian countries. Breeding lines with the desirable traits have been selected, and are being tested, beginning in 1994, in replicated yield trials at IRRI. Pest resistance and improved grain quality characteristics are being incorporated into the new plant type lines.

This new plant type in association with direct seeding could increase rice yields by 20–25 percent per hectare. Using the new plant types in hybrid rice breeding could give an additional 20–25-percent yield advantage (Khush, 1995). It would appear that the yield gains in farmers' fields from these approaches are likely to be 10–15 years away.

It is still too early to conduct an adequate economic assessment of the new plant type and its potential impact on production and productivity. It may, however, be useful to speculate on the potential of the new plant type in alleviating (or aggravating) the post-Green-Revolution problems identified in this book: diminishing returns to intensification, increased resource competition, and declining productivity of the paddy resource base.

One of the first issues that needs to be established, when experimental data are available, is whether the shift in the productivity frontier with the new plant type relative to the current modern varieties is a parallel shift or a shift skewed towards higher input levels. In other words, would the productivity gains be proportionately the same for all input levels (an intercept shift) or would high-input-using farmers have a proportionately higher yield gain relative to lower-input-using farmers (a slope shift)? A related question is whether farmers at the low end of input use obtain higher or lower yields with the new plant type relative to the current modern varieties. (See Fig. 10.4 for a presentation of the above scenarios with respect to the productivity frontier.)

A parallel shift in the productivity frontier is unlikely because of: (i) grain filling and grain weight being highly responsive to nutrient input levels; (ii) the anticipated interaction between high nitrogen application and fungicide use; and (iii) the suitability of the new plant types for direct seeding which is intrinsically high-input-using, especially herbicides (Kropff *et al.*, 1993). The characteristics just described indicate the likelihood of a productivity frontier that is skewed towards high input levels. The gains in absolute terms from

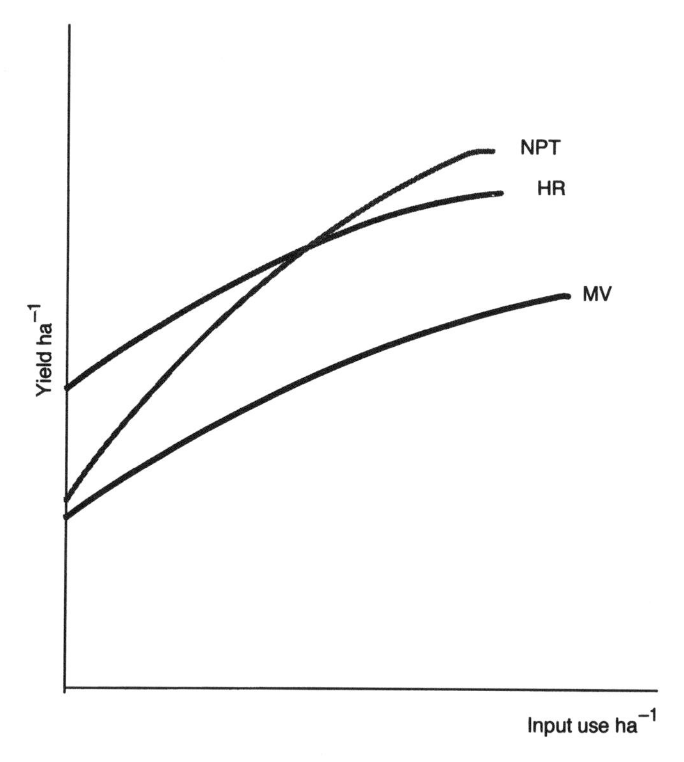

Fig. 10.4. Input responsiveness of modern varieties (MV), hybrid rice (HR) and the new plant type (NPT).

this plant type can definitely be expected to be higher in the high-potential irrigated lowland environments than in the less favorable environments.

For the irrigated environments, the steeper the slope of the productivity frontier, the lower the unit costs of production with the new plant type relative to current modern varieties. Productivity growth can be expected as irrigated farmers move towards and then along the new productivity frontier, until diminishing returns set in again at a future date. The extent of productivity gains, and the period for which they last, depend on complementary investments in new irrigation infrastructure, the maintenance of current infrastructure, and land investments necessary for the switch to direct seeding.

A switch to the new plant type is less likely in the rainfed lowlands because their inability to effectively control water would make it difficult to adopt direct seeding techniques (using pre-germinated seeds). In environments where transplanting is expected to persist, current modern varieties could be expected to dominate the new plant type, both in terms of yields and profitability. In other words, the spillover benefits of the new plant type from the irrigated to the rainfed environments are likely to be smaller than those experienced with the current modern varieties. The exception could be the deepwater environment where the opportunities for a dry-season irrigated crop using groundwater are high, and direct seeding could be easily adopted. Pumped water has intrinsically better control than gravity-flow irrigation water.

It could be expected *a priori* that, in the irrigated environments with good water control, the new plant type could lead to a reversal of the current trends in diminishing returns to further increases in input use. What are the possible effects of the new plant type on competition for agricultural resources – specifically land, labor and water?

If a quantum leap in productivity is expected from the favorable, irrigated lowlands, then it could reduce intensification pressures on the more fragile environments, especially the uplands. With increased competition, high cost (in terms of human labor) production of subsistence rice in the uplands will be less desirable than importing lower-cost rice for consumption from the lowlands. The uplands would be increasingly diverted to non-rice activities. The rainfed lowlands have fewer options to switch out of a dependence on rice production, especially in the wet season (Chapter 8). These areas could be worse off due to the output-price-depressing effects of productivity growth in the irrigated lands. The favorable rainfed lowlands that currently use modern rice varieties would of course find the switch to the new plant type profitable.

Demand for hired labor would be significantly reduced with the adoption of direct seeding and the associated increase in herbicide use. Transplanting and weeding labor requirements in particular would be reduced. The higher output potential of the new plant type implies an increased demand for harvest and post-harvest labor. Possibilities for mechanization could alleviate the labor constraint for these operations (see below). In the case of family labor, especially management and supervision time, the consequences of adopting the new plant type are harder to predict. If the new plant type can be adopted without significant changes in crop management techniques, e.g. similar nitrogen management practices as for the current varieties, then management labor requirements would not increase significantly. On the other hand, if productivity gains can be sustained only with the concurrent adoption of knowledge- or management-intensive technologies, such as matching nitrogen supply to plants' need, then management labor requirements could rise substantially (see below).

The new plant type could be expected to be neutral with respect to competition for water. It could, however, require more effective management, control and timely release of irrigation water in the context of direct seeding. Improved water management would require high levels of farmer management time both at the farm and at the system level. Farmer participation in system-level water-allocation decision making becomes crucial for ensuring adequate and timely water supply. Given the increasing cost of farmer time (Chapter 3), adequate water management could become a constraint to sustaining the productivity gains from the new plant type.

Finally, what can be said *a priori* about the impact of the new plant type on sustaining the paddy resource base? Both positive and negative consequences could be expected. On the positive side, increasing lowland productivity could reduce the pressure on the uplands, as mentioned above. Also, improved water control investments made for improving the productivity of the new plant type could allow for seasonal diversification of ricelands, thus allowing for a break in the monoculture cycle.

On the negative side, there could be a substantial increase in herbicide use associated with direct seeding, which could have significant pollution effects. Intensive production systems have in the past led to adverse changes in the pest complex (Chapter 4) and these problems could be aggravated. Areas of particular concern for the future are increased disease buildup due to interaction with high levels of nitrogen use, changes in the weed complex, and increasing weed resistance to herbicides. The environmental consequences of introducing the new plant type have to be adequately identified and assessed in the context of increased productivity benefits. Fortunately, the processes that lead to resource base degradation with intensification would be similar to those observed with the Green Revolution technologies, and therefore it could be easier to conduct a thorough ex-ante (forecast) impact analysis.

Contributions of Biotechnology to Seed Improvement

Any discussion of potential germplasm improvements in rice is not complete without an assessment of the likely contributions of biotechnology. It is important to note, however, that recent advances in shifting the yield frontier for rice have come from the use of conventional breeding techniques rather than biotechnology. Biotechnology would be most useful in generating rice plants with improved resistance to pests and tolerance to abiotic stresses. Modern biotechnology tools can enhance conventional breeding efforts in two ways: (i) by increasing the efficiency of conventional breeding; and (ii) by using genetic engineering to bring non-rice genes into the rice gene pool. Breeding efficiency is improved by reducing the trial and error of conventional screening techniques, through the use of molecular markers, and reducing the time period involved in obtaining a stable breeding line, through the use of anther culture techniques. 'Biotechnology's most novel contribution will probably be in adding alien genes to the rice gene pool through genetic engineering,' (IRRI, 1993, p. 118).

The application of biotechnology to rice germplasm improvement is based on recombinant DNA technology, and cell and tissue culture (Khush and Toenniessen, 1991). Recombinant DNA technology is a series of techniques that permit the manipulation of DNA, the essential genetic material in the cells. From an understanding of how cells function and how genetic material can be transferred from a cell of one species to another, techniques that permitted the manipulation of processes in the cell were developed. Genetic engineering refers to a process which includes the identification and isolation of suitable genes to transfer the delivery systems needed to introduce the desired gene into the recipient cells, and the expression of the new genetic information in the recipient cells. Cell and tissue culture offers novel approaches to plant production, propagation and preservation.

This section provides a brief summary of the available biotechnology tools and their potential contributions. In the short to medium term the contributions of biotechnology are likely to be most significant in improving the durability of pest resistance, through the use of molecular markers, tissue culture and genetic engineering. In the longer term, biotechnology tools could contribute to the yield and grain quality improvements.

In assessing the usefulness of genetic engineering techniques it is important to understand possible negative consequences. An example is the incorporation of a gene for herbicide resistance into the rice plant (Datta *et al.*, 1992). A herbicide-resistant plant could sustain rice productivity in the face of increasing weed competition, especially in direct-seeded systems. Widespread use of such plants could, however, increase the risk of cross-pollination of the gene into weed and wild rice species, thus aggravating rather than reducing the problem of weed competition. Also, a herbicide-resistant rice plant could encourage high levels of herbicide use and thereby contribute to increased environmental pollution. A case study is required to monitor and evaluate these concerns of negative approach.

Biotechnological inputs in conventional breeding

The advent of molecular marker technology has helped breeders select plants with useful economic traits and use them in the breeding process. The probability of success in breeding increases with the use of molecular markers as described below.

Over 20 single genes for disease and insect resistance have been located relative to the restriction fragment length polymorphism (RFLP) markers. Among them are several genes for resistance to rice blast fungus, and major genes for resistance to bacterial blight and three insects (brown planthopper, whitebacked brown planthopper, and gall midge). A number of other useful genes such as those determining photoperiod sensitivity and aroma have also been identified. Gene mapping results can be immediately applied to breeding programs (IRRI, 1993).

The product of a genetic cross receives half its chromosomes, and thereby half its genes, from one parent and half from the other. The product of a cross thus results in the inclusion of many undesirable genes along with the few desired ones. Further rounds of crossing and selection are necessary to eliminate undesirable genes while retaining the desired ones. Selection is often a rate-limiting step in plant improvement. This is because conventional methods have not been able to directly select for the presence of the DNA that makes up a gene, or for primary products of gene action. Instead, the only way to select plants containing the desired genes has been to grow a number of plants from a cross and to observe the effect of genes on the plant's morphology, its physiological characteristics, its resistance to an insect/pathogen or some other aspect of the plant's phenotype.

Molecular techniques make it possible to introduce a desired rice gene into rice by facilitating its selection. Genetic maps are likely to be used as general purpose tools with a variety of plant-breeding applications. A genetic map shows the relative position of genes on a chromosome and the distance between genes. The phenomenon underlying genetic mapping is that the farther apart genes are from one another on a chromosome, the more often they will segregate and form new combinations with other genes in the process called recombination. Since the precise number of genes present in any complex organism is unknown but suspected to be as large as 50,000 in crop plants, identifying the exact location of all the genes is a complex process. The position of useful and important genes on the chromosomes is roughly identified or tagged by the use of molecular markers. Tagging allows plant breeders to use indirect selection for genes of interest. Instead of selecting progeny plants containing the gene of interest by checking for the action of the gene, marker-aided selection directly selects for the plants having the gene through the presence of the marker in the DNA of the plant.

The use of these new tools requires increased knowledge about the plant genome. Although some progress has been made in mapping major and minor rice genes and applying mapping results to breeding, two constraints remain in marker-aided selection (N. Huang, IRRI, 1995, personal communication). The speed of gene mapping remains slow and the techniques need to be refined to facilitate indirect selection of desired genes. The procedure is expensive and requires highly trained personnel. Although these new molecular techniques allow traditional plant-breeding problems to be resolved, the challenges facing plant improvement programs remain essentially the same. The gene pool still needs to be screened for desirable genes which then have to be transferred to the crop and, finally, the performance of the improved plant has to be tested in the field.

Unconventional approaches to plant breeding

In conventional breeding the only method of introducing genes into a plant is by crossing it with another plant containing the desired gene or genes. A plant breeder searching for desirable genes would have to limit the search to plants that could be successfully crossed with rice. Advances in tissue culture and genetic engineering allow genes from wild relatives of rice and non-rice genes to be introduced into rice plants.

Two tissue culture techniques, embryo rescue and anther culture, have already made significant contributions to rice improvement. Embryo rescue enables breeders to make crosses between varieties that could not be crossed before – for example, a cross between cultivated rice and wild rice species. Anther culture allows faster stabilization of breeding lines. Stabilization over a series of generations of the rice plant is done in the laboratory rather than through actual field cultivation. A salt-tolerant rice variety is about to be released based on conventional breeding and anther culture (Alejar *et al.*, 1995).

Hybrids between cultivated rice and 12 wild species have been produced using embryo rescue. Genes for resistance to brown planthopper (BPH), whitebacked brown planthopper (WBPH), bacterial blight, and blast have already been transferred from wild species into elite breeding lines of rice. Cold tolerance is one of the traits introduced into cultivated Indica varieties by anther culture (IRRI, 1993).

Non-rice genes can be introduced into rice through genetic engineering – a gene transfer not possible with conventional breeding techniques. Relatively few useful foreign genes have been introduced into rice because reliable transformation protocols have only recently emerged. Novel genes such as the *Bacillus thuringiensis* (*Bt*) gene for insect resistance and coat protein genes for tungro resistance are likely to become available within the next five years. When introduced into rice, they should impart high levels of resistance. Coat-protein-induced virus resistance is directed against the rice tungro virus (Beachy *et al.*, 1989), which is the most important disease problem for rice (Herdt and Riely, 1987). A coat-protein-mediated resistance to rice stripe virus (an insect-transmitted virus) has been introduced in Japonica rice (Hayakaya *et al.*, 1992). The *Bt* genes are targeted against stem-borers (Joos and Morrill, 1989), which is the most important insect pest problem for rice (Herdt and Riely, 1987). Insect-resistant rice generated by introducing modified δ-endotoxin gene or *Bt* has been shown to produce *Bt* protein, and enhance insect resistance in Japonica rice plants (Fujimoto *et al.*, 1993).

Genes for resistance to viral, bacterial, fungal and other diseases have been identified and successfully transferred into many crops (Persley, 1990; Ahl Goy and Duesing, 1995). The first genetically engineered Indica rice based on protoplast system (Datta *et al.,* 1990) now provides sheath blight resistance in Indica rice (Lin *et al.,* 1995). Several chitinase genes in combination with *Bt* genes have been introduced in rice to further enhance protection against pathogenic fungi, bacteria and insects. The identification of many other genes that may enhance the genetic yield potential of rice or confer resistance to biotic and abiotic stresses is still at an early stage. After their identification these genes have to be isolated and constructed in a manner suitable for their introduction and expression into the rice genome.

Potential impact of biotechnology

Biotechnology calls for an evaluation of the impacts of technology with little knowledge about the final form of technological advance. However, if the output of social science research is information that could lead to a new and improved product (Norton and Schuh, 1981, as cited in Lindner, 1987), the potential and limitations of biotechnology need to be understood so as to generate the kind of information needed for biotechnological products to have maximum impact.

Biotechnology offers innovative possibilities for increasing crop production and protecting the environment by the reduced use of agrochemicals. Expectations of an increase in yield potential through biotechnology are too optimistic at this stage. Yield in rice, as in other plants in general, is believed to be controlled by many genes working in synchrony but not enough is known yet about which genes specifically contribute to yield so that they can be moved together successfully, at the same time without introducing undesirable genes that have to then be weeded out. With time, yield gains may be achieved, but in the nearer future substantial gains from crop protection can be expected. Many potentially useful genes to combat biotic stresses from insects, fungi, bacteria and viruses have been identified and transferred to rice and other crops. Greater precision and speed of genetic manipulation will enhance the rate of these developments.

Since biotechnology offers a number of ways to modify the genetic composition of a living organism to suit local conditions, all its potential and limitations stem from that capacity. As in the case of the new plant types and hybrid rice, genetically engineered varieties of rice will have their own special characteristics that have an impact on their utility and adoption. Some researchers argue that biotechnology products are not unlike seed-based technologies of the Green Revolution period and therefore easy to adopt, while others believe biotechnological products are knowledge intensive and make demands on farmer knowledge and management, which in turn may require greater investments in research, effective extension, and infrastructural capacity. Knowledge-intensive technologies are harder to disseminate and therefore adopt. If the trait for resistance in *Bt* rice is not inherited, farmers need to purchase transgenic rice seeds every season. The price and availability of seed will be important factors determining rates of adoption. In order to predict the success of adoption of the new technology, the analyst will have to consider differences in managerial skills, economic circumstances and resource endowments of farmers.

The commercial or widescale application of biotechnology products in general may still, however, be somewhat problematic. A major policy issue is the regulatory climate governing the release of genetically engineered products to ensure public health and environmental safety. This is a particularly thorny issue related to biotechnology because of growing public concern about the use of genetically modified organisms. The history of regulation of biotechnologies is short and governments and regulatory bodies require information on the likely costs and benefits of biotechnological products. The deployment of transgenic plants is still uncertain as many countries have stringent legislation that regulates the use of agricultural products derived through biotechnology. In a brief report on the use of *Bt* products in India, Tichon (1994) observes that no *Bt* products were registered for use in cotton until June 1992. Even though high resistance levels to conventional pesticides used in India were recorded, an impediment to registration was an erroneous perception held by officials that *Bt* could be more damaging to silkworm production than other pesticides.

Even in countries where the existing legislation is sufficient to allow the use of biotechnology products, the requirement is for guidelines that cover field testing of these products and the assessment of any risk associated with their release into the environment. Countries where transgenic rice may be deployed need to establish review bodies and institutional biosafety committees to develop guidelines to monitor and regulate the use of the biotechnology products. The benefits to developing *Bt* rice may be severely diminished if countries where yellow stemborer causes extensive damage do not allow the use of transgenic products.

A contentious issue is that of property rights and genetic resources. The term 'genetic resources' refers to germplasm that encodes any useful or potentially useful characteristic of living material. The argument for patenting biotechnology is that the processes, substances and organisms that fall under the nomenclature of biotechnology do not occur naturally or, even if they do, the process used to produce it in commercial quantities is done by people. The case again stems from the view that the building blocks of the processes, the naturally occurring germplasm, are part of the common heritage of mankind (Persley, 1990). The debate includes distributional issues: should the country (people) where the genetic resource originated receive compensation? The role of social scientists is not yet well defined in these areas of concern.

Few discussions on crop biotechnology are complete without raising the issue of environmental and health hazards associated with the release of genetically engineered organisms. For instance, should scarce research funds be used to develop herbicide-resistant plants, when there is the fear that the escape of herbicide-resistant genes might result in the production of an over-competitive, herbicide-resistant weed? Basta and glyphosate belong to a new generation of relatively benign herbicides. However, their use has been limited because of their nonselective mode of action. It is proposed that the development of Basta and glyphosate-resistant crop plants will encourage the phasing out of the highly toxic herbicides currently in use. In the face of imminent field release of herbicide-resistant transgenic plants, general concern has been expressed about the risks associated with the escape of recombinant genes from transgenic plants into populations of unmanaged and/or wild, related species through pollen movement and subsequent hybridization and selection (Meadows, 1993). The trait most commonly evaluated in transgenic plants is herbicide resistance (34 percent of all field trials). Every year, in every geopolitical area except the Pacific Rim, herbicide-resistant plants represented the largest proportion of trials (Ahl Goy and Duesing, 1995). A risk assessment, when risk concerns the escape of genes through pollen movement into wild relatives, would require information on whether pollen from the genetically modified crop can hybridize with wild relatives and whether any of these wild relatives occur nearby, and under what ecological conditions (Manasse and Kareiva, 1990). This sort of information is hard to obtain, but before investing in herbicide-tolerant rice the risks of creating ecologically deleterious plants need to be minimized.

Biotechnology products need to be integrated into existing research efforts and agricultural development programs of the individual countries. There is a need for ready access to information on how best this may be achieved. Institutional and management arrangements have to be worked out in order to successfully deploy transgenic products and retrieve information that may serve as early warning signals of any potential negative effects of biotechnology at the farm, commodity or country levels. The critical linkage for effective disciplinary collaboration between molecular biologists, entomologists, pathologists, breeders, agronomists and social scientists must be established. Not least, information flows need to feedback to institutions so that the identification of current problems where biotechnology may be helpful will come in use for priority setting at the research level. Priority setting – modified by probabilities of success, the probability of adoption, potential impact on production and consumption, and potential impact on the welfare of the poor – will be the basis for investment in biotechnology research.

Conclusions

Recent advances in shifting the yield frontier for rice indicate that the prospects for sustaining productivity growth in the intensively cultivated irrigated lowlands of Asia are high. In the short to medium term, yield growth would come from the adoption of hybrid rice and in the longer run through the use of 'super rice' and second-generation hybrids between modern Indica varieties and the new plant type. While the prospects are good for the generation of new seed technologies, research and adaptation work is by no means completed, and continued high levels of research investments are needed in order to make widespread farmer adoption a reality. In order to secure productivity growth of rice, the new seed technologies ought to be complemented by continued investments in irrigation and other rice-related infrastructure. Profitable adoption of new seed technologies would require high-level input use, especially fertilizers and herbicides, to achieve the yield potential of the new varieties. The implications of input intensification are discussed in Chapter 11.

Much has been said about the role of biotechnology in securing the food supplies of the future. This chapter took a cautious view of the potential of biotechnology. The role of biotechnology in shifting the yield frontier for rice is limited in the foreseeable future. The greatest contribution of biotechnology tools would be in improving yield stability through generating rice plants with improved resistance to pests and tolerance to abiotic stresses. Even where transgenic rices are available, their farm-level adoption may be constrained by regulatory impediments that may restrict their access. Intellectual property protection and plant breeders' rights may further prevent the easy access of materials developed through the use of biotechnology tools. Substantial policy reforms ought to be in place before the fruits of biotechnology are available to the farmers of Asia.

Notes

1. Rice is a self-fertilized crop. This means that farmers can save seeds for planting from one crop to the next. To grow a commercial hybrid rice crop, however, farmers need fresh seed every season. If seeds produced in the commercial crop are saved and planted, the resulting crop will not be uniform, will show mixed grain types, and will not have a yield advantage (Virmani and Dedolph, 1994).
2. Significant heterosis, commonly observed for vegetative vigor and root characteristics, suggests that the use of hybrid rice should also be explored for certain stress environments (rainfed, lowland, drought-prone) where transplanting is practiced or some seeding equipment is available for direct seeding with a reduced seed rate (Virmani, 1996).

11

Fertilizers and Pesticides: Higher Levels versus Improved Efficiencies

Chapter 6 discussed the need for increasing rice productivity growth in order to meet the unabated growth in rice demand. Meeting the long-term requirements for rice would require a shift in the yield frontier for rice (as discussed in Chapter 10), and also fundamental changes in the way fertilizers and pesticides are used. In order to sustain rice productivity growth while at the same time conserving the resource base would require rice production increases to be achieved with less than proportionate increase in chemical input use. Recent advances in the generation of efficiency-enhancing fertilizer and pesticide technology could help in meeting the dual goals stated above.

Changes in fertilizer application, especially in terms of timing and method of application, could contribute significantly to reduction in nutrient losses through volatilization and seepage, and improve plant nutrient uptake. Efficiency gains made through improvements in fertilizer management may contribute to a reduction in the overall fertilizer requirements for sustaining productivity growth. In the case of insecticides, recent evidence indicates a very modest impact of insecticides on rice production. Farm-level experiences with integrated pest management (IPM) indicate that judicious decision making on insecticide use could lead to substantial reductions in its use without a reduction in rice yields.

While the growth in fertilizer and insecticide use can be managed through the adoption of efficiency-enhancing technologies, herbicide use is expected to increase dramatically across Asia for the foreseeable future. Growth in herbicide use is closely linked to increasing wage rates, and therefore to the substitution out of manual weeding to chemical control. Rapidly rising wages also contribute to the shift from the labor-intensive transplanting operation to direct seeding, a switch that cannot be accomplished without the complementary use of herbicides. Growing water scarcities in irrigated rice systems also contribute to the increasing trends in herbicide use. In wetland rice cultivation systems, water is used as a means of weed control. Standing water in the field deters weed growth, but as water becomes scarce herbicides begin to be substituted for water as a means of preventing weed buildup and weed growth. Opportunities for reducing herbicide use through efficiency improvements are limited in tropical rice systems.

While efficiency-enhancing technologies are available for insecticides and fertilizers, and savings in input costs per unit of output are possible, it ought to be recognized that the adoption of these technologies comes at a significant cost in terms of farmer time. Most efficiency-enhancing technologies are knowledge-intensive and require substantial farmer time for learning, decision making and supervision. The monetary cost savings associated with the adoption of the above technologies ought to be weighed against the requirements for farmers' time which is becoming increasingly expensive. Finally, the profitability of adopting efficiency-enhancing technologies will be lower in countries where the relative prices of inputs are kept lower through explicit or implicit subsidies.

This chapter provides the following:

1. An assessment of farm-level fertilizer use and productivity, and the opportunities for enhancing fertilizer use efficiency.
2. A review of herbicide use trends in rice farming and the outlook for sustaining rice productivity growth while minimizing herbicide use.
3. An evaluation of insecticide productivity and the prospects for rice production with zero or minimum insecticide use.
4. Concluding remarks on the challenges involved in the transfer of knowledge-intensive technologies to Asian rice farmers.

Fertilizer Use and Response in Intensive Irrigated and Rainfed Rice Systems

Fertilizer use in irrigated rice systems was very low prior to the Green Revolution and during the first decade after the adoption of modern rice varieties. Average nitrogen applications in the rice bowls of Asia prior to the Green Revolution were around 9–15 kg ha^{-1} (IRRI, 1995). The amounts of chemical fertilizers applied increased dramatically over the decades of the 1970s and the 1980s. David and Otsuka (1994) estimated fertilizer demand functions for rice cultivation with farm household survey data for a number of countries in Asia. The estimates of demand elasticity for fertilizer with respect to the adoption of modern varieties show that a 10 percent increase in the area under modern varieties increased fertilizer use by 24 percent for the Philippines, 14 percent for Indonesia, 13 percent for Thailand and 10 percent for India. In the 1990s, nitrogen application rates for the irrigated rice systems of South and Southeast Asia are typically from 80 to 150 kg ha^{-1}. Fertilizers account for 20–25 percent of total production costs in irrigated rice systems in Asia (Rosegrant and Pingali, 1994).

Global rice requirements by the year 2020 are expected to be 60 percent higher than in 1995 (IRRI, 1995). It is expected that a large part of the incremental rice production will have to come from the irrigated lowlands of Asia. There are essentially two avenues for increasing rice production in the irrigated environments, the first being an exploitation of existing yield potential at the farm level, i.e. a movement along the current yield frontier, and the second a shift in the yield frontier through the new plant types discussed in Chapter 10. In the absence of technologies for improving nutrient use efficiency at the farm level, either of the above avenues for increasing productivity per unit land area would require higher levels of chemical fertilizers relative to current use. Recent estimates indicate a doubling of total fertilizer used for rice by the year 2020 (Cassman and Pingali, 1995).

The question that needs to be asked is whether farmers growing irrigated rice are efficient in their exploitation of the fertilizer-responsiveness of modern varieties. In other words, have the irrigated rice farmers moved along their fertilizer response function to the point of economic optimum? Table 11.1 presents estimates for fertilizer response functions using farm-level data from Bangladesh, Thailand, the Philippines and Indonesia. Response functions were estimated separately for the irrigated and rainfed environments and for the wet and dry seasons (Fig. 11.1). Economic optimum levels of fertilizer were determined by deriving the marginal productivity of fertilizer and equating it to the ratio of fertilizer to rice prices.

It is important to note from the results that the intercept and slope of the fertilizer response functions vary by season and by rice ecosystem. The yield response to fertilizer is higher in the dry season relative to the wet season and for the irrigated environments relative to the rainfed environments. At the sample average, fertilizer use is less than the economic optimum for all locations, across ecosystems and seasons, although average fertilizer use in the irrigated environments in Indonesia is very close to the economic optimum level (Table 11.2). Comparisons at the high end of fertilizer use (mean plus one standard deviation) indicate, however, that the opportunity for profitably

Table 11.1. Regression estimates of fertilizer response using intensive farming household survey, selected Asian countries.

Country	Intercept	Regression coefficients of			
		Labor ha^{-1}	NPK ha^{-1}	(NPK ha^{-1})^2	Dummy variable for rainfed ecosystem
BANGLADESH (1987 WS)					
Irrigated (n=277)	2221.34***	–0.0093 (0.0786)	12.3419*** (1.5937)	–0.0078*** (0.0012)	
Rainfed (n=111)	1509.84***	0.0651 (0.1083)	33.1249*** (9.0146)	–0.0789** (0.0362)	
THAILAND (1987 WS)					
Irrigated (n=87)	2402.69***	1.7249 (1.0668)	11.4859* (6.7574)	–0.0112 (0.0369)	
Rainfed (n=151)	1099.66***	4.3516** (1.7387)	9.0815*** (3.2234)	-0.0071 (0.0245)	
PHILIPPINES					
1985 WS (n=231)	1940.06***	4.0274** (1.8506)	16.7338 (14.1340)	–0.0084 (0.0778)	-156.1844 (132.6496)
1986 DS (n=132)	1832.32***	10.9864*** (2.0674)	19.6232*** (6.7195)	-0.0425** (0.0202)	
INDONESIA (1988 DS)					
Irrigated (n=71)	2579.97*	14.0915** (5.7019)	31.6715* (18.3169)	–0.0822*** (0.0499)	

***,**,*, Significant at the 1, 5 and 10 percent levels, respectively.
Figures in parentheses are standard errors of estimate.
Data for this analysis came from various projects of the Social Sciences Division, IRRI, Philippines.
Source: Pingali *et al.* (1996a).

increasing fertilizer use is limited for the intensively cultivated rice bowls of Asia. For the Philippines and West Java, Indonesia, high-end users are applying fertilizers beyond the economic optimum. The dry-season economic optimum for irrigated rice in the Philippines and Indonesia is 200 kilograms and 190 kilograms respectively. For countries with a large proportion of intensively irrigated ricelands, the

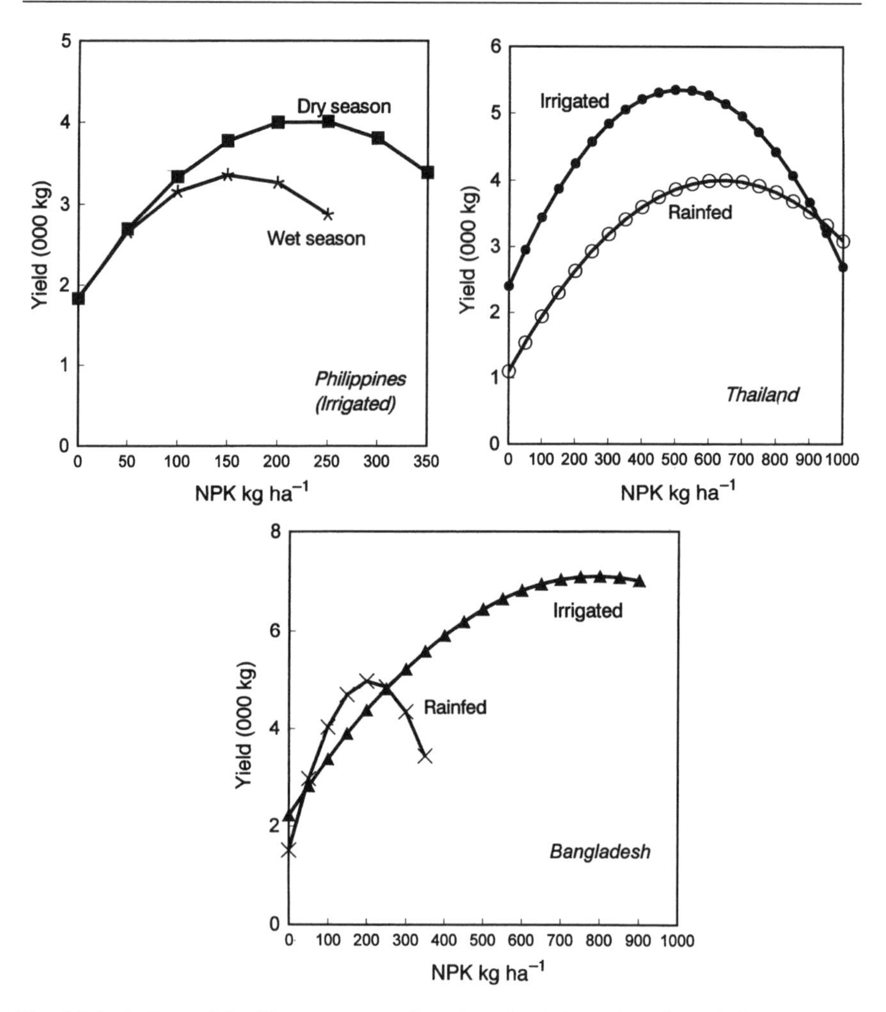

Fig. 11.1. Estimated fertilizer response functions for irrigated and rainfed environments, selected Asian countries.

opportunities for profitably increasing fertilizer use are limited and rice productivity enhancement would have to come from an improvement in the technical efficiency of chemical fertilizers. For the less intensively cultivated irrigated rice environments, such as in Bangladesh and Thailand, fertilizer application rates are below the economic optimum, and for these countries the opportunities for enhancing rice productivity through an increase in fertilizer use continues to be high.

Rice productivity improvements in the favorable rainfed lowlands can also be achieved through an increase in fertilizer use from the

Table 11.2. Marginal productivity of fertilizer use in different rice production environments, selected Asian countries.

	Mean NPK (kg ha^{-1})	Standard deviation (SD)	Marginal productivity at mean NPK	Marginal productivity at mean + 1 SD NPK	Fertilizer/ paddy price ratio
BANGLADESH (1987 WS)					2.30
Irrigated	86.54	141.30	10.99	8.79	
Rainfed	76.94	40.37	20.98	14.61	
THAILAND (1987 WS)					3.79
Irrigated	71.67	41.91	9.88	8.94	
Rainfed	16.15	26.62	8.85	8.47	
PHILIPPINES					4.08
1985 WS	84.20	24.46	15.32	13.90	
1986 DS	133.64	58.05	8.26	3.33	
INDONESIA (1988 DS)					1.70
Irrigated	176.38	53.62	2.67	–6.14	

The values were computed using the regression results presented in Table 11.1.
Source: Pingali *et al.* (1996a).

current level to the economic optimum level. Despite a rapid growth in the use of chemical fertilizers in the wake of the Green Revolution in Asia, farmers in rainfed environments still apply a relatively low dose of chemical fertilizers. Average chemical fertilizer usage in irrigated rice in India is around 170 kg ha^{-1} of NPK as compared to only 32 kg ha^{-1} in rainfed environments (Hossain and Singh, 1995). The yield response of nitrogen similarly is lower in rainfed rice ecosystems. The low rate of application and low nitrogen use efficiency in rainfed environments are due to a host of abiotic stresses which plague rice production. The adoption of nutrient-responsive high-yielding varieties in most rainfed environments is still limited (Kshirsagar and Pandey, 1995). Production is risky and farmers have to contend often with flood or drought and sometimes both flood and drought in the same field during the same production season.

District-level data from Eastern India indicate that the average level of fertilizer application is negatively correlated with the magnitude of yield risk indicating that, in more variable environments, farmers apply smaller average quantities of fertilizers (Hossain and Laborte, 1994). Similar results have been obtained from a micro-level study of fertilizer application in rainfed rice in Central Luzon, Philippines (Pandey *et al.*, 1995). Based on an analysis of field-level panel data for four years, Pandey *et al.* (1995) also report that the temporal variability in fertilizer application is higher in fields with more variable environmental conditions. Apparently, farmers are varying fertilizer dosage from year to year by a greater magnitude in fields with higher temporal variability in hydrological conditions in order to benefit in favorable years and avoid losses in unfavorable years.

The effect of risk on fertilizer use depends also on other options available for reducing risk. If farmers are concerned about the variability of consumption, risk-reducing strategies like adjustments in cultural practices and participation in land, labor and capital markets can help smoothen consumption stream. This dampens the perceived risk associated with fertilizer use, thus encouraging a higher level of application (Anderson and Hazell, 1994). The possibilities for such adjustments, however, depend on the agroclimatic conditions and institutional and policy factors (Binswanger and Rosenzweig, 1986; Walker and Ryan, 1990).

Meeting the food demands of the future would require a shift in the yield frontier for irrigated rice and a movement along the current yield frontier for the favorable rainfed rice production systems. In either case, in the absence of techniques for increasing the technical efficiency of fertilizer use, the absolute quantities of chemical fertilizers used can be expected to increase substantially. The profitability of future rice production systems, at current relative prices for rice and fertilizers, would depend on the proportionate increase in fertilizer use relative to the yield gains achieved. Improved technical efficiencies in fertilizer use could contribute significantly to improved profitability of rice production systems. The environmental consequences of fertilizer use, in terms of losses to the atmosphere and increased risk of groundwater pollution, would be further aggravated if significant improvements in fertilizer efficiency did not emerge. The challenge for the research system today is to find cost-effective methods for increasing the partial factor productivity of nutrient inputs in irrigated rice systems. An equally important challenge is to understand the conditions under which farmers would be interested in acquiring and using technologies for increasing nutrient use efficiency.

Knowledge-intensive Nutrient Management for Irrigated Rice Systems

The technical efficiency of fertilizer use can be increased in two ways: firstly by matching nutrient supply to crop demand (better timing of fertilizer application), and secondly by switching to improved fertilizer technologies such as controlled-release fertilizers and deep placement technologies. The second set of issues are not pursued in this section; for a state-of-the-art discussion on them, see De Datta and Buresh (1989) and Singh (1995). Organic fertilizers are often seen as a means of sustaining long-term soil fertility and as a means of enhancing the efficiency of chemical fertilizers. Box 11.1 provides a discussion on the farm-level profitability of using organic fertilizers.

For the purposes of this section we restrict our discussion on technologies for enhancing nutrient use efficiency to those that are designed to match nutrient supply to crop demand. In other words, the quantity and timing of nutrient applications at the farm level is determined by the farmers' perceptions of native soil fertility and the crop's demand for additional nutrients at a particular stage of crop growth. Increasing fertilizer use efficiency at the farm level requires a greater level of farmer knowledge in the acquisition and use of new fertilizer management technologies. It also requires farmers to have an in-depth understanding of their soil resource base. Tools such as the chlorophyll meter can complement farmer knowledge and help in making informed decisions on fertilizer use. The chlorophyll meter is used to assess nitrogen deficiency in the rice plant at any point in its growth phase and to guide fertilizer applications. Such knowledge-based applications enhance plant uptake and reduce nutrient losses, thereby increasing the efficiency of fertilizers applied. Technical details on the use of the chlorophyll meter for irrigated rice systems are provided in Peng *et al.* (1996). This section discusses the conditions under which knowledge-intensive nutrient management systems will be adopted at the farm level.

Farm-level profitability of knowledge-intensive nutrient management technologies is determined by the following factors: (i) the extent of the rise in partial factor productivity of fertilizer relative to conventional fertilizer management systems; (ii) the magnitude of fertilizer savings relative to the cost of farmer time required for making informed decisions on fertilizer use; and (iii) the cost of acquiring and/or using decision aids such as the chlorophyll meter. In this section, we work with the presumption that knowledge-intensive nutrient management systems do not shift the technological yield frontier for irrigated rice, rather they increase the partial productivity of fertilizer nutrients.

Box 11.1. Economics of organic fertilizers.

Farmers in rainfed environments have traditionally relied on organic sources such as farmyard manure, compost, straw and green manure crops for maintaining soil fertility. A closer integration of livestock with cropping and low opportunity costs of land and labor in traditional rainfed systems have made the use of organic sources of nutrients economically viable.

To the extent that nutrients provided by organic sources are perfect substitutes for nutrients from inorganic sources, the effective relative prices of nutrients from these two sources determine their usage. A long-term decline in the real price of inorganic fertilizers has made their use economically more attractive to farmers. Organic fertilizers require land, labor and other inputs for their production and application and the effective price of nutrients from this source has been found to be higher than that from inorganic sources (Rosegrant and Roumasset, 1987; Garrity and Flinn, 1987; Ali and Narciso, 1994).

Agronomic evidence indicates that organic fertilizers improve soil characteristics and could result in additional yield gains (Eaqub *et al.*, 1988). Possible long-term adverse effects of inorganic fertilizer on soil properties and the environment could be ameliorated by using organic sources. These long-term and environmental benefits are not priced in the market and organic fertilizers may socially be more beneficial than they appear in terms of the short-term benefit–cost criterion. However, currently available information on this is very limited (Ali and Narciso, 1994). Further research on quantifying long-term benefits under farmers' conditions in a range of environments and management practices is warranted.

Where organic fertilizers do in fact improve soil quality and enhance sustainability, it does not necessarily mean that they should be produced *in situ*. These fertilizers could be produced *ex situ* by other producers and sold to farmers who use them. Where opportunity costs of labor and land are rising, *in situ* production may not be economical. Recent trends in vegetable production in peri-urban areas in the Philippines using organic manure imports from poultry and other sources attest to this. Often, discussions on organic fertilizers confuse issues related to the economics of their production with the economics of their use.

At the current state of knowledge, the long-term prognosis is that organic fertilizers are likely to play a minor role in the management of nutrients in more favorable rainfed environments. Rising opportunity costs of land and labor in these environments raise the cost of organic fertilizers making them less competitive relative to inorganic sources. In less favorable rainfed environments with limited access to purchased inputs, farmers may remain more reliant on organic fertilizers. Research in improving the productivity of organic fertilizers and lowering the cost of their production may be more beneficial in these environments.

Source: Pingali *et al.* (1996a).

Figure 11.2 provides a conceptual model of fertilizer response functions for knowledge-intensive (R′) versus conventional nutrient management (R). The switch from conventional to knowledge-intensive nutrient management makes the slope of the fertilizer response function steeper, and it peaks earlier than the response function for conventional nutrient management. In theory, for a given level of fertilizer application, knowledge-intensive nutrient management systems ought to provide a higher level of output relative to conventional management systems, as long as the marginal productivity of fertilizer remains in the positive domain. The technical (agronomic) optimum level of fertilizer use for knowledge-intensive systems (N′) is lower than that of the conventional system (N). At zero nutrient costs, there ought to be no difference in yield per hectare between the two fertility management systems, although the knowledge-intensive system requires a lower level of fertilizer input. Therefore, at zero nutrient costs, knowledge-intensive nutrient management systems,

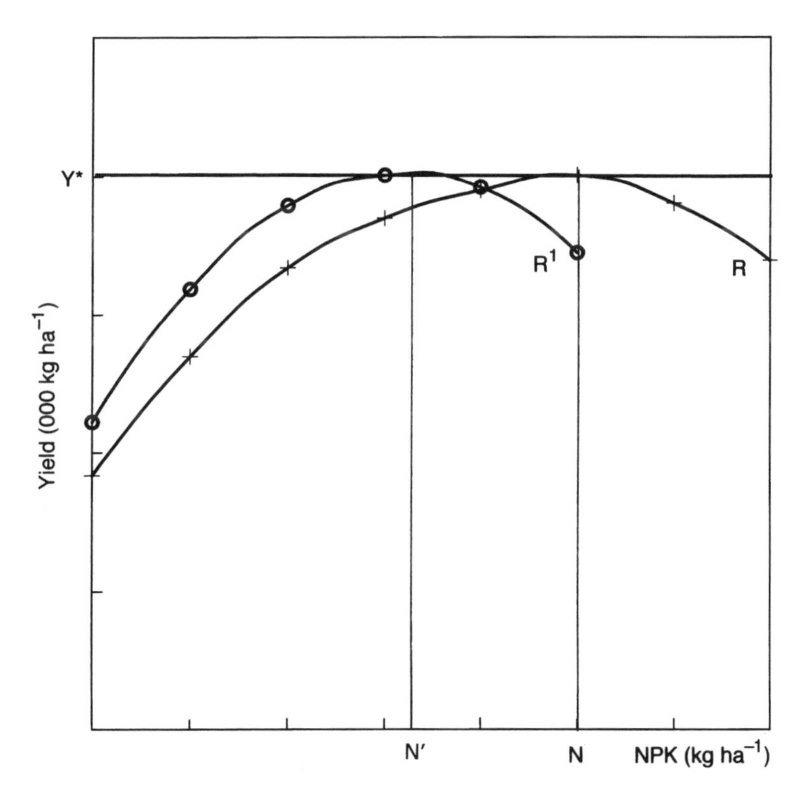

Fig. 11.2. Fertilizer response functions with knowledge-intensive (R′) vs conventional (R) nutrient management.

with their higher technical efficiency, will always dominate conventional systems.

Farmers, as discussed earlier, operate at the economic optimum rather than the technical optimum; optimal fertilizer use is determined by the point where the marginal productivity of fertilizer equals the ratio of fertilizer price to output price. The dominance of knowledge-intensive systems is not as clear when input and output prices are explicitly considered. Let us look at two scenarios, the first in which the costs of monitoring and decision making are zero and the second in which they are not. When the costs of monitoring and decision making are zero, economic optimum is determined as before: the fertilizer level at which the marginal product equals the fertilizer to rice price ratio. If the slope of the response function of the knowledge-intensive system is steeper than that of the conventional system then the former will dominate the latter except at very low ratios of fertilizer to rice prices. This conclusion does not hold when monitoring nutrient requirements, and decision making on nutrient applications has a cost associated with it.

Monitoring soil nutrient status takes time. We need to know how much, at what frequency, and whether farmers can do it on their own, or whether they need to seek expert judgments. Knowledge-based decisions on fertilizer timing and quantities to apply require judgments on crop demand for nutrients and how they can be effectively supplied. Costs associated with decision making include farmer time for learning and decision making, and the cost of acquiring and using decision tools such as the chlorophyll meter. When decision costs are non-zero, fertilizer savings through the use of knowledge-intensive technologies ought to be greater than the costs of acquiring the tools plus the time cost of making the decision. Farmer adoption of knowledge-intensive nutrient management systems will be less likely where the cost of knowledge acquisition and decision making is high.

Reducing decision costs associated with the use of knowledge-intensive technologies is a major challenge for the scientific and extension communities. A key to understanding how decision costs can be reduced is to assess the scale at which nutrient management decisions need to be made. The question we have to ask is whether such decisions need to be made specifically for each farm or even for particular parcels within a farm, or whether decisions can at a point in time be generalized across farms within a particular area and over particular cropping systems and varieties. Decision costs are lower when the geographic domain over which a decision can be generalized is larger. Can an extension agent, for example, make a recommendation each week on the need for additional fertilizer application in a particular area based on monitoring fields using a chlorophyll meter? If such generalizations are possible, the costs of acquisition and use of decision tools, such as the chlorophyll meter, become smaller for individual farmers.

If nutrient management decisions are highly farm- and/or parcel-specific, then the farm-level costs of acquiring decision tools and using them can be a major deterrent to their adoption. Significant fertilizer savings, on a sustainable basis, are required before farmers would be interested in adopting knowledge-intensive techniques that are highly farm specific. In this case, other opportunities for reducing decision costs that ought to be explored are whether it is possible to rent (or borrow) decision tools, and/or whether it is possible to purchase the decision (such as from a consultant who tells the farmer if and how much fertilizer to apply). These are questions that need to be asked; we do not believe answers are currently available.

Where nutrient application decisions are highly farm specific, the costs of learning improved fertility management methods can be quite high for the farmer. We ought to consider the costs associated with transmitting such complex information to the farmers and the high degree of variability in their assimilation of the information provided. Are there other cost-effective methods that would allow the farmers to achieve the same end of increasing nutrient use efficiency without having to bear the high cost of learning? For example, would some simple rules on the timing of nitrogen application help enhance nitrogen efficiency, albeit at a lower level than with a chlorophyll meter? Devising a simple message, even one that is based on complex scientific principles, could increase the rate of adoption of knowledge-intensive technologies. Designing effective systems for the transmission of scientific knowledge to the farmer in a way that is useable by him/her is an extremely important research activity in the pursuit of enhanced fertilizer efficiency.

From Manual Weeding to Herbicide Use

Weeds are one of the most important sources of yield loss in tropical rice. In the Philippines, 30 percent of the yield gap between the yield attainable on farmers' fields and actual yield has been attributed to poor weed control (De Datta and Flinn, 1986). Worldwide loss in rice yield due to weeds has been estimated to be around 10 percent of total production (Moody, 1991). Taking 1993 rice production to be 520 million tons, this amounts to a loss due to weeds of US$ 14 billion.

Weed management practices used by rice farmers could be classified into direct and indirect methods. Direct methods refer to those methods that are designed mainly to control weeds. These are herbicides and mechanical or manual weeding. Current trends are towards increased use of chemical weed control in place of manual and/or mechanical methods. Indirect methods include general crop management practices which in addition to improving crop productivity also reduce weed infestations.

Land preparation, method of crop establishment, puddling, water management, date of planting, plant density, cultivar choice and fertility management impact indirectly on weed population. Farmers manipulate these components of rice culture to achieve a certain degree of weed control. Land preparation, the choice of crop establishment method, and water management have the most significant impact on weed growth and weed-related yield losses. Weed populations can be substantially reduced by increasing the intensity of land preparation. Weeds are generally kept in check under transplanted conditions with good water control. In contrast, dry- or wet-seeded rice suffers from more severe weed problems. The technical aspects of these indirect methods of weed control have been comprehensively discussed by Moody (1991). Growing water scarcity and increasing wage rates tend to reduce the profitability of relying on indirect weed-control methods, and bias the production system towards increased reliance on herbicides.

Evolving trend towards increased herbicide use

The use of herbicides to control rice weeds has increased over time (Table 11.3). The Philippines, Thailand and India have more than doubled their herbicide imports in the 1980s. The total value of herbicide sales in rice-producing countries was estimated to be around

Table 11.3. Change in the value of herbicide imports in selected Asian countries.

Country	Period	Percentage change in the value of herbicide imports
Philippines	1984–1988	165
Thailand	1980–1990	217
India	1987–1990	272
China	1987–1990	23
Indonesia	1984–1988	94

Source: FAO (1988) as cited in Pandey and Pingali (1996).

US$ 1.2 billion in 1993 (Table 11.4). Over 50 percent of this total is used in Japan. The share of developing countries is about 20 percent (Fig. 11.3). Farm-level data for the Philippines and Vietnam indicate that there has been an upward trend on expenditure on herbicides (Fig. 11.4). Several factors have contributed to this increased popularity of herbicides for weed control, and these can be classified into technological and economic factors.

The traditional varieties of rice, although low yielding, were naturally competitive with weeds due to their early vegetative growth, tall stature and drooping leaves. Modern high-yielding varieties are relatively less competitive. Thus compared with traditional varieties, modern varieties tend to be more heavily infested by weeds. Due to the higher yield potential of modern varieties, yield gains obtained from a given intensity of weed control are also higher.

Table 11.4. Herbicide use in rice.

Country	Value of herbicide use[a] (million US$)	Herbicide use[b] ha^{-1} (US$ per ha)	Percent rice area irrigated[c]	Wage rate[c] (US$ per day)
India	23	0.55	44	1.03
China	80	0.41	94	1.03
Bangladesh	5	0.47	19	1.42
Indonesia	6	0.60	81	1.14
Thailand	18	1.8	27	1.83
Philippines	11	3.14	59	2.10
Japan	659	314	100	45.00
South Korea	110	92	100	22
USA	100	91	100	—
Other countries	158			
Total	1170			

Sources of basic data:
[a] Ciba-Geigy, Personal Communication, 1994.
[b] Rice area obtained from IRRI (1991).
[c] IRRI (1991).
Source: Pandey and Pingali (1996).

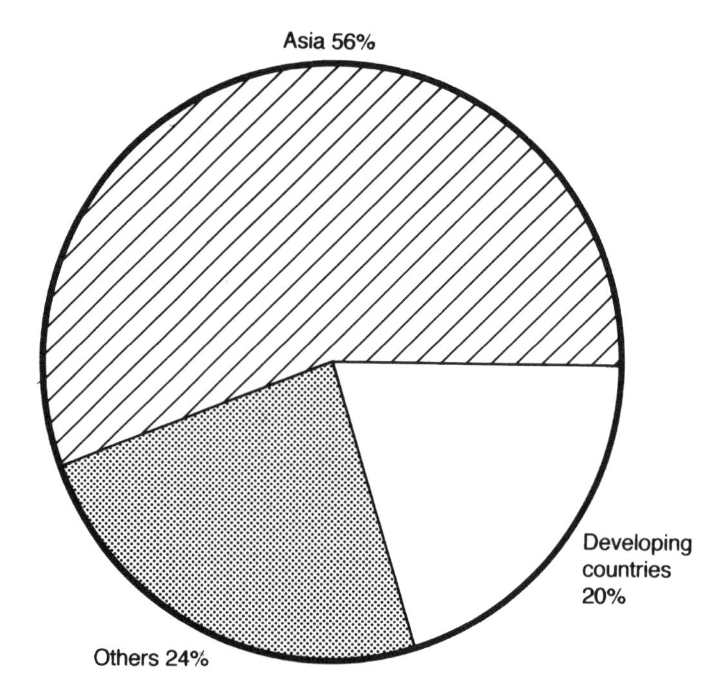

Fig. 11.3. Global herbicide use in rice, 1993.

Widespread adoption of high-yielding varieties thus created economic incentives for farmers to reduce weed infestations. Simultaneously, cheap herbicides became available while wage rates increased. In the Philippines, the ratio between the price of herbicide and the wage rate declined steadily over time, making herbicide use economically more attractive (Fig. 11.5). In Iloilo province, Philippines, the cost of weed control by herbicide in wet-seeded rice is less than one-fifth of the cost of a single handweeding (Moody, 1996). Similarly, in West Java and Mekong Delta, the cost of handweeding is three to five times the cost of herbicides. Economic analysis of weed control practices in the Philippines indicates that the marginal benefit–cost ratio associated with herbicide usage is as high as 16 (Table 11.5).

The prospects are for further increases in wage rates (Fig. 11.6) and increasing profitability of labor-saving technologies in rice production. A major response to increasing agricultural wages has been a shift from transplanting to the direct-seeding method of rice establishment. Direct seeding consists of sowing rice seeds directly in the soil mostly by broadcasting. Thus direct-seeding methods substantially save on labor cost. Most of the irrigated rice in Malaysia is wet seeded. Wet

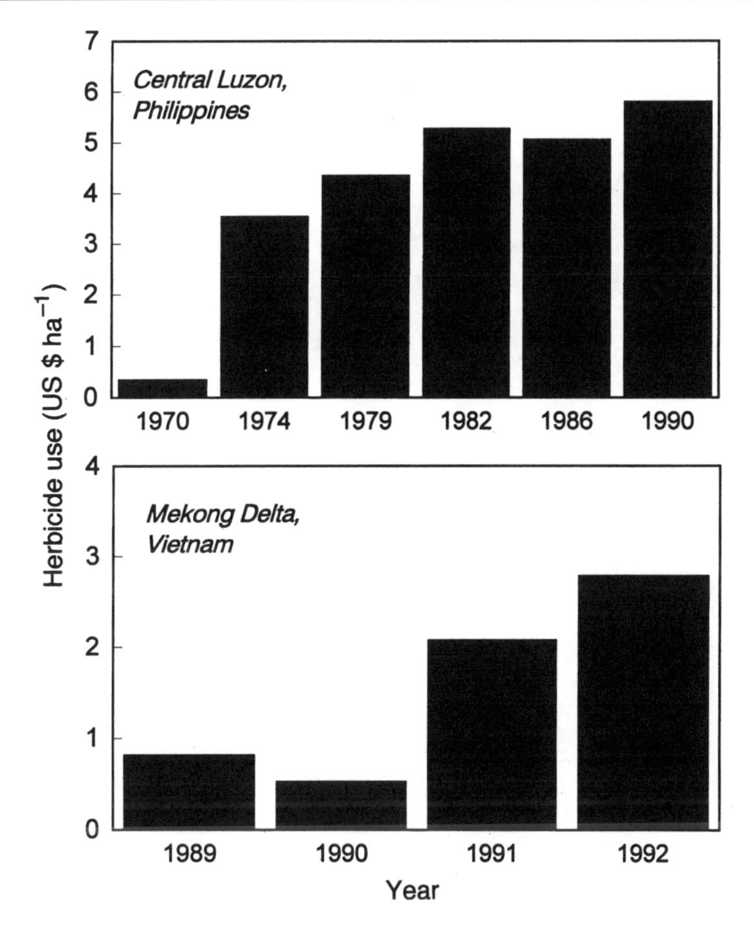

Fig. 11.4. Farm-level herbicide use in Central Luzon, Philippines, and Mekong Delta, Vietnam. (Source of basic data: Social Sciences Division, IRRI.)

and dry seeding methods are becoming increasingly popular in other rice-producing countries of Southeast Asia. Even though chemical weed control costs are higher for direct-seeded rice, savings in labor cost have been more than the additional cost of herbicide (Table 11.6).

Weed problems are, however, more serious in direct-seeded rice than in transplanted rice. As weeds and rice emerge simultaneously in direct-seeded fields, the competitive effect of weeds on rice is much greater. The success of the direct-seeding method depends critically on weed control. Empirical studies in the Philippines and Vietnam indicate that farmers who used wet seeding relied more heavily on herbicides than farmers who transplanted (Table 11.7). Similarly in Peninsular Malaysia, 95 percent of farmers who direct seeded applied

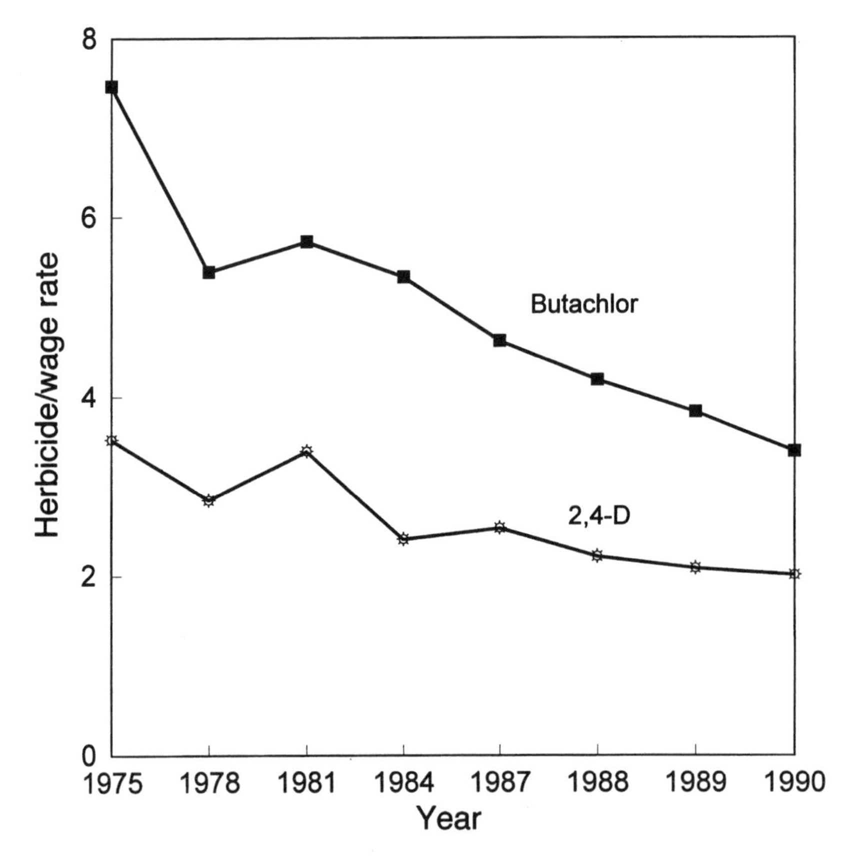

Fig. 11.5. Trend of herbicide and wage rate ratios in the Philippines, 1975–1990.

herbicides, with an average herbicide cost of US$ 30 per hectare. On the other hand, only 46 percent of the farmers who transplanted used herbicides, with an average herbicide expenditure of US$ 4 per hectare (Moody, 1996).

Herbicide use is increasing even in areas which are transplanted. Increasing unreliability of water, due to a host of factors including deterioration in irrigation infrastructures (Rosegrant and Pingali, 1994), has made weed control via water management somewhat uncertain. In addition to the general increase in wage rates, seasonality of weeding often creates labor scarcity during the optimal time for weed control. The peak of labor demand for weeding is likely to be accentuated in more intensive production systems. Thus, demand for chemical methods can be expected to increase as rice production is intensified.

Table 11.5. Economics of weed control in irrigated transplanted rice[a], Philippines, 1990.

Control method	Grain yield ($t\ ha^{-1}$)	Total variable cost[c] (US\$ ha^{-1})	Total return (US\$ ha^{-1})	Return above variable cost (US\$ ha^{-1})	Marginal benefit–cost ratio[d]
No weeding	1.5	0	263	263	—
2 handweedings[b]	3.7	90	648	558	3.3
Thiobencarb/ 2,4-D (1.0 + 0.5 kg ha^{-1})	3.3	19	578	559	16

[a] Data on rice prices and yields were obtained from World Rice Statistics 1990 (IRRI, 1991). Data on physical inputs and yields were obtained from Ampong-Nyarko and De Datta (1991), using analysis from previous IRRI studies.
[b] Assumptions:
Labor = US\$ 2.09 day^{-1}; 2 handweedings = 43 day ha^{-1}; Cost of herbicides = US\$ 19.
[c] Total variable costs reflects costs for weed control only.
[d] Marginal benefit–cost ratio = change in returns above variable cost over change in total variable cost.
Source: Naylor (1996) as cited in Pandey and Pingali (1996).

Equity, environmental and health consequences of herbicide use

Although economic and technological factors are likely to lead to increased substitution of labor by herbicides, such a substitution may result in short-term adverse social consequences. Due to the need to complete weeding within a short time, farmers rely on hired labor for weeding (Table 11.8). As small and marginal farmers are the main source of hired labor, their income and employment will be adversely affected unless they can find an alternative use of their time. Increased cropping intensity made possible by improved technology has in the past been able to absorb most of the displaced labor (David and Otsuka, 1994). The extent to which substitution of labor by capital in agriculture will occur without imposing welfare costs on certain groups depends mainly on government policies on exchange rate, pricing of inputs and outputs, and macroeconomic policies. To the extent that herbicides are made cheaper relative to labor due to distortionary price policies, substitution of herbicides for labor is socially undesirable. Negative environmental and health effects of herbicide use also entail additional social costs.

Herbicide residues have been detected in the ground and surface water systems in rice production areas where herbicides are used (Bhuiyan and Castañeda, 1995). Although the levels detected are

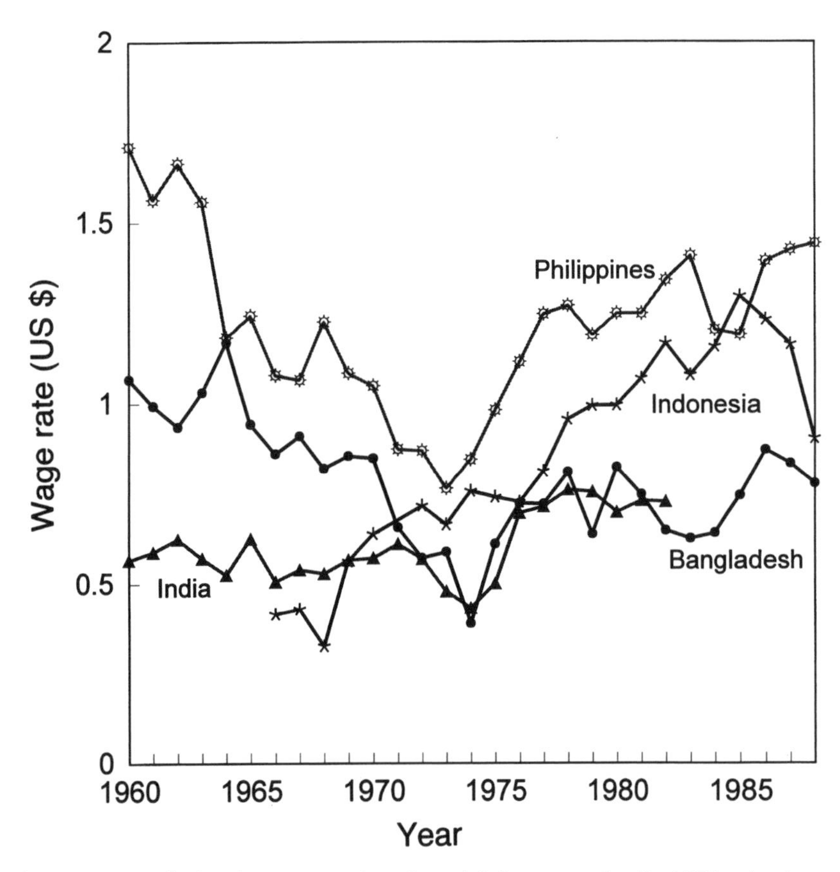

Fig. 11.6. Trend of real wage rates in selected Asian countries (in 1975 prices).

Table 11.6. Herbicide and labor costs associated with wet seeding and transplanting methods of rice establishment.

	Herbicide cost (US$ per ha)			Pre-harvest labor cost (US$ per ha)		
Country	Trans-planted	Wet-seeded	Addi-tional	Trans-planted	Wet-seeded	Addi-tional
Philippines[a]	7	9	2	36	8	–28
Malaysia[b]	6	20	14	205	114	–91

Sources of basic data as cited in Pandey and Pingali (1996): [a] Erguiza *et al.* (1990). [b] Ho (1994).

Table 11.7. Changes in weed control practices over time in Vietnam.

Method of crop establishment		% of farmers reporting[a]			
		1989	1990	1991	1992
Dry-seeded	Herbicide + manual	48	47	67	71
	Herbicide only	12	16	16	16
	Manual only	38	36	15	12
	None	2	1	2	1
Transplanted	Herbicide + manual	23	33	59	71
	Herbicide only	2	0	10	18
	Manual only	71	67	27	12
	None	4	0	5	0

[a] Some columns may not total 100 due to rounding.
Source: Pandey and Pingali (1996).

Table 11.8. Labor use (mandays ha^{-1}) by operation by method of crop establishment for three locations in the Philippines, wet season 1981–1982.

	Bulacan–N. Ecija		Iloilo		Camarines Sur	
	WSR	TPR	WSR	TPR	WSR	TPR
Land preparation	6	8	8	8	7	14
	(57)	(48)	(38)	(38)	(71)	(43)
Seeding/transplanting	1	20	1	13	1	18
	(12)	(100)	(0)	(92)	(0)	(100)
Handweeding[a]	3	3	7	19	5	1
	(23)	(29)	(29)	(74)	(40)	(54)
Other pre-harvest	8	13	1	2	1	1
	(32)	(38)	(0)	(50)	(0)	(0)
Total pre-harvest	18	44	17	42	14	34

WSR = wet-seeded rice; TPR = transplanted rice.
[a] Includes replanting and handweeding.
Figure in parentheses indicate proportion of hired labor.
Source: Cordova *et al.* (1984) as cited in Pandey and Pingali (1996).

currently small, the potential for a rapid increase in water contamination with an increase in herbicide use is a matter of serious concern. The health effects of pesticide use in the Philippines, even at current levels of use, are quite substantial (Rola and Pingali, 1993; Antle and Pingali, 1994; Pingali *et al.*, 1994). Eye, skin, pulmonary and neurologic problems are significantly associated with long-term pesticide exposure. The majority of the pesticides, the highly hazardous Category I and II chemicals, which might be linked to these impairments, while commonly available in the Philippines, are banned or severely restricted in the developed world. Herbicides have been particularly implicated in the high incidence of skin diseases, polyneuropathy, and gastrointestinal problems (Pingali *et al.*, 1994).

Rola and Pingali (1993) found that the net benefits of insecticide use are negative when health effects are explicitly accounted for. Antle and Pingali (1994) found that insecticide-related health impairments cut farm household productivity. When insecticide use is reduced, the value of the positive health benefit of reduced exposure is invariably greater than the value of the crop lost to pests (Chapter 5). In the case of herbicides, the health–productivity trade-off is less clear. In the case of a switch from transplanting and manual weeding to direct seeding with herbicide use, the positive productivity benefits are substantially greater than the value of the adverse health effects (Pingali and Marquez, 1996). On the other hand, the productivity benefits of a switch to herbicides in transplanting systems are smaller and may not outweigh the health costs associated with herbicide exposure. Table 11.9 provides a comparison of yields and net benefits for alternative weed management strategies with and without considering health costs. For details on the methodology used for the above ranking see Pingali and Marquez (1996).

Aspects of integrated weed management

Given the trend towards increased herbicide use and the likely environmental and health consequences of such a trend, integrated weed management involving a combination of several methods is desirable. In fact, most rice farmers adopt some form of integrated weed management practice (De Datta and Baltazar, 1996).

Table 11.9. Mean net benefits of weed control strategies incorporating health cost (US$) for Nueva Ecija rice farmers, Philippines.

		Mean net benefits		
Weed control strategy	Yield (kg ha^{-1})	Estimated health costs (US$)	Without health costs (US$)	With health costs (US$)
Transplanting + handweeding	5043	0	666	666
Transplanting + handweeding + herbicide	4927	5	668	663
Direct seeding + herbicide	5131	5	818	813

Source: Pandey and Pingali (1996).

Long-term management of weed population without excessive reliance on only one method is the main purpose of integrated weed management. Reducing the buildup of a weed seed bank is a major aspect of such a long-term strategy. A good understanding of the population dynamics of weeds helps identify intervention points which may be more effective in regulating weed populations than the more conventional approach of killing weeds.

The seed bank of a self-contained weed population can be manipulated by altering recruitment, seedling mortality, seed viability and fecundity. Herbicides and handweeding reduce weed populations by increasing seedling mortality. Plowing, harrowing and other preplanting operations help deplete the seed bank by altering recruitment. Fecundity is reduced by removing late-emerging weeds before their seeds are shed. By putting weed seeds at varying depths, deep plowing can alter seed viability.

In the case of rice, two major methods of weed management have been the manipulation of recruitment and of seedling mortality. The population dynamics of rice weeds have not been widely enough studied to be definite about the importance of these two interventions relative to regulation of fecundity. However, in the case of wild oats, economic returns to the regulation of fecundity have been estimated to be high (Pandey *et al.*, 1993). When this tactic was combined with post-emergence herbicide treatments, the seed bank of wild oats declined rapidly and a long-term control of weed was made possible (Pandey and Medd, 1990).

A tactic that is relevant to rice farmers is to use clean certified seeds. Contamination of rice seeds with weed seeds is one of the major factors causing the dispersal of weeds of rice (Moody, 1993). In the Philippines, approximately 24,000 weed seeds were found in a kilogram of rice (Rao and Moody, 1990). When farmers keep their own seeds, contaminated seeds can add substantially to the seed bank, despite their attempts to clean the seeds (Moody, 1993).

Using more competitive varieties of rice is another tactic of reducing the effect of weed competition. However, in the case of rice, there appears to be a trade-off between the plant's ability to compete with weeds and yield (Moody, 1991). Research on varietal improvement for weed management in rice is at a very early stage (Khush, 1996).

The challenge for weed management research is to develop control strategies that sustain and/or enhance farm profits while safeguarding the environment and human health. Strategies that are sustainable over the long term are more likely to be holistic in approach and well integrated with other crop management practices. Emphasis ought to shift from quick fix solutions to problem prevention.

Farmer incentives for adopting a holistic weed management

approach depend upon the relative prices of rice, labor and herbicides. Where herbicide prices do not reflect the true social costs, indiscriminate use will persist. Regulatory policy could help change the relative price structure in order to reflect social costs associated with injudicious herbicide use. A three-pronged regulatory approach aimed at producers would be: (i) cancellation of the highly hazardous and/or persistent chemicals; (ii) selective taxation to make the liberal use of all chemicals expensive; and (iii) farmer training in the judicious and discriminate use of chemicals.

From Indiscriminate Insecticide Use to Judicious Pest Management

While herbicides are a growing phenomenon in rice production, insecticide use is well established having been promoted intensively during the early years of the Green Revolution. Intensive rice monoculture systems popularized by the Green Revolution created an environment that was conducive to pest growth. Although pest-related yield losses were small in percentage terms, they were highly visible and led to apprehensions of major outbreaks. The policymakers' aversion to large-scale pest outbreaks resulted in policies that made pesticides easily accessible and affordable at the farm level. Farm-level promotion of pesticide use was accompanied by little, if any, training in judicious use, farmer safety and adverse ecological and health consequences. Frequent and indiscriminate pesticide use is the norm in Asian rice production systems today.

The adverse effects of indiscriminate and unsafe insecticide use on the paddy ecosystem, the environment and human health were discussed in Chapter 5. Insecticides provide a clear case of the potential for knowledge as a substitute for physical inputs in crop production. Farmers with improved knowledge of pest–predator ecology and pest-control decision making have invariably used lower amounts of insecticides.

Some early signs already indicate that the global insecticide market has peaked and aggregate use is beginning to decline. The decline in the global insecticide market in recent years can be attributed to: (i) the withdrawal of government subsidies for insecticides in most Asian countries; (ii) the widespread use of second-generation modern rice varieties that are resistant to a wide range of insect and disease pressures; and (iii) the growing recognition, by farmers as well as policymakers, of the need for judicious insecticide applications as promoted by the growing number of integrated pest management (IPM) programs across Asia.

The Green Revolution, insecticide use and pest outbreaks

The global insecticide market for rice was valued at US$ 1.0 billion in 1993. Wood Mackenzie Consultants Ltd. (1993) reported that Japan has the highest percent use of insecticides (40 percent); the other large users of insecticides are China, South Korea and India (Fig. 11.7). Southeast Asia accounts for less than 10 percent of the global rice insecticide market. In the Philippines, for example, insecticide sales in 1993 amounted to US$ 15 million, about half of which was used for rice. Across Asia, insecticide use increased from US$ 347 million in 1980 to a peak of US$ 1.079 billion in 1990, an average annual increase of US$ 66.5 million, or almost 20 percent.

At the farm level, Asian rice farmers commonly make two to three applications of insecticides, although six or seven applications are not uncommon (Rola and Pingali, 1993; Warburton *et al.*, 1995). While trends are towards increased acceptance of judicious pest control, as in the case of IPM, the vast majority of farmers still continue to use prophylactic measures for insect control. Insecticides used include the organochlorine endosulfan; organophosphates such as methyl parathion, monocrotophos and chlorpyrifos; carbamates such as BPMC, carbaryl and carbofuran; and pyrethroids such as cypermethrin and deltamethrin. The vast majority of insecticides used in developing Asian countries are the highly hazardous Category I and II chemicals, as classified by the World Health Organization, many of which have

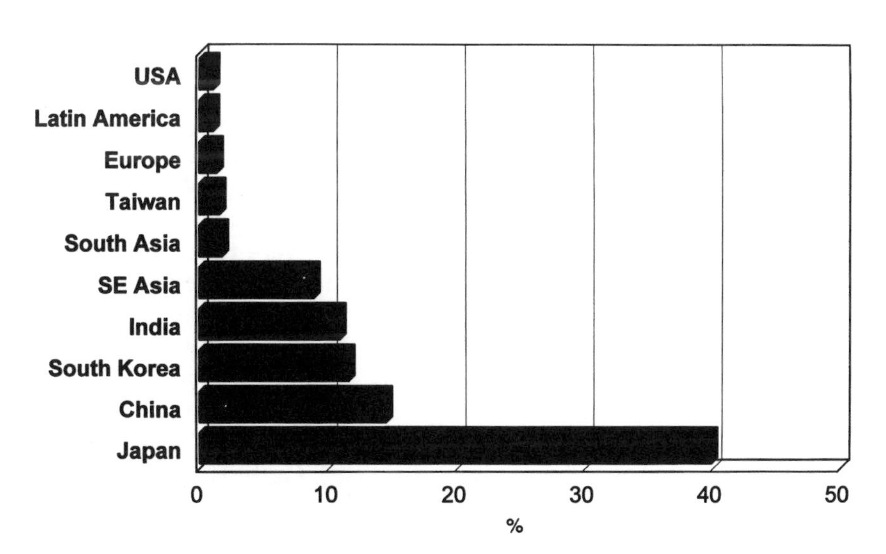

Fig. 11.7. Global rice insecticide market (total of US$ 1105 million).

been banned for agricultural use in most developed countries (Warburton *et al.*, 1995). They, however, continue to be widely used in tropical Asia because their costs are substantially lower than the safer alternatives.

Unsafe pesticide storage, handling and disposal practices subject the farmer to high levels of health hazards and contaminate the paddy ecosystem (see Pingali and Roger, 1995, for a detailed documentation). Safe spray equipment and protective clothing, suitable for tropical conditions, are not available for Asian rice farmers. Farmers also forage the rice paddies for food such as fish and frogs, and feed such as aquatic plants and rice straw, that could be contaminated by pesticides (Warburton *et al.*, 1995). Current pesticide pricing and regulatory structure plus inadequate storage, unsafe handling practices, short re-entry intervals, and inefficient sprayer maintenance taken together provide an environment of greater accessibility and exposure to chemicals not only by the farmer applicator but by the farming household as well (Pingali and Roger, 1995).

The rapid growth in insecticide use despite its adverse health and environmental effects can be attributed to the perception, of both the policymakers and the farmers, that insecticides were a necessary input for modern rice production. Rola and Pingali (1993) have argued that policymaker and farmer perceptions of pest-related yield losses were in general substantially greater than actual losses. Perceptions of pest-related yield losses were, however, driven by observed plant damage and not by actual yield losses. For example, leaf-feeding insects are common pests of rice. They damage the rice crop by feeding on the leaves, reducing leaf area, and causing highly visible damage symptoms. Studies at IRRI have shown that leaf damage, even as much as 50 percent, has insignificant effects on rice yields. The incidence of leaf-feeding insect infestations that cause a 50 percent damage to leaves is extremely rare, yet 80 percent of the insecticide applications for rice in Asia are targeted towards leaf-feeding insects (Heong *et al.*, 1994).

In the early years of the Green Revolution when rice varieties with no resistance to insect and disease pressures were used, widespread fear of pest epidemics resulted in policies that were conducive to the rapid adoption and use of insecticides. In reality, however, the large number of pest outbreaks in tropical Asia have been associated more with injudicious insecticide applications, rather than with the use of modern high-yielding cultivars, high cropping intensity and/or high chemical fertilizer use (Heinrichs and Mochida, 1984; Kenmore *et al.*, 1984; Joshi *et al.*, 1992; Schoenly *et al.*, 1996). Outbreaks of secondary pests, such as the brown planthopper (BPH) *Nilaparvata lugens*, and tungro virus vectored by the green leafhopper (GLH) *Nephotettix virescens*, began to occur in regions adopting modern varieties and concomitant use of agrochemicals (Table 11.10). In particular, insecticide use preceded outbreaks of secondary pests, notably the BPH, that were previously of minor significance (Kenmore, 1991; Way and Heong, 1994). Across time, the number of leaf-damaging insect pest outbreaks excluding BPH had been consistently highest (Table 11.11). This group of pests included leaf- and planthoppers, leafrollers, leaf-folders, case- and armyworms, whorl maggots and locusts.

When little or no insecticides are used, tropical irrigated ricefields possess a rich diversity of pest and predator populations; in most instances, the species richness and abundance of predator populations may be greater than those of the pest populations (Way and Heong, 1994). However, pest (especially planthopper) population development following insecticide applications has been enhanced by several factors: higher recolonization rates due to greater mobility; higher reproduction

Table 11.10. Most frequently reported insect pest outbreaks in Asia.

Pest	Frequency of occurrence (%)	Year first reported	Location
Brown planthopper (*Nilaparvata lugens*)	27	1967	Malaysia
Armyworm (*Mythimna separata*)	20	1967	Malaysia
Whitebacked planthopper (*Sogatella furcifera*)	14	1963	Bangladesh
Locusts (*L. migratoria*)	8	1967	Malaysia
Stemborer (*S. incertulas*)	7	1967	India
Leaffolder (*Cnaphalocrosis medinalis*)	5	1966	Malaysia
Hispa (*Dicladispa armigera*)	2	1975	Bangladesh
Gall midge (*Orseolia oryzae*)	2	1986	Bangladesh
Black bugs (*Scotinophara coarctata*)	2	1966	Malaysia

Sources of basic data:
Frequency data from Litsinger *et al.*, 1987.
Years of first occurrence from various volumes of the International Rice Research Newsletters (IRRN).
Source: Pingali and Gerpacio (1996).

Table 11.11. Number of reported pest outbreaks and area affected (in hectares), by pest group, all Asia, 1967–1993.

	Area affected (ha)						No. of reports w/ area affected	Total no. of pest reports
Pest group	< 100	101–1000	1001–5000	5001–10000	10001–50000	> 50000		
BPH	2	9	3	1	2	2	19	213
Root-damaging	—	1	—	—	—	—	1	24
Stem-damaging	1	8	8	3	3	2	25	169
Leaf-damaging	9	14	16	3	5	9	56	439
Grain-damaging	1	2	2	—	—	—	5	11

Source: Pingali and Gerpacio (1996).

rates because of high fecundity, short life span, and eggs embedded in plant tissues escaping insecticide effects; and reduction in egg parasitism and predation because of parasite and predator mortalities.

In the late 1970s, rice varieties with resistance to several important pests were developed and rapidly released across the intensively cultivated rice areas in Asia (Rola and Pingali, 1993). With the advent of rice varieties resistant to pest pressures, the importance of pesticides for reducing yield variability has declined. However, the dissemination of pest-resistant varieties was not accompanied by extension messages on the reduced need for insecticides. Consequently, high and injudicious insecticide applications continued as before. Continued prophylactic applications of broad spectrum insecticides soon caused the breakdown in varietal resistance; the pests' ability to evolve genetically over successive generations make the plants' resistant characteristics benign. New varieties generated to replace varieties with resistance breakdown were subsequently overcome through further biotype changes in pest populations. It was recognized that the above 'breeding treadmill' could only be overcome through dramatic changes in crop management practices, especially in the use of insecticides. The recognition of, and the impetus for, integrated pest management was largely the result of the urgency created by the increasingly frequent resistance breakdowns and the subsequent pest outbreaks (Litsinger, 1989).

The productivity impact of insecticide use

On-farm experiments and examination of farmers' yields do not suggest any positive yield or profitability response to insecticide applications. One of the earliest studies to identify the farm-level productivity effects of insecticides was done by Herdt *et al.* (1984) as part of a broader analysis on yield constraints. The constraints study conducted between 1973 and 1979 across several Asian countries identified and quantified the factors contributing to the yield gap between the experiment station and farmer fields. The study concluded that rice yields in farmers' fields could be raised by an additional 0.5–1 t ha^{-1} with prophylactic insecticide applications, as compared to farmer yields of 4–5 t ha^{-1} at low insecticide applications. However, achieving the full yield potential of a variety at the farm level is often not cost effective. Herdt *et al.* (1984) concluded that the expected returns to rice production are lower for farmers applying insecticides on a prophylactic basis rather than not applying insecticides at all. This result was validated by on-farm trials of alternative pest control practices conducted by Litsinger (1989) and Waibel (1986). Both Litsinger and Waibel observed no significant yield differences between the insecticide treated and untreated plots in more than half the cases.

Rola and Pingali (1992, 1993) estimated an insecticide response function and evaluated the returns to four alternative pest management strategies for the Philippines, using both farm panel data and data from on-farm experiments conducted over several growing seasons. Response function estimates indicate very modest effects of insecticides on the mean and variance of the yield distribution (Table 11.12). Comparing the returns to prophylactic control, economic threshold levels, farmers' practice and natural control, Rola and Pingali found natural control to be the economically dominant pest management strategy. Natural control, in association with varietal resistance, proved to be consistently more profitable in an average year than prophylactic treatment and economic thresholds. The dominance of natural control was found to hold even when farmer risk aversion is accounted for. The dominance of natural control becomes even greater when the health costs of exposure to insecticides are explicitly accounted for (Chapter 5). The positive production benefits of applying insecticides are overwhelmed by the increased health costs. The value of crop loss to pests is invariably lower than the cost of pesticide-related illness (Rola and Pingali, 1993) and the associated loss in farmer productivity (Antle and Pingali, 1994).

In recent years, village-level experiments were conducted in several Asian countries to assess the impact of natural control or zero insecticide strategy (see Box 11.2). The results across these countries are similar: the production response to insecticide applications is modest at best, while profits are generally lower when compared to farmers not applying any insecticides (Rice IPM Network, 1995).

Table 11.12. Impact of insecticides on the mean and variance of yield distribution.

	Percent change in	
	Mean	Variance
Experimental data	0.007	–0.002
Farm-level data		
Laguna	0.06	–0.12
Nueva Ecija	0.06	–0.40

Sources of basic data:
Rola and Pingali (1993).

Under low pest-infestation levels, natural control is the economically dominant pest management strategy (Rola and Pingali, 1993). Natural control relies on predator populations to control pest infestations under normal circumstances. Pesticides may have to be used as a last resort in the rare instance of high pest infestations. Natural control does not imply 'do nothing'; it is based on the premise of in-depth farmer knowledge of the pest–predator ecology and frequent monitoring of field conditions by the farmer. In this regard, natural control can be considered the ultimate goal of an IPM program and farmers who are well-versed in IPM techniques would converge towards it. Therefore, continued investments in IPM training would be essential for the successful adoption of zero insecticide practices.

The above discussion on a zero-insecticide strategy is not meant to imply that it is not without any risks of major pest losses at the farm level. As similar farm-level experiments are conducted over time and over several locations that are geographically and agroclimatically differentiated, more information will become available on the probability of pest losses when zero or minimum insecticides are used. Issues that require particular attention in future research are: (i) the probability of large-scale infestations when insecticides are not applied, and the strategies for dealing with them, including crop insurance; (ii) the probability of migratory infestations and the factors that cause them; and (iii) the consequences of indiscriminate insecticide use on neighboring farms on pest buildup and pest losses on farms practicing a zero-insecticide strategy.

Integrated pest management (IPM): implementation issues

IPM as a technique in crop protection recognizes (Bottrell, 1979) that: (i) potentially harmful species will continue to exist at tolerable levels

Box 11.2. Zero insecticide strategy: community-level experiments.

In the Philippines, community-level experiments were conducted by Heong, Pingali, Palis and other staff at IRRI to determine the effect of judicious insecticide use on farm-level yields and profits. Three communities that were spatially separated but similar in agroclimatic conditions, cropping intensity and production practices were part of the experiment. In the first community (T1) farmers chose not to apply any insecticides through the entire crop cycle (although they were asked not to apply insecticides only in the first 40 days after the crop was planted). In the second community (T2) farmers were taught Integrated Pest Management (IPM) and used its principles in deciding on pest management practices. The third community (T3) was held as a control and monitored through the study period. The study initiated in 1993 has been carried out over six crop seasons.

In T1, the natural control village, the percentage of farmers using insecticides dropped from a baseline level of about 80 percent to less than 10 percent in the wet season within the first year of the experiment (Fig. 11.8). For T2, the IPM village, the percentage of farmers using insecticides dropped from a baseline level of close to 100 percent to less than 20 percent in the same period. In T3, the insecticide group, wet-season insecticide use dropped slightly from a baseline level of 100 percent of farmers in 1993 to about 85 percent in 1994. Despite the dramatic drop in insecticide use in T1 and T2, there has been no significant effect on yields over the six seasons of the study. For a particular location there was no significant change in average yields over time. Comparison across locations shows that there is no significant difference in the mean and variance of both wet- and dry-season yields over the study period (Figs 11.9 and 11.10). Comparison of dry-season profits across locations indicate that there was no significant difference between T1 and T2 over the study years, but their profits were significantly higher than those of T3. Price and Luis (1995) found that the natural control community also witnessed a reappearance of wild aquatic food, such as fish, other vertebrates and edible wild plants for food and feed. The reappearance of wild food was also seen, at a smaller extent, in the IPM community. There was no change in wild food availability in the control community over the period of the study.

In Vietnam, Heong *et al.* (1994) conducted experiments on around 460 farmers' fields to evaluate the necessity of insecticide use in rice production. Farmers left approximately 500 m^2 of their field as control, on which no insecticide was sprayed throughout the 1992 wet season, and compared the yields with the rest of the field which was treated normally. There was no significant difference in the mean and variance of yields between the treated and untreated fields. Similar community-level experiments to determine the need for insecticides are now under way in other Asian countries, such as Thailand, Malaysia, China and Indonesia.

Source: Barangay IPM Project, Social Sciences Division, Entomology and Plant Pathology Division, IRRI, 1995.

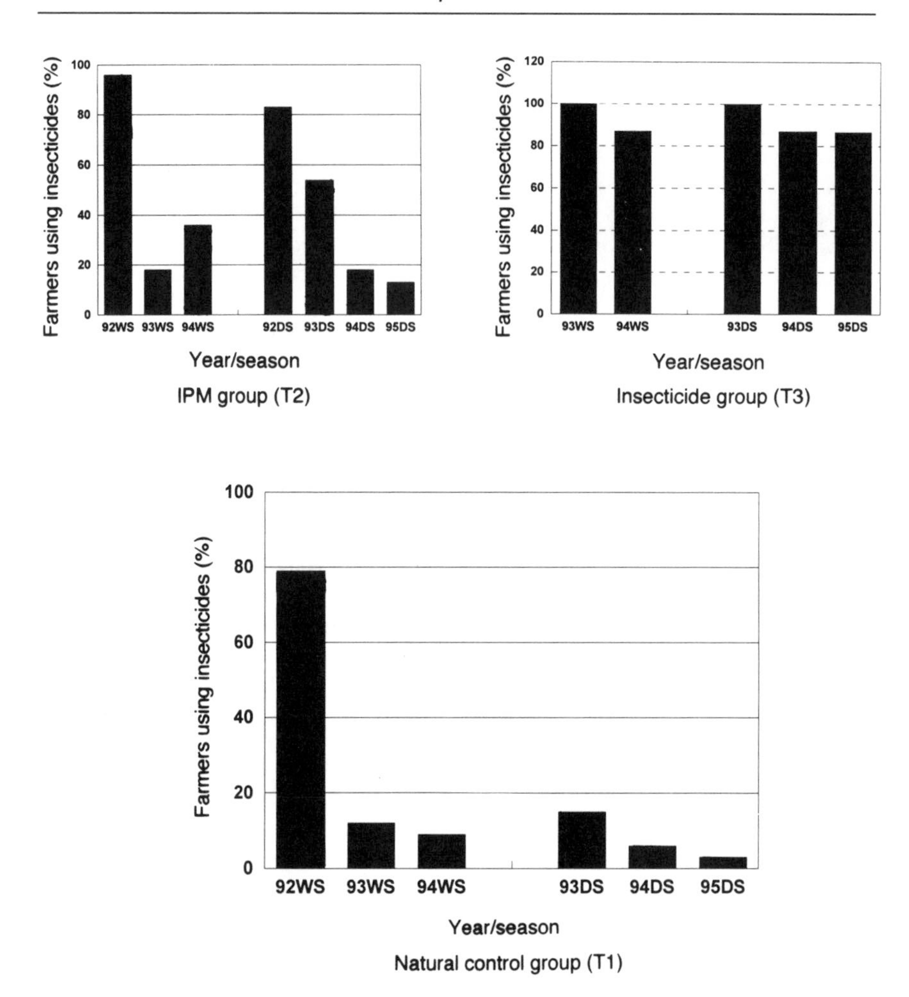

Fig. 11.8. Proportion of farmers using insecticides, by treatment group. (Source of basic data: Barangay IPM Project, Social Sciences Division, IRRI, 1995.)

of abundance, and low-level infestations of some pests may in fact be desirable; (ii) the ecosystem is the management unit, which is manipulated to hold pests to tolerable levels while avoiding disruptions of the systems; (iii) the use of natural control agents is maximized; (iv) any control procedure may produce unexpected and undesirable effects; and (v) an interdisciplinary systems approach is essential in collecting the information and formulating the management strategy. In essence, it integrates multidisciplinary methodologies in developing agroecosystem management strategies that are practical, effective, economical, and protective of both public health and the environment (Rola

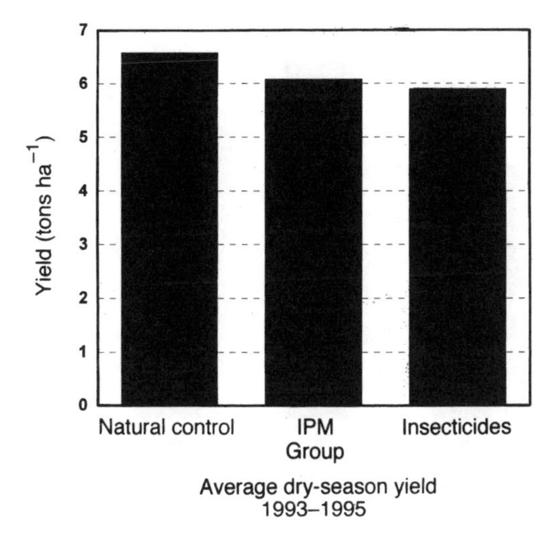

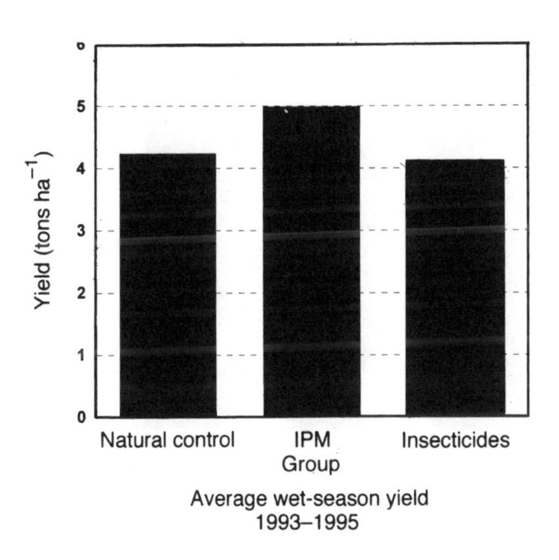

Fig. 11.9. Average wet- and dry-season yield, by farmer group. (Source of basic data: Barangay IPM Project, Social Sciences Division, IRRI, 1995.)

1.6
1.4
1.2
1
0.8
0.6
0.4
0.2
0
Standard deviation
Natural control
IPM
Insecticides
Group
Standard deviation in dry-season yield
1993–1995

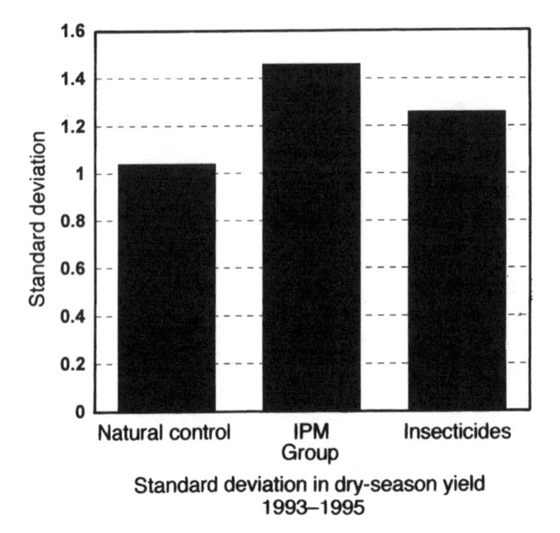

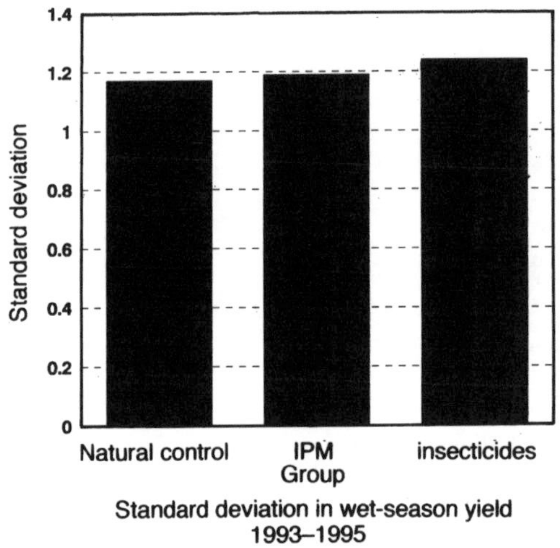

1.4
1.2
1
0.8
0.6
0.4
0.2
0
Standard deviation
Natural control
IPM
insecticides
Group
Standard deviation in wet-season yield
1993–1995

Fig. 11.10. Standard deviation in wet- and dry-season yield, by farmer group. (Source of basic data: Barangay IPM Project, Social Sciences Division, IRRI, 1995.)

and Pingali, 1993). Promoting sustainable pest management for tropical rice within an IPM framework requires improved research-extension linkages, effective farmer training methods, community action, and an undistorted price structure.

The IPM concept is holistic; it requires farmers to take a systems view of the farm enterprise and understand the interlinkages among the various components of the system. Disseminating such a holistic message requires new sets of skills of the extension system that has traditionally been geared towards promoting component technologies, such as improved seeds, fertilizer, etc. Pest and pesticide problems are intrinsically local in nature. National policy ought to nourish the rural community's capacity to handle pest problems effectively, profitably and equitably. To be successful, the IPM concept needs to be adapted to particular local situations. Such adaptation has to be done with close collaboration of researchers, extension personnel and farmers. Most IPM success stories have been preceded by research done in farmers' fields with the farmer actively participating in all stages of the research process (Escalada and Heong, 1993).

The wider dissemination of such locally validated research results requires a decentralized extension system where the formulation of the extension message is determined at the sub-provincial or municipality level. In such a system, the extension worker would act as a local-level researcher adapting research and technology to suit local agroecological conditions, rather than act merely as a transmission agent of messages barely understood and often inappropriate to the situation at hand. Such a decentralized extension set-up would only come about with a paradigm shift from its current emphasis on top-down information transfer.

Farmer training in sustainable pest management is an essential component of a strategy towards zero or minimum insecticides. The eventual goal is to build farmer capacity in problem-identification and problem-solving based on a thorough understanding of paddy field ecology. The experience of the FAO Farmer Field Schools has shown that trained farmers use significantly lower levels of pesticides than untrained farmers (Kenmore, 1991). Trained farmers are also more likely to experiment with other components of sustainable production systems, such as improved fertilizer management and more efficient water management.

There are as yet several unresolved research and policy issues related to intensive farmer training in IPM. The most important one is of the costs and benefits of farmer training. The costs of training the 120 million farmers in Asia are enormous and these costs need to be clearly justified in terms of farm-level benefits, both of reduced expenditure on insecticides and in terms of reduced social costs of pesticide use. Attention ought to also be focused on the opportunities for reducing the overall costs of training. There are essentially two options, not mutually exclusive, for reducing training costs. The first is to train a core group of farmers within a geo-political unit, such as a municipality, and then rely on farmer-to-farmer training for disseminating the IPM message to a wider group of farmers. There are definite scale economies to the farmer-to-farmer training approach provided the quality of the transmitted message does not deteriorate as it gets passed down the line.

The second approach for a cost-effective transfer of the IPM message is to condense the complex message into several simple rules that are easily implemented by the farmer. An example of such a rule is: 'Do not spray insecticides against leaf-feeding insects for the first 40 days of crop growth'. The rule is based on detailed pest ecology studies which have shown that the predominant insect pests during the first 40 days of crop growth are leaf-feeding insects, and even very high levels of infestations of these insects rarely leads to any yield loss (Heong, 1990). Leaf-feeding insects are very visible and farmers tend to attach great importance to controlling them (Heong *et al.*, 1995). Controlling leaf-feeding insects, however, comes at a cost that substantially exceeds the value of yield savings, if any. Early-season insecticide applications tend to wipe out the leaf-feeding insects as well as the beneficial predator populations that are building up in the paddy. Rice paddies receiving one or two insecticide applications within the first 40 days tend to be susceptible to secondary pest infestations, especially the brown planthopper, which build up unchallenged due to the lack of natural controls. Controlling the growth of secondary pests requires further insecticide applications, thus spiralling the use of insecticides. By not applying insecticides early in the crop season farmers can reduce the need for them later on due to the abundance of predator populations in the paddy fields.

The 'no spray for 40 days' rule is an example of a simple message, distilled from in-depth scientific investigation, that can be easily transmitted to farmers. A set of such mutually consistent rules could go a long way towards improving insect pest management in tropical rice. Simple rules ought not to be seen as a substitute for farmer training but rather a strong complement to a training program. Simple messages can be transmitted rapidly even while investments for intensive training continue to be made towards the ultimate goal of reaching all tropical rice farmers.

Even with a well-established IPM program, pesticides may have to be kept as a technology-of-last-resort. Essentially, the idea of pesticide use in IPM is to spray only when imperative, using the smallest amount possible to do the job. To make IPM more attractive, pesticides ought not to be subsidized. In fact, farmers would bother to learn and apply IPM techniques only when the cost savings associated with insecticide application make it worth their while. The highly acclaimed IPM training in Indonesia was preceded by a comprehensive reform of pesticide regulation and the removal of all subsidies for rice insecticides (Rola and Pingali, 1993). In addition to the removal of subsidies, a sustainable pest management program may warrant the taxation of agrochemicals to account for the social costs associated with indiscriminate pesticide use. Antle and Pingali (1994) have shown that a 100 percent tax on insecticides for rice could actually have a positive productivity effect associated with improved farmer health due to reduced exposure and thereby increased labor productivity.

Conclusions

Meeting the dual goals of maintaining rice productivity growth while at the same time sustaining the agricultural resource base would require substantial improvements in input use efficiencies. Improving the partial factor productivity of chemical inputs would directly contribute to enhanced environmental protection. Knowledge-based management of chemical fertilizers could lead to an increase in output per unit of input applied. In the case of insecticides, the widespread adoption of integrated pest management could lead to a dramatic drop in chemicals applied for rice. The one problem area continues to be herbicides; in this case, competing demand for labor and water make alternatives to chemical management of weeds relatively less profitable. There is an urgent need for identifying sustainable weed management systems for rice production systems. With the exception of herbicides, the long-term prospects are for greater use of knowledge and judgment in chemical input use decisions.

The ultimate decision maker in all input use decisions is the farmer, and good decision making is costly especially in terms of the time required for monitoring crop performance and deciding on appropriate amendments. Profitable adoption of knowledge-intensive input management technologies would depend on the value of fertilizer savings relative to the cost of additional time required for learning and for decision making. Where explicit decision aids are utilized, the cost of acquiring and using the decision aids has also to be considered in the above equation.

The challenge for the research system is in finding cost-effective methods for transferring knowledge-intensive crop management technologies to farmers. In order to effectively use efficiency-enhancing technologies, farmers are required to acquire complex knowledge about modern production systems and use some very intricate decision rules. Knowledge-intensive technologies call for a complete change in the way recommendations are formulated, and in the concept of recommendation domains. The ultimate goal is to have farmers who are capable of deriving the appropriate recommendation for each farm or field using 'if–then' type analyses.

The current practice of providing blanket recommendations on fertilizer applications to farmers over large areas is no longer a valid mode of operation for the extension service. Extension services of the future would have to transfer knowledge and decision-making skills to farmers rather than make input use recommendations. In order to effectively transfer knowledge and to change decision-making paradigms, the scientific community and the extension service would need to have a better understanding of the farmer knowledge base. As mentioned earlier, any new information provided to farmers will be processed through the existing farmer knowledge base. This requires a rather dramatic change in the way information is transferred to farmers, from a top-down approach, with the presumption that farmers have limited knowledge, to a bottom-up approach, where farmer knowledge is given the legitimacy it deserves.

Unlike input use recommendations, knowledge transfer requires that farmers adapt scientific principles to their own particular circumstances and derive farm-specific practices. Therefore, to be effective, any system of knowledge transfer ought to encourage farmer learning through experimentation and adaptation. Farmer training, in this context, is open-ended and participatory, with the emphasis being placed on learning by doing. One certainly should not expect all farmers to become experts and adept at transforming scientific principles to usable recommendations. A few farmers will be successful in acquiring and using the new skills, and they would then become the source of expertise for other farmers within the same geographic and agroecological domains. Such farmer-to-farmer training would be the primary mechanism for transferring input use recommendations within an extrapolation domain. Extrapolation potential may be greatest in the intensive cropping areas that not only have more homogeneous environmental conditions but have large linguistically and culturally homogeneous populations.

The complexity and costs associated with training programs that try to incorporate the above features are very high and often act as impediments to action on the part of the extension service. Complex scientific knowledge, however, need not always result in a complex message. In many instances, complex knowledge can be transmitted in terms of simple rules that farmers can gradually assimilate into their knowledge base. With the use of simple rules, farmer-experimenters are apt to have a high probability of achieving expected results with the new practice. The simple rule approach allows the alteration of a component of the traditional knowledge system without endangering the entire system itself. The change allows for farmer experimentation to develop the best fit in the system for the new component, which may lead to a synergistic reaction throughout the system. Farmers are left to make their own empirical observations and incorporate these observations into their existing knowledge systems as appropriate to their view of the world and natural phenomena.

We ought to recognize, however, that the profitable adoption of knowledge-intensive technologies that enhance input efficiency will only occur when the price is right. Holding input prices low through the use of subsidies or output prices high through price support programs will reduce the likelihood of adopting input-efficient technologies.

12

Dealing with Labor Scarcity: Mechanical Technologies

Over the past three decades Asia has seen an unprecedented level of mechanization of agricultural operations, especially in the high-potential irrigated environments. Intensification of rice production systems created power bottlenecks around the land preparation, harvesting and threshing operations. Alleviating the power bottlenecks with the adoption of mechanical technologies helped enhance agricultural productivity and lowered the unit cost of rice production even in the densely populated countries of Asia. Mechanization of agricultural operations was very selective; power-intensive operations such as land preparation, threshing and milling were readily mechanized, while control-intensive operations such as transplanting and weeding continued to be manually performed.

Economic growth and the commercialization of agricultural systems is leading to further mechanization of rice systems in Asia. The advanced countries of East Asia have a completely mechanized rice production system, while the rapidly-growing countries of Southeast Asia are moving in that direction. In the 1990s, Southeast Asia, especially Thailand, has begun the process of mechanizing the harvest operation using combine-harvesters. Early signs indicate that the commercialization of rice production systems is also leading to a shift from small to larger machines for land preparation, harvesting and threshing. Vertical integration of the post-harvest processing industry is also leading to the replacement of small village-based post-harvest facilities with large-scale processing plants.

The critics of the Green Revolution have argued that the widespread mechanization of rice production systems has had serious equity consequences in terms of the displacement of labor and tenant farmers. Existing evidence indicates, however, that the equity consequences have not been as severe or as widespread as they are presumed to be. The mechanization of power-intensive operations have had minimal equity effects even in the labor-surplus economies of Asia. The switch from manual labor to mechanical or chemical technologies for control-intensive operations, such as weeding, has had adverse equity effects in low-wage countries. However, where markets have been allowed to function with minimal government intervention, control-intensive operations continue to be performed by human labor until wages rise due to increased labor withdrawal from the agricultural sector.

This chapter provides the following:

1. A documentation of the process of agricultural mechanization with rice intensification and agricultural commercialization.
2. An assessment of evidence on the productivity and equity impact of mechanization.
3. An agenda for mechanization research and policy.

Intensification, Economic Growth and the Adoption of Mechanical Technologies

The need for increased energy requirements with the intensification of agricultural systems has been extensively documented in the literature (Boserup, 1965; Pingali and Binswanger, 1987). The movement from land-extensive cultivation systems, such as shifting cultivation systems, to the land-intensive permanent cultivation systems leads to an increase in the tasks performed and the intensity with which these tasks are performed. In the early stages of the intensification process, energy requirements for performing the additional tasks are met through increased human labor use. Over time, however, there has been a gradual substitution of human labor for mechanical power. In the absence of labor-saving technical change, the movement to permanent agricultural systems would not have been profitable. The progressive substitution of mechanical energy for human energy has helped sustain labor productivity in Asian rice systems even prior to the Green Revolution.

Chapter 3 discusses the classification of agricultural operations in terms of their requirement for power versus the requirement for judgment and control. Operations such as land preparation, pumping, milling and transport are power intensive, while weeding, crop care and harvesting operations require less power and more judgment for effective performance. The mechanization of power-intensive operations has taken place rapidly even in countries with high population densities and low wages, while the mechanization of control-intensive operations has occurred only under high-wage conditions (Binswanger, 1978; Herdt, 1987). In the Asian rice systems the transition from animal draft power to tractor use for land preparation started in the late 1940s and the early 1950s. The use of mechanical mills, diesel pumps and large threshers also started around that time (Herdt, 1987). The advent of the Green Revolution accelerated the process of mechanization across Asia, although the pace of transformation varied by regions, being fastest in the rapidly-growing economies of East Asia and slowest in South Asia.

Investments in irrigation infrastructure and the spread of modern variety use resulted in a shift in cropping patterns in Asia, from a system of a single rice crop per year to one in which two or three crops of rice are grown on the same field each year. Higher cropping intensities resulted in an outward shift in the demand for labor as well as an increase in the demand for timely completion of operations. The shift from traditional varieties to modern varieties had an independent positive effect on the demand for labor. The absolute quantity of labor required for transplanting, weeding, harvesting and threshing increased significantly with modern rice cultivation systems, because of more systematic and intricate planting methods, and higher quantity of output. Chapter 3 presents data on the increased labor requirements for modern variety cultivation as compared to traditional rice varieties. The resulting rise in real wages for agricultural operations, even in densely populated regions, resulted in a substitution out of human labor and animal draft power for power-intensive operations.

With the spread of intensive rice production systems and the higher quantities of rice output processed and marketed, power-intensive operations were rapidly mechanized. The operations commonly mechanized were land preparation, water pumping, threshing and milling. Japan and South Korea led the rest of Asia in the speed and extent of mechanization and set a pattern that other countries followed. In Japan by 1960, the mechanization of pumping and threshing had already been completed and the use of power tillers had just started to take off. The number of power tillers on Japanese farms grew from 750,000 units in 1960 to 2.5 million units in 1965 (Kisu, 1983). By 1965 there was one power tiller for every 2 hectares of crop land in Japan (Herdt, 1983); by 1989, it had more than one power tiller per hectare of rice paddy land (Mizuno, 1991). The Korean experience was similar: the power tiller numbers rose from a little over 1000 in 1965 to around 290,000 by 1980 (Cho, 1983). By 1970, South Korea had approximately one power tiller for every 10 hectares of paddy land (Herdt, 1983), and by 1990, one power tiller for every 2 hectares of paddy land (Chang,

1991). The process of decollectivization in China has led to rapid mechanization of farm operations. By 1992 it had around 220 power tillers per 1000 hectares of ricelands (China Statistical Office, 1993).

In Southeast Asia, mechanization of rice paddy lands took off in the early 1980s. Power tiller use rose from 26 and 14 per thousand hectares of paddy land in Thailand and the Philippines, respectively, in 1980 (Herdt, 1983), to approximately 56 and 20 per thousand hectares by 1990 (APO, 1991). Indonesia, Myanmar and Vietnam have been slower in the switch to power tillers for paddy land preparation with less than one power tiller per 1000 hectares of ricelands. For Thailand and the Philippines, national average figures do not indicate substantial variability in power tiller adoption by rice environments. The irrigated rice-bowl provinces of the two countries tend to be highly mechanized, with over one power tiller per 10 hectares of paddy land, while the less favorable rice environments continue to rely on animal power.

South Asia is very diverse in terms of the mechanization of the land preparation operation, both across countries and within countries. Although a superficial look at national average figures would seem to indicate that these countries continue to rely on animal draft power, the intensively cultivated rice-bowl provinces tend to be on the same mechanization pathway as similar rice-growing regions in Southeast Asia.

Sri Lanka is the most advanced in the mechanization of land preparation, while Nepal is the least. Unlike the countries of Southeast Asia, the mechanization of land preparation in South Asia has emphasized four-wheel tractors rather than power tillers. The larger tractors are more conducive to the rice–upland crop rotations that are common in South Asia, more suitable for operating rental markets over a larger geographic area, and more easily amenable for use as transport vehicles.

Mechanical threshers closely followed power tillers throughout the intensively cultivated rice bowls of Southeast Asia. With the advent of double and/or triple cropping of rice and the increased quantities produced, the period around harvesting and threshing one crop and land preparation for the next crop became extremely time-bound and hence resulted in a major labor peak. Small axial flow threshers (around 10 hp) developed at IRRI were widely disseminated in Southeast Asia around 1975 (Khan, 1986). The small threshers quickly replaced alternative means of threshing rice grain, such as human labor, animal treading and tractor treading. The advantage of the small threshers was their mobility and amenability to contract operations. During harvest time, one can see the small threshers moving from farm to farm providing contract threshing services. In Central Luzon, Philippines, small threshers were introduced in 1975, and by

Table 12.1. Changes in cropping intensity and mechanical technology (percent) used by rice farmers in Central Luzon, Philippines, 1966–1990.

	Central Luzon						
	1966	1970	1974	1979	1982	1986	1990
Area with two crops per year	17	15	20	54	54	50	44
Parcels using power tillers	0	5	19	49	57	67	85
Parcels using portable threshers	0	0	5	21	75	97	100

Source of basic data: Social Sciences Division, IRRI.

1982, 73 percent of the rice farmers were using them (Herdt, 1987); by 1990, all other forms of threshing disappeared from the area (Table 12.1). Similar rapid adoption of threshers has been observed in the other intensively cultivated rice-bowl provinces of Southeast Asia, such as the Central Plains of Thailand, West Java, Indonesia and the Mekong Delta, Vietnam. By the 1990s, most of the irrigated rice in Southeast Asia was threshed by mechanical threshers and thresher use is expanding to the rainfed environments. South Asia still lags behind in thresher use; tractor or animal treading of grain is still a common practice, although high-potential areas such as the Indian Punjab have shifted to thresher use.

Economic growth, agricultural commercialization and the resulting factor scarcities are leading to long-term changes in the organization of agricultural production. The most prominent changes are the consolidation of landholdings to create larger farms and the rapid withdrawal of labor from the agricultural sector (Chapter 9). Such changes are most evident in the developed countries of East Asia and are beginning to emerge in the rapidly growing countries of Southeast Asia, particularly Malaysia and Thailand. The long-term changes in the organization of farm production have a significant impact on the choice of technologies for power-intensive operations. The use of small machines for land preparation and threshing declines and larger four-wheel tractors and combine-harvesters are rapidly adopted. Larger machines become economical as farm sizes increase and as wage rates rise.

Figure 12.1 shows the evolution of agricultural machinery use in Japan from 1927 to 1987. The switch from small to larger machines with economic growth is evident from the Japanese experience. Up until 1963 very few four-wheel tractors were used in paddy land preparation in Japan; by 1987, there were over 2 million of them, approximately 500 per 1000 hectares of arable land. During the same time period, power tiller use declined steadily after reaching a peak of

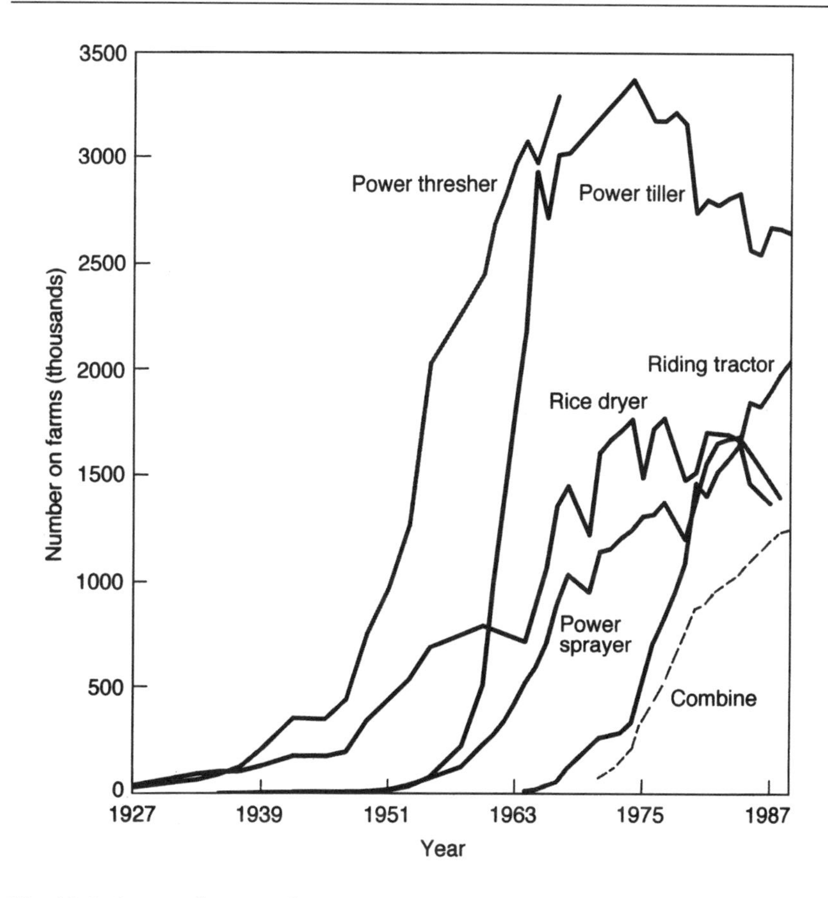

Fig. 12.1. Japan's farm machinery, 1927–1989. (Source of basic data: APO, 1991.)

around 3.5 million units in 1970. Since the early 1970s, Japan has seen a switch from the use of threshers to combine-harvesters. By 1990 there were around 300 combines per 1000 hectares of arable land in Japan (Mizuno, 1991). Korea has been going through a similar transition from small to larger machines for land preparation and harvest operations since 1980 (Chang, 1991). Among the Southeast Asian countries, Malaysia had concentrated on larger machines from the start because of larger farm sizes and low agricultural labor supply (Mohamed, 1991), while Thailand began to make the transition in its irrigated ricelands in 1990. With continued growth in the other Southeast Asian countries, similar transitions ought to be expected starting in the next decade. South Asia will continue to be slow in the transition to large machines, although once again, high potential areas such as the Indian Punjab can be expected to move more rapidly than the rest of the region.

The above discussion on the transition in agricultural machinery for power-intensive operations is not meant to imply that there is no role for power tillers, small threshers and other small machines in rice production systems with economic growth. In areas where plot sizes continue to remain small due to biophysical constraints or where farm consolidation is prevented by legislation, small machines will continue to be used for tillage, harvesting and threshing. In the rainfed environments, plot sizes are generally small in order to maximize water capture and retention through the growing season. Moreover rainfed paddies tend not be uniformly shaped, since they tend to mimic the topography of the location, hence consolidation of rainfed plots into larger fields is often infeasible. Small machines could be expected to play an important role in rainfed agriculture as labor withdrawal from the agricultural sector leads to rising wages. In the irrigated environments, increases in plot sizes although physically possible may be constrained by land tenure legislation designed decades ago that may still persist. Legal restrictions are often circumvented by consolidating operational holdings while retaining ownership on small farms to the ceiling limit. In countries where the consolidation of holdings is not feasible due to legal restrictions, the use of small machines would persist.

Mechanization of control-intensive operations

Several attempts have been made to mechanize control-intensive operations in rice production systems since the Green Revolution (IRRI, 1986). Mechanical transplanters, weeders and fertilizer applicators were some of the technologies released by the rice research systems in Asia. With the exception of mechanical transplanters (machine-driven) in East Asia, most attempts at mechanizing crop establishment and crop care activities have failed. The failures can be attributed to the higher cost of using the mechanical technology relative to the alternatives available, as discussed below.

Mechanical transplanters have been widely adopted in high-wage countries where the opportunity for direct seeding is limited, such as Japan, South Korea and Taiwan. The use of transplanters in Japanese rice production took off in 1970, and by 1989 there were 2.2 million transplanters in use nationwide (Mizuno, 1991), approximately one transplanter for every 2 hectares of arable land. For South Korea, by 1989, approximately 66 percent of the rice planted area was mechanically transplanted (Chang, 1991). Taiwan has similarly seen a rapid growth in transplanter use since the late 1970s (Peng, 1983). As wages rise, manual transplanting becomes very expensive; the alternative to mechanical transplanting, as a means of saving labor in crop establishment, is to broadcast pre-germinated rice seedlings into the paddy. Direct seeding is, however, not generally possible in the temperate environments because the cold spring temperatures are not conducive to seedling establishment and growth in the paddy. Rice seedlings are grown under controlled temperature conditions and transplanted into the paddy when they are old enough to resist cold stress and, being later in the season by a few weeks, the temperatures are warmer. Mechanical transplanting has not taken off in the humid tropics of Southeast Asia even under high-wage conditions because of the possibility of direct seeding. Chapter 9 has shown the switch from transplanting to direct seeding as wages rise in Southeast Asia. Manual transplanting of rice, using mainly female labor, continues to persist in South Asia. Where hydrological and other physical constraints prevent the adoption of direct seeding, there is potential for mechanical transplanter adoption.

Mechanical weeders were introduced into rice systems in the mid-1970s, but their record of adoption was very poor primarily because of the availability of herbicides that were a substantially cheaper source of weed control. Discussion on the productivity and environmental implications of herbicide use is provided in Chapter 11. Attempts to enhance the efficiency of chemical fertilizer use through mechanical technologies for fertilizer deep placement have also generally not been successful. The long-term decline in global fertilizer prices has resulted in low farmer interest in improving fertilizer use efficiency and hence investment in equipment for efficiency enhancement. The poor performance of fertilizer placement technology has been observed even in the high-income countries of East Asia. Given the proliferation of non-mechanical options for improved crop management, the future for the adoption of machines for control-intensive operations seems rather bleak.

Demand for quality and mechanical technologies

As incomes grow, consumers prefer to purchase higher-quality rice, the price differential for quality increases, and the market responds with increased supplies of high-quality rice (Unnevehr *et al.*, 1992). In general, urban/high-income consumers tend to pay higher premiums for quality and for a larger number of quality characteristics than rural/low-income consumers. Several studies in Asia have documented the quality preferences of rice consumers and the relationship between income, urbanization and the demand for rice quality (see for example, Damardjati and Oka, 1992 for Indonesia; Abansi *et al.*, 1992 and Maranan *et al.*, 1992 for the Philippines; Choudhury *et al.*, 1992 for Bangladesh; and Wong *et al.*, 1992 for Malaysia).

Improved quality characteristics can be in terms of taste, cooking quality and physical appearance of the rice grain. While improved taste and cooking quality can be achieved through varietal selection and breeding programs, the physical appearance of the grain can be improved by engineers through better post-harvest processing technologies. Physical quality is determined by the percentage of head rice (unbroken grains), whiteness and the absence of foreign materials. Threshing, drying and milling operations have direct effects on the physical quality of the rice grain. As incomes rise, the choice of harvest and post-harvest machines would be determined by the desire for quality enhancement in addition to labor savings.

As economies grow, the switch from manual harvesting and mechanical threshing to the use of combine-harvesters is motivated by quality enhancement as much as by labor savings. Grain harvested using combines is immediately loaded on trucks and transported to millers, thus minimizing the time spent on the field and in farm-level handling, and reducing grain spoilage and the chance of foreign material content in the grain. The switch to combines also helps reduce grain cracking, thereby increasing the proportion of head rice in marketed grain. The drying operation that immediately follows harvesting was traditionally done by spreading the unmilled grain under the sun for a day, often by the side of a road. Sun-drying does not effect a uniform moisture content for the grain, and it causes an increase in the percent of foreign matter and broken grains in the final product. Faced with increasingly quality-conscious consumers, millers in the advanced countries of Asia use large mechanical dryers that can turn around several tons of grain of uniform moisture content. Such dryers are becoming common among the commercial milling companies of Southeast Asia that specialize in supplying rice to the urban market (Tetter, 1986). Attempts to introduce small dryers in rural areas have generally failed because of small farm-level price differences between wet and dry paddy (Bonifacio *et al.*, 1990). When the harvest operation is mechanized, commercial rice dealers would find it more profitable to buy wet paddy in bulk and go through the entire post-harvest processing operations themselves. For subsistence consumption purposes, small-scale mechanized dryers cannot compete against sun-drying, except in areas with a bi-modal rainfall distribution where it rains around harvest period.

The final post-harvest operation is milling, the de-husking of rice to produce white grains ready for cooking. Mills have preceded the Green Revolution by several decades. These are often small mills designed to handle small quantities of paddy per hour (500 kg) and are used mainly for providing subsistence requirements. IRRI has recently introduced an even smaller mill known as the micro-mill with milling capacity of 50 kg paddy hr^{-1} (Paris *et al.*, 1995). In subsistence societies, small village mills and micro-mills have been successfully adopted with minimal equity consequences (see next section). As economies grow and agriculture becomes commercialized and the majority of the rice is produced for the market, small mills give way to large commercial mills. Substantial economies of scale can be expected from the vertical integration of all post-harvest processing activities: transport of paddy from the farm, drying, milling, storage and the transport to urban centers being all done by the same commercial enterprise. A vertically integrated post-harvest processing system is well established in the advanced countries of East Asia, Malaysia and in Thailand, and it is beginning to emerge in the other Southeast Asian countries as incomes grow. Even in South Asia, Basmati rice producers in India and Pakistan face an integrated post-harvest processing system, especially for the portion that is exported. The rest of South Asia can be expected to be slower in the transition to a completely commercialized large-scale post-harvest processing system.

Productivity and Equity Impact of Mechanization

Productivity impact of the switch to mechanical technologies for agricultural operations is measured in terms of changes in yields, labor savings, area expansion (in terms of increases in cropping intensities) and quality enhancement of the marketed output. Equity impact, on the other hand, is measured in terms of labor displacement and income distributional effects, particularly for the landless labor households and for women. The productivity and equity impact of mechanization varies by power intensity of the operation that is being mechanized, by the land and labor endowments of the region, and by level of economic development of the country. Since the mechanization of power-intensive operations has been well under way throughout Asia, several studies have documented the impact, and evidences from these studies are presented below. Few studies are available in the case of control-intensive operations and quality enhancement technologies, since the introduction and adoption of these technologies has been sparse. The power-intensive operations considered here are land preparation, threshing, small harvesters and small mechanical mills. The evidence presented below indicates that, for power-intensive operations, the productivity benefits of mechanization are mainly in terms of labor savings, and the equity implications are minimal even in labor-abundant, low-wage economies.

Land preparation

The movement from using animal-drawn plows to tractors or power tillers is considered efficient if yield per hectare increases and/or if labor hours required for land preparation per hectare are reduced. Yield increases are possible only when mechanization improves tillage quality. Existing evidence, however, indicates that in general no significant yield difference exists between animal draft and tractor tillage. Herdt (1983) found that yield differences between animal draft and tractor farms were negligible after differences in fertilizer use were considered (see Table 12.2). This is consistent with results from South Asia, where more than 50 percent of farms using tractors had significantly higher yields. However, almost all those cases were associated with an increase in fertilizer use (Binswanger, 1978). If we find no yield differences between animal draft and tractor farms, we must conclude that the transition to tractor-drawn plows is rarely motivated by improvement in tillage quality. Area expansion and/or labor saving must be the driving forces for such a transition.

Pingali *et al.* (1987) reviewed 24 studies on labor use by operation of animal draft and tractor farms in Asia, and found changes in labor use by operation, in total labor use, and shifts in the levels of labor use between operations. Twenty-two of the 24 studies reviewed reported lower total labor use per hectare of crop production for tractor farms compared to animal draft farms. Twelve studies reported reductions in labor use of 50 percent or more.

The greatest reduction in labor use was for land preparation, with all studies reporting reduction in labor input that exceeded 75 percent. Let us consider cases where there was a 50 or more percent reduction in labor use for land preparation, and trace through the effect on other operations. Consider weeding first. Of the 14 cases with 50 percent or more reduction in land preparation labor, only two reported a reduction in weeding labor greater than 25 percent. Ten reported reduction in weeding labor smaller than 25 percent, and two reported increases in weeding labor relative to animal traction farms. The situation is very similar in harvesting. Of the 14 cases with 50 percent or more reduction in land preparation labor, only one resulted in an equal reduction in harvesting labor, nine reported labor reduction less than 25 percent, and one found increased labor requirements.

The above results indicate that labor savings resulting from the transition to tractors are confined mainly to land preparation where one observes a substantial reduction in labor peak. However, where aggregate area expansion effects exist (including an increase in cropping patterns), one could expect a rightward shift in the demand curve

Table 12.2. Summary of studies comparing rice yields on farms that used animal or hand for land preparation with farms that used machinery.

Author	Area	Comparison	Reported yield (t ha^{-1})	Fertilizer (urea) (kg ha^{-1})	Adjusted yield (t ha^{-1})
Pudasaini	Nepal (without pumps)	Bullock vs tractors	1.7 2.1	16 164	1.7 1.4
Pudasaini	Nepal (with pumps)	Bullock vs tractors	2.1 2.3	214 264	2.1 2.1
Sinaga	West Java, Indonesia (1979–1980 wet season)	Animal vs tractors	4.9 4.9	323 323	4.9 4.9
Sinaga	West Java, Indonesia (3 seasons, 1978–1980)	Manual vs tractors	3.8 3.9	285 308	3.8 3.8
Tan and Wicks	Nueva Ecija, Philippines (1979 wet season)	Water buffalo vs tractors	2.6 4.0	89 129	2.6 3.8
Anuwat	Central Thailand (irrigated–transplanted)	Bullock vs tractors	2.6 2.8	32 48	2.6 2.6
Anuwat	Central Thailand (rainfed broad-cast)	Bullock vs tractors	0.2 0.2	3 2	0.2 0.2
Alam	Bangladesh	Bullock vs power tiller	1.5	n.a.	1.5
Deomampo and Torres	Central Luzon, Philippines	Before vs after tractor and tillers	2.2 2.6	57 79	2.2 2.1
Antiporta and Deomampo	Philippine provinces	Animals vs tractors tillers	2.6 2.8	86 117	2.6 2.5

Source: Herdt (1983).
n.a., Not available.

for weeding and harvesting labor, despite the lower per hectare requirements.

What about area expansion possibilities? Where the land frontier is not exhausted and the opportunities for expanding domestic or international trade exist, the switch to tractors can lead to an expansion in rice area. In land-scarce economies, however, the introduction of tractors has rarely resulted in an expansion of ricelands (Binswanger, 1978). Rice cropping intensities have increased significantly from the joint mechanization of land preparation and threshing, especially in humid tropical Southeast Asia (Juarez and Pathnopas, 1983). With the introduction of modern rice varieties and irrigation infrastructure in Southeast Asia, two to three crops of rice can be grown on the same plot of land per year. With the first crop usually harvested during the rainy months, the danger of grain spoilage was very high if threshing was not done soon after harvest. Also, fields should be cleared of the previous crop before irrigation water is released. Without the use of power tillers and threshers, high intensity cropping could not have been sustained in much of Southeast Asia.

The following generalization can be made from the preceding discussion: efficiency gains in the transition to tractors for land preparation occur mainly due to per hectare labor savings rather than yield improvements. The exception to the above generalization is Southeast Asia, where high rice-cropping intensities are sustained through the mechanization of land preparation and threshing.

What are the equity consequences of a shift to tractors from animal-drawn plows for land preparation? The answer depends on the following subquestions: (i) Did tractor owners expand the size of their operation by displacing tenant farmers? (ii) Was there a power bottleneck around the land preparation before arrival of tractors? (iii) Has total labor use on tractor owners' farms increased or decreased since the advent of tractors?; and (iv) Has there been a net transfer of income from tractor owners to agricultural labor?

Aggregate area expansion

Most comparisons of farms with and without tractors conclude that tractor-using farms are larger than non-tractor-using farms. These findings suggest that the transition to tractors is associated with an expansion in the private area cultivated, and therefore with an increase in production on tractor farms. Whether there is also an aggregate area expansion and production effect is generally not known from these studies.

Aggregate area expansion occurs only when private farms extend into fallow or reclaimed land or when cropping intensity is increased on a given plot of land. Where private area expansion occurs at the cost of other farmers (mainly through the displacement of tenant farmers), there are no aggregate area expansion benefits. One observes, rather, adverse equity consequences in this case. Any inference on the aggregate expansion in agricultural output through an increase in land area cultivated due to tractors would require additional information on where the additional area comes from.

Lockwood *et al.* (1983) found in Faisalabad, Pakistan, that 70 percent of the area expansion on tractor farms came from tenant displacement, the remaining 30 percent came from increasing area rented in, leading to further tenant displacement. Eighty-eight of the original 105 tenants lost their land when the landowners bought their first tractor. The average tenant farm size declined from 4.4 to 3.4 hectares. This study confirms the earlier findings in the same area of McInerney and Donaldson (1975). Jabbar *et al.* (1983) reported that 81 percent of power tiller owners in Mymensingh, Bangladesh, increased their cultivated area by expanding into previously rented-out land.

Where uncultivated or fallow land was not available, the promotion of tractors for land preparation has led to high levels of tenant displacement. The exception as discussed above is the growth in rice cropping intensity due to mechanization, which did not lead to tenant displacement and resulted in positive employment benefits (David and Otsuka, 1994).

Power bottleneck

Mechanization of the land preparation can lead to significant productivity gains with minimal equity effects in areas with a power bottleneck during land preparation. Such a bottleneck could occur either due to a shortage of labor or of draft animals. Labor shortages are most common in sparsely populated areas and in areas with good non-farm employment opportunities. Peak season labor shortages have also occurred, with an increase in rice cropping intensities, around the harvest operation for one crop and land preparation for the next (see Chapter 3 for details). Draft power shortages could occur in sparsely or densely populated areas. In the latter, it occurs because draft animal maintenance is relatively expensive due to the high cost of fodder. Significant productivity benefits accrue to the alleviation of labor and power bottlenecks, through mechanization.

The promotion of tractors (power tillers) in areas with a power bottleneck during the land preparation could lead to general improvement in income levels.

Total labor use

We concluded earlier in this section that tractor use results in reduced labor use per hectare, but it also leads to an increase in area cultivated by tractor farms. Total labor use on tractor farms could increase despite per hectare reduction in labor use as indicated by the following generalization: where uncultivated or fallow land is available, or where cropping intensities increase, the use of tractors (or power tillers) for land preparation could lead to an increase in labor employment.

Income transfer

Income distributional effects of mechanized land preparation depend on the nature of final demand and the extent of output growth (Table 12.3). Let us first consider areas with an inelastic demand for food output produced. These are typically small, closed economies, where neither exports nor imports of food (rice) are allowed. In such economies the total demand for food is determined only by domestic demand, and any increase in output will lead to a decline in price.

Where final demand is inelastic, aggregate area expansion caused by mechanized land preparation results in a transfer of income from tractor-owning households to landless-labor households. This happens for two reasons: (i) expansion into previously uncultivated areas (or an increase in cropping intensities) results in increased employ-

Table 12.3. Income distribution effects of mechanical technology[a].

	Income distribution effects		
	Tractor plowing	Mechanical threshing	Milling (small-scale only)
A. Inelastic demand for food			
1. Output expansion	+	+	(+)
2. Labor savings only/ no output expansion	–	–	+
3. Subsidies for machines and equipment	–	–	(+)
4. Guaranteed support price	(+)	–	0
B. Elastic demand for food			
1. Output expansion	+	–	+
2. Labor savings only	–	–	+
3. Subsidies for machines and equipment	?	–	(+)

[a] The effect considered here is the transfer of income from the landowners to the landless labor resulting from the mechanization of the above operations. The positive sign implies expected redistribution of income from landowners to the landless labor, a negative sign implies the reverse, while 0 implies no change. Parentheses imply very small change.

ment opportunities; and (ii) real price of food output declines as a result of output expansion, and since the landless labor are net purchasers of food the effect of this price decline is analogous to increasing their income[1].

The direction of income transfer is reversed where the opportunities for aggregate area expansion are limited. Here, tractor farms expand into land previously cultivated by tenants and the result is net labor displacement rather than an increase in aggregate output.

Where final demand is elastic (open economies, or economies with large domestic demand), aggregate area expansion does not lead to a decline in prices, but a limited amount of income transfer to labor is observed due to expanded employment opportunities. If opportunities for aggregate area expansion are not available, the effects are the same as discussed in the preceding paragraph.

Mechanization of post-harvest operations

Next to land preparation, harvesting, threshing and milling are the most arduous operations in rice production. Consequently, where mechanical technologies for these operations exist and can be profitably used, they tend to be adopted fairly rapidly. Small mechanical mills, for instance, spread spontaneously across the world without any government program promoting them. What have been the productivity and equity consequences of mechanizing post-harvest operations?

Threshers

The late 1970s and 1980s have seen the rapid spread of mechanical threshers in parts of Southeast Asia and parts of India (Walker and Kshirsagar, 1981; Duff, 1986). In any given area, the private profitability (efficiency) of using a mechanical thresher over handbeating and animal or tractor treading is determined by yield benefits of mechanized threshings, marketable surplus generated on the farm, and by labor wages and availability during the harvesting–threshing period.

Proponents of the thresher technology usually argue that the mechanical thresher presents a significant increase in realized yields due to: (i) a more complete threshing of grain than manual or treading techniques; (ii) a reduction in losses caused by repeated handling of both threshed and unthreshed materials; and (iii) an increase in cropping intensity resulting from a lower turnaround time with mechanical thresher use. On-farm experiments comparing manual and mechanical threshing have shown that mechanical threshers reduce grain loss by 0.7–6 percent of total yield (Toquero and Duff,

1985). However, there have been no studies of actual farmer thresher use to see if such savings are observed in practice. Cropping intensity effect has been discussed above, and to the extent that threshers contribute to sustaining high cropping intensities, thereby allowing more timely completion of operations, they have a positive effect on yields.

The existing evidence indicates that the primary motivation for mechanical thresher use is the labor-saving benefits. Table 12.4 compares labor use and productivity by threshing technique in Thailand and the Philippines. The adoption of a portable axial-flow thresher in the Philippines resulted in a decline in threshing labor requirements from 7.69 labor days per ton of grain for foot treading to 0.81 labor day per ton. In Thailand, the adoption of a large axial-flow thresher resulted in labor savings of 3.50 labor days per ton of grain relative to the traditional threshing by buffalo treading. Consequently, large gains in labor productivity were observed in both cases.

Whether the above labor savings actually lead to efficiency gains or not depends on the real wage of threshing labor, the timely availability of labor and other sources of power for threshing, and the nature of final demand. The following generalization is possible: mechanical threshing is economically efficient where the farmer faces an elastic demand curve for his/her output and a power bottleneck exists during the harvesting–threshing period either due to high land–labor ratios or due to opportunities for non-farm rural and urban

Table 12.4. Labor requirements and labor productivity for rice threshing in Thailand and the Philippines.

Method	Time (labor t^{-1})	Amount (kg labor d^{-1})
Thailand		
Buffalo treading	4.85	206
Tractor treading	2.40	415
Large axial-flow thresher	1.33	752
Philippines		
Foot treading	7.69	130
Hand beating	5.49	182
Portable axial-flow thresher	0.81	1230
Large axial-flow thresher	0.94	1070

Source: Duff (1986).

employment. It is important to remember that a combination of elastic demand and a threshing power bottleneck is required for mechanical threshers to be profitable. Equity considerations in thresher adoption center around the source of the labor that is displaced. Surveys in Nueva Ecija, Philippines, showed that post-harvest labor on mechanized farms was 25 percent lower than on farms in which rice was manually threshed. Disaggregated into family and hired labor, the data reveal that much of the labor savings came at the cost of landless households whose labor services declined by 31 percent (Duff, 1986). A similar decline in the use of hired labor was observed in Laguna and Iloilo, Philippines (Juarez and Pathnopas, 1983).

Ahammed and Herdt (1981) used a general equilibrium model to estimate the nationwide employment implications of thresher use (among other production methods). They found that a 1 percent increase in rice production would generate an employment effect of 16,000 labor years if manual threshing was used. The employment effect would be 22 percent lower (12,400 labor years) if portable threshers were used. Taking all sectors of the economy into account, substituting mechanized threshing for manual threshing would reduce the employment-generating potential of increased rice production by 7 percent. This overall adjustment depends on urban employment opportunities, the ability of labor to migrate between regions and between sectors, and industrial policies.

A similar analysis in Thailand found that all the labor savings came from family labor and not hired labor (Sukharomana, 1983). This is because Thailand has a very favorable land-to-labor ratio and, therefore, a very small landless labor class. Traditionally, the threshing operation is done by family labor with buffalo or tractor treading. The use of large axial-flow threshers on a contract basis has resulted in the release of family labor for other activities.

For both family and hired labor, the thresher had a differential impact on men, women and children. Mechanical threshers considerably reduced the arduousness of post-production tasks. The lighter nature of the work made it possible for women and children to substitute for men in the threshing operation (Ebron, 1984). Where off-farm employment opportunities exist, the thresher use can result in increased incomes of labor households since male workers may be released for other income generation while women and children provide threshing labor.

Milling

The use of small mechanical mills for dehusking paddy is perhaps the least controversial of all forms of mechanization. Handpounding of grains, the traditional alternative to mechanical milling, is an extremely labor-intensive task usually performed by women. The efficiency gain in the shift to mechanical mills comes from the resulting labor savings. Because much of the rural milling is done piecemeal for home consumption only and because traditional handpounding is done by female household members, switching to small mechanical mills results in substantial gains in leisure time for women. Consequently, one observes the widespread use of small rural mills for dehusking rice on a contractual basis.

With one significant exception, small mechanical mills have increased efficiency of food production without adverse equity effects. The exception is Bangladesh where significant displacement of hired labor has occurred as a consequence of mechanical mills (IBRD, 1987). Traditional rice milling in Bangladesh is done by a foot-operated mortar and pestle known as a 'dheki', which is usually owned by large landowners and operated on a contractual basis by women from landless and low-income households. The ratio of milling costs with the 'dheki' to those with the mill is about 12:1, not counting the transport costs of bringing the rice to the mill. Thus, the mill owners can charge much lower rates for milling than 'dheki' operators.

Currently, more than 40 percent of all rice milling in Bangladesh is mechanized. The rapid spread of mechanized milling has benefited large landowner households, subsistence-farming households, and urban consumers, including the urban poor, whose rice prices have been reduced. Female members of large surplus farms have more leisure time, because they no longer have to supervise hired 'dheki' operators. Female members of subsistence farms who previously operated the 'dheki' for their home consumption are relieved from time-consuming and physically-demanding labor. Nonetheless, women from landless families who work for wages have suffered as a result of the mills because of absent alternative remunerative employment.

Rice processing is one of the few income-earning opportunities available for rural women in Bangladesh, because it can be accomplished within the confines of the 'purdah' system (Cain *et al.*, 1979). With an estimated 10,000 rice mills in rural areas, between 1.4 million and 2 million women have lost a traditional source of part-time employment. Each year, with approximately 700 new mills coming into use, an estimated 100,000 to 140,000 women are being displaced (IBRD, 1987). The policy challenge in Bangladesh is to find alternative employment opportunities for sustaining the incomes of the women of landless and low-income households, without slowing down the growth in small mechanical mills.

Small harvesters

While combine-harvesters have been in use in East Asia and in Malaysia for several decades now and are emerging in Thailand, their spread in the rest of Asia has been limited due to low harvest wages, and small plot sizes. Harvest wages, though relatively low in comparison to the developed countries of East Asia, are rising and the prospects are that over the next decade mechanization of the harvest operation would be demanded in Southeast Asia.

Small harvesters are seen as an intermediate step in the transition to combine-harvesters in much of Southeast Asia. In the absence of land consolidation and the re-design of paddy lands to form large contiguous fields, the prospects for large-scale adoption of the combine-harvesters is limited. The recently introduced stripper–harvester machine is seen as a viable technology to overcome the problem of small farm sizes.

The stripper–harvester consists of a simple stripper rotor that spins in the crop as the machine moves forward and combs or strips the grain from the plant (Douthwaite *et al.*, 1993). The stripped material is threshed and cleaned using existing threshing machines. The stripper weighs 240 kilograms, light enough to be moved across bunds and can easily operate in small fields. It can harvest 0.80 hectares of rice per day and needs around 7 person days per hectare for harvesting and threshing. The switch to the stripper from manual harvesting leads to labor savings of approximately 12 person days per hectare for the harvesting and threshing operation. The machine costs around US$ 2500 and is ideal for contract operations. Moya *et al.* (1996) report that the stripper, if used for harvesting 16 hectares per season, will reduce harvesting costs by 40 percent compared to manual harvesting. If the machine is used for harvesting 24 hectares per season, the cost savings would be over 50 percent.

What are the prospects for stripper–harvester adoption in Southeast Asia? Considering the irrigated environments first, the immediate prospects for widespread adoption are limited. Harvesting wages in Southeast Asia, with the exception of Thailand and Malaysia, are not high enough to make the switch to the stripper–harvester profitable, although as the economies grow one could expect increased profitability of mechanized harvesting. Over the next decade, Indonesia, Philippines and Vietnam may gradually move towards using the stripper in the irrigated environments, with the rate of adoption depending on the rate of withdrawal of labor from the agricultural sector. The stripper, however, ought to be seen only as an intermediate solution to the labor scarcity problem in the irrigated environments. Rapid economic growth and the commercialization of the agricultural sector could lead to the consolidation of small farms, making larger combines more profitable than the smaller stripper–harvester in the irrigated environments.

The long-term prospects for small harvesters are better for the rainfed environments. In the rainfed environments, even with economic growth and withdrawal of labor out of the agricultural sector, plot size will continue to remain small. The size and shape of rainfed rice paddy is determined by topography and hydrological conditions, the objective being to optimize rainwater harvest and retention in the paddy. Therefore, in the rainfed environments, consolidation of farms will not necessarily lead to larger paddy fields. Small harvesting machines would be profitable in the rainfed environments as economies grow.

Research and Policy Implications

The movement of labor out of the agricultural sector and the consequent rise in agricultural wages is an inevitable phenomenon of economic development. The speed with which the above transformation proceeds will vary by region, with China and Southeast Asia being faster than South Asia, although the direction of change will be similar across Asia. The search for labor-saving innovations will be a very important phenomenon in Asian agriculture over the next two decades. In promoting agricultural mechanization, it is important to understand the evolutionary patterns in mechanization and the conditions under which different operations are shifted from human and animal power to mechanical power.

In the developed countries of East Asia, the process of mechanization has been completed. As farm consolidation occurs in these countries, one ought to expect small machines to be replaced by larger machines, especially for land preparation, harvesting and threshing. In Southeast Asia, the irrigated environments have almost completed the process of mechanizing power-intensive operations; harvesting is the one remaining operation that could be mechanized in the short to medium term. In areas where direct seeding is not possible, one could also see increased interest in mechanical transplanting machines. As incomes grow in Southeast Asia and as consumer preference for quality increases, large-scale post-harvest facilities can be expected to replace small-scale (often village-based) systems. The rainfed environments in Southeast Asia have been slower than the irrigated environments in the adoption of mechanical technologies, but they ought to be expected to catch up as labor withdrawal from the agricultural sector proceeds.

South Asia has lagged behind the other regions in the adoption of mechanical technologies, although the extent of adoption has been higher in the high-potential irrigated environments. In real terms, wages are also rising in South Asia, and the mechanization of power-intensive operations ought to be completed over the next two decades. The speed with which South Asia mechanizes its agriculture would depend on the extent to which the governments of the region liberalize their economies and create an environment conducive to international capital flows and investment in non-agricultural job creation. The rainfed environments of South Asia can be expected to have minimal levels of mechanization over the foreseeable future.

The public sector has been active in the provision of mechanical technologies for the rice Green Revolution; both the international and the national agricultural research systems in Asia have been involved in technology design and generation. Active public sector research investments in mechanization have been justified on the ground that Asian rice farms are small and the farmers need low-cost machines that are small in size. It was presumed that the large multinational corporations would not find it profitable to meet the unique needs of rice farm mechanization. Also, the private sector in Asia was in its infancy during the 1960s and 1970s and hence did not have the capacity to provide the R&D necessary for developing small machines. The public sector was able to reduce the research costs and provide the industry with prototype technologies that could be manufactured and disseminated widely at prices lower than competing options from the multinational corporations based in Japan and other developed countries.

Public sector investment in engineering research provided intermediate solutions to breaking the power bottlenecks that emerged as a result of rice intensification. Small machines such as the power tiller and thresher were widely disseminated throughout Asia, and their adoption has been rapid, particularly in the irrigated environments of Southeast Asia. Public sector attempts to mechanize control-intensive operations, such as transplanting and weeding, have generally not been successful. Most Asian societies were not yet at a stage in economic development where mechanizing these operations would have been profitable. The question that needs to be asked now is whether there is a role for continued public sector involvement in agricultural mechanization research and development. The post-Green-Revolution role of the public sector in engineering research can be expected to be limited to very specific situations where the private sector's presence is small.

During the Green Revolution period, the public sector created a successful niche for itself in the development of small machines. One needs to question the relevance of a miniaturization approach in high-potential environments in the face of rapid economic growth and the commercialization of agriculture. High opportunity cost of labor and the anticipated consolidation of landholdings for the favorable irrigated environments of the future imply that small machines may not remain economically efficient, and that a gradual upscaling of agricultural machinery ought to be expected. Continued research and development efforts in small-machinery development run counter to the anticipated growth in the operational size of farmholdings. For the irrigated high-potential environments, small machines ought to be seen as an intermediate stage in agricultural mechanization. In the case of the rainfed lowlands, however, small machines will continue to dominate even with economic growth. Hydrologic constraints prevent the increase in field sizes in the rainfed environments and hence the unique niche for small machines. International and national research systems would have a comparative advantage in concentrating on machinery development for the rainfed environments, while the private sector provides for the high-potential irrigated environments.

Public sector engineering research has in recent years begun to devote an increasing share of its resources to rice quality enhancement through improved post-harvest technology, especially for drying and milling. With the vertical integration of a harvest/post-harvest processing system anticipated in much of Southeast Asia over the next decade, operations currently performed by small machines will increasingly be done in an integrated manner using large-scale infrastructure. Private industry would have a distinct advantage in developing and setting up such post-harvest facilities. The role of the public sector would once again be limited to finding intermediate solutions for harvest and post-harvest operations for the rainfed evironments. Wage rates are expected to rise in the rainfed environments at a rate similar to that in the irrigated environments, and the demand for labor-saving technology is expected to increase sharply over time, especially in the rapidly growing economies of Southeast Asia.

One final area where public sector research may play an important role, once again for the rainfed environments, is mechanized transplanting. Earlier chapters of this book have argued that farmers will switch from the labor-intensive transplanting operation to direct seeding as wages rise. It ought to be noted, however, that lands with poor drainage and water control will find it difficult to switch to direct seeding; transplanted rice systems will persist in these lands despite rising wages. For the irrigated environments, the transplanting bottleneck can be overcome by adapting transplanters used in Japan and South Korea to Southeast Asian conditions. For the rainfed environments, further research and development may be necessary before a suitable and low-cost mechanical transplanting system is available.

Is there a policy trade-off in the promotion of agricultural mechanization?

In foodgrains (rice), the policy trade-off can be stated as follows: agricultural mechanization increases the efficiency of foodgrain (rice) production, and could therefore lead to a decline in foodgrain prices. Lower foodgrain prices provide net welfare benefits to urban consumers and other net purchasers of food such as landless labor and small farm households. On the other hand, increases in the level of mechanization could lead to the displacement of agricultural labor and tenant farmers. The trade-off between productivity and equity has become less severe with economic growth and the consequent increase in farm wages and the anticipated consolidation of farmholdings. In the developing countries of Asia, particularly South Asia, policy debates on the equity consequences of mechanization can be expected to continue over the next two decades. The discussion below is directed towards the debate on mechanization in the South Asian countries.

The extent of the policy trade-off varies by resource endowment, operation mechanized, level of economic development, and macroeconomic policies. The trade-off is severe in labor-abundant land-scarce economies with no non-farm employment opportunities. In this case, the mechanization of land preparation and threshing will almost always result in the displacement of labor and/or tenant farmers. The mechanization of milling on the other hand could have positive welfare effects even in labor-abundant areas because most rural small-scale mills substitute for the hand-pounding labor of female family members rather than hired labor.

Where alternative employment opportunities exist, such as urban employment, a shortage of agricultural labor for land preparation, weeding and harvesting–threshing could be observed even in densely populated areas. Here, the trade-off is smaller and mechanization of the operations for which a labor bottleneck is observed could have net welfare benefits for all groups concerned. Rice intensification has caused power bottlenecks around the land preparation, harvesting and threshing operations even in areas with few alternative employment opportunities. In areas where output expansion is constrained by power bottlenecks for the above operations, mechanization could lead to positive equity benefits.

Expansion of agricultural output (where markets exist) in land-abundant areas is generally constrained by labor scarcity. In such economies, operation-specific mechanization is efficient and generally leads to positive equity effects. Of course, mechanization is not always efficient in land-abundant, labor-scarce areas (see Pingali *et al.*, 1987, for a detailed discussion of the conditions under which it would be efficient).

The efficiency–equity trade-off is also affected by a government's macroeconomic policies. Consider the case of a closed foodgrain market (foodgrains are neither imported into nor exported out of the country). Mechanization of land preparation and threshing could have positive welfare effects in this case, if opportunities for expanding into uncultivated or fallow land, or for increasing cropping intensities, are available. Where this is true, mechanization leads to both an increase in area cultivated and an increase in output. More area cultivated results in an increase in employment for weeding, interculture, harvest and post-harvest operations, and higher output results in a decline in the price of food. Both these consequences have direct benefits for the landless poor. If opportunities for area expansion are not available, food prices may still go down but net displacement of tenant and landless labor occurs and high welfare costs are imposed on these groups.

Some governments combine a closed foodgrain market policy with a support price program to protect farmers' incomes. These policies, when combined with an aggressive promotion of mechanization, result in equity costs for the rural poor (in areas without a land frontier), since they lose their income source without the compensation of lower food prices. This situation can be further exacerbated by providing subsidies for machinery purchase or cheap credit.

Over-valued exchange rates, in addition to support prices and subsidies, distort the efficiency benefits of mechanization and thereby impose significant equity costs on society. Over-valued exchange rates also reduce the real cost of imports, thereby providing an advantage for imported inputs relative to domestic inputs in production decisions. For our purposes, over-valued exchange rates result in the choice of imported machines, fertilizers and pesticides over domestic, labor-intensive technologies.

The situation changes somewhat when the economy is involved in exporting foodgrains. In this case, an increase in output does not lead to a decline in domestic food prices, and hence the beneficial effects of lower food prices do not accrue to the rural poor. In the open-economy scenario, mechanization would only be profitable and equitable if either of the following existed: opportunities for area expansion or opportunities for non-farm employment. Even if these conditions hold, adverse equity effects could be induced by government policies on imports, credit, machinery supply and land acquisition.

Even under the best situation where equity effects are minimal, there is bound to be labor that is displaced and that needs to be relocated. Government policies could help in relocating, and helping the displaced labor acquire new skills for employment elsewhere in the economy. Sometimes, as in milling in Bangladesh, the equity costs are high, but the overall benefits of the technology are so large that the total transition from hand-pounding is inevitable. In such cases, the government has a large role to play in minimizing the equity costs for displaced labor through readjustment and skill-acquisition programs.

Notes

1. See Hayami and Herdt (1977) for an analytical discussion on the price effects of technological change on income distribution.

References

Abansi, C.L., Duff, B., Lantican, F.A. and Juliano, B.O. (1992) Consumer demand for rice grain quality in selected rural and urban markets. In: Unnevehr, L.J., Duff, B. and Juliano, B.O. (eds) *Consumer Demand for Rice Grain Quality.* International Rice Research Institute and International Development Research Centre, Los Baños, Laguna, Philippines.

Abrol, I.P. (1987) Salinity and food production in the Indian Sub-continent. In: Jordan, W.R. (ed.) *Water and Water Policy in World Food Supplies.* Texas A&M University Press, College Station, pp. 109–113.

ACIAR – Australian Centre for International Agricultural Research (1984) *Soil Erosion and Management.* Paper presented in a workshop held at the Philippine Council for Agricultural Resources Research and Development, Los Baños, Laguna, Philippines, 3–5 December.

ADB – Asian Development Bank (1987) *A Review of Forestry and Forest Industries in the Asia-Pacific Region.* ADB, Manila, Philippines.

Agcaoili, M. and Rosegrant, M.W. (1994) World rice trade: prospects and issues for the future. In: *Proceedings of the Third Workshop of the Rice Supply and Demand Project,* 24–26 January, Bangkok, Thailand.

Ahammed, C.S. and Herdt, R.W. (1981) A general equilibrium analysis of the effects of farm mechanization in the Philippines. Paper presented at the joint ADC/IRRI Workshop on Consequences of Small Rice Farm Mechanization in Asia. International Rice Research Institute, Los Baños, Laguna, Philippines, 14–18 September (mimeo).

Ahl Goy, P. and Duesing, J. (1995) From pots to plots: genetically-modified plants on trial. *Bio/Technology* 13, 454–458.

Ahmed, N. (1985) Fertilizer efficiency and crop yields in Pakistan. *Phosphorus in Agriculture* 89, 17–32.

Ahmed, R. and Rustagi, N. (1987) Marketing and price incentives in African and Asian countries: a comparison. In: Elz, D. (ed.) *Agricultural Marketing Strategy and Pricing Policy.* World Bank, Washington DC, pp. 104–118.

Alagcan, M.M. and Bhuiyan, S.I. (1992) Excess water management for growing upland crops in rice irrigation systems. In: Maglinao, A.R. (ed.) *Irrigation Management for Rice-Based Farming Systems in the Philippines.* PCARRD, IIMI, IRRI, Los Baños, Laguna, Philippines.

Alejar, M.S., Zapata, F.J., Senadhira, D., Khush, G.S. and Datta, S.K. (1995) Utilization of anther culture as a breeding tool in rice improvement. In: Terzi, M., Cella, R. and Falavigna, A. (eds) *Current Issues in Plant Molecular and Cellular Biology.* Kluwer Academic Publishers, Dordrecht, Boston, London, pp. 137–142.

Ali, M. and Narciso, J.H. (1994) Economic evaluation and farmers' perception of green manure use in rice-based farming systems. In: Ladha, J.K. and Garrity, D.P. (eds) *Green Manure Production Systems for Asian Ricelands.* Selected papers from the International Rice Research Conference. International Rice Research Institute, Los Baños, Laguna, Philippines.

Allan, W. (1965) *The African Husbandman.* Oliver and Boyd, Edinburgh.

Anderson, J.R. and Hazell, P.B.R. (eds) (1994) *Variability in Grain Yields: Implications for Agricultural Research and Policy in Developing Countries.* International Ford Policy Research Institute, Johns Hopkins University Press, Baltimore, London.

Anderson, K. and Hayami, Y. with associates (1986) *The Political Economy of Agricultural Protection: East Asia in International Perspectives.* Allen and Unwin, London.

Anderson, K. and Tyers, R. (1990) How developing countries could gain from agricultural trade liberalization in the Uruguay Round. In: Goldin, I. and Knudsen, O. (eds) *Agricultural Trade Liberalization: Implications for Developing Countries.* OECD/World Bank, Paris, Washington DC.

Antle, J.M. and Pingali, P.L. (1994) Pesticides, productivity and farmer health: a Philippine case study. *American Journal of Agricultural Economics* 76(3), 418–430.

APO – Asian Productivity Organization (1991) *Utilization of Farm Machinery in Asia.* Report of APO Multi-Country Study Mission, 19–29 June 1990. Tokyo, Japan.

Astor, V. and Rowntree, B.S. (1946) *Mixed Farming and Muddled Thinking – An Analysis of Current Agricultural Policy.* MacDonald, London.

Avery, D. (1988) China. *World Agriculture Trends Strategic Outlook* 1(4), 1–22, illus. November.

Bardhan, P.K. (1970) Green Revolution and agricultural laborers. *Economic and Political Weekly* 5 (29–31) (Special Number, July), 1239–1246.

Barker, R. and Cordova, V. (1978) Labor utilization in rice production. In: *Economic Consequences of the New Rice Technology.* International Rice Research Institute, Los Baños, Laguna, Philippines.

Barker, R. and Duff, J.B. (1986) *Constraints to Higher Rice Yield in Seven Rice Growing Environments in South and Southeast Asia.* (mimeo) International Rice Research Institute, Los Baños, Laguna, Philippines.

Barker, R., Herdt, R.W. with Rose, B. (1985) *The Rice Economy of Asia.* Resources for the Future, Inc., Washington DC and International Rice Research Institute, Los Baños, Laguna, Philippines.

Beachy, R.N., Subba Rao, B.L., Gough, K., Shen, P. and Kaniewska, M. with Hibino, H.H. (1989) *Characterization of the Viruses Associated with the Tungro Disease.* Paper presented at the Third Annual Meeting of the Rockefeller Foundation International Program on Rice Biotechnology, Colombia, Missouri, 8–10 March.

Bhuiyan, S.I. and Castañeda, A.R. (1995) The impact of ricefield pesticides on the quality of freshwater resources. In: Pingali, P.L. and Roger, P.A. (eds) *Impact of Pesticides on Farmer Health and the Rice Environment.* Kluwer Academic Publishers, Norwell, Massachusetts.

Bhuiyan, S.I. and Tabbal, D. (1996) *Competition Between Agriculture and Other Sectors for Water from Rice Irrigation Projects.* SWS Water Science Internal Document No. 96–1, Soil and Water Sciences Division, International Rice Research Institute, Los Baños, Laguna, Philippines.

Binswanger, H.P. (1978) *The Economics of Tractors in South Asia.* Agricultural Development Council and the International Crops Research Institute for the Semi-Arid Tropics. Hyderabad, India.

Binswanger, H.P. and Pingali, P.L. (1987) The evolution of farming systems and agricultural technology in Sub-Saharan Africa. In: Ruttan, V.W. and Pray, C.E. (eds) *Policy for Agricultural Research.* Westview Press, Boulder.

Binswanger, H.P. and Rosenzweig, M.R. (1986) Behavioural and material determinants of production relations in agriculture. *The Journal of Development Studies* 22(3), 503–539.

Bonifacio, E.P., Quick, G.R. and Pingali, P.L. (1990) Deriving design specifications for mobile dryers: an end-user perspective. In: *Proceedings of the First International Agricultural Engineering Conference for Asia-Pacific.* Asian Institute of Technology, Bangkok, Thailand. 3–6 December.

Boserup, E. (1965) *Conditions of Agricultural Growth.* Aldine Publishing Company, Chicago.

Bottrell, D. (1979) Putting the integrated pest control package together. In: *Documents, Regional Training Seminar on Rice Integrated Pest Control for Irrigated Rice in South and Southeast Asia.* Manila, 1978, 13 p.

Bouis, H.E. (1994) The effect of income on demand for food in poor countries: are our food consumption databases giving us reliable estimates? *Journal of Development Economics* 44(1), 199–226 (June).

Brabben, T.E. (1979) *Reservoir Sedimentation Study. Korangkates, East Java, Indonesia.* Report No. OD 22. Hydraulics Research Station, Wallingford. 19 p.

Brando, A.S.P. and Martin, W.J. (1993) Implications of agricultural trade liberalization for the developing countries. *Agricultural Economics* 8, 313–343.

Bray, F. (1986) *The Rice Economies: Technology and Development in Asian Societies.* Oxford, Blackwell.

Budowski, G. (1980) *Agroforestry in Managing Tropical Forests.* Paper presented at the International Symposium on Tropical Forests, Yale University, 15–16 April.

Bulatao, R.A., Bos, E., Stephens, P.W., and Vu, M.T. (1990) *World Population Projections, 1989–90 Edition.* World Bank, Johns Hopkins University Press.

Burton, S., Shah, P.B. and Schreier, H. (1989) Soil degradation from converting forest land into agriculture in the Chittawan District of Nepal. *Mountain Research and Development* 9(4), 393–404.

Byerlee, D. (1987) *Maintaining the Momentum in Post-Green Revolution Agriculture: A Micro-Level Perspective for Asia.* International Development Paper No. 10. Michigan State University, East Lansing, Michigan. 57 pp.

Byerlee, D. (1992) Technical change, productivity, and sustainability in irrigated cropping systems of South Asia: emerging issues in the post-Green Revolution era. *Journal of International Development* 4, 477–496.

Byerlee, D. (1993) *Modern Varieties, Productivity, and Sustainability: Recent Experiences and Emerging Challenges.* Paper presented to the AAEA Pre-Conference Workshop on 'Post-Green Revolution Agricultural Development Strategies: What Next?' Orlando, Florida, 30–31 July.

Byerlee, D. and Pingali, P.L. (1994) Agricultural research in Asia: fulfillments and frustrations. *Proceedings of the XXII Conference of the International Association of Agricultural Economists* Harare, 22–29 August.

Byerlee, D. and Siddiq, A. (1994) Has the Green Revolution been sustained? The quantitative impact of the seed-fertilizer technology in Pakistan revisited. *World Development* 22(9).

Cain, M.L. *et al.* (1979) *Class, Patriarchy, and the Structure of Women's Work in Rural Bangladesh.* Center for Policy Studies Working Paper 43. Population Council, New York.

Carlson, G.A. (1989) Pest resistant varieties, pesticides and crop yield variability: a review. In: Anderson, J.A. and Hazell, P.B.R. (eds) *Variability in Grain Yields: Implications for Agricultural Research and Policy in Developing Countries.* International Food Policy Research Institute, Johns Hopkins University Press, Baltimore, pp. 242–250.

Cassman, K.G. and Pingali, P.L. (1995) Extrapolating trends from long-term experiments to farmers' fields: the case of irrigated rice systems in Asia. In: Barnett, V., Payne, R. and Steiner, R. (eds) *Agricultural Sustainability: Economic, Environmental and Statistical Considerations.* John Wiley and Sons Ltd, New York.

Cassman, K.G., De Datta, S.K., Olk, D.C., Alcantara, J., Samson, M., Descalsota, J. and Dizon, M. (1994) Yield decline and nitrogen balance in long-term experiments on continuous, irrigated rice systems in the tropics. *Advances in Soil Science*, Special Issue.

Cassman, K.G., Peng, S.B., Olk, D.C., Ladha, J.K., Reichardt, W., Dobermann, A. and Singh, U. (1995) Opportunities for increased nitrogen use efficiency for improved resource management in irrigated rice systems. *Proceedings of the Workshop on Nutrient Use Efficiency of the Cropping System.* International Rice Research Institute, Los Baños, Laguna, Philippines. 13–15 December.

Castañeda, A.R. and Bhuiyan, S.I. (1988) Industrial pollution of irrigation water and its effects on riceland productivity. *The Philippine Journal of Crop Science* 13(1), 27–35.

Chalamwong, Y. and Feder, G. (1986) *Land Ownership Security and Land Values in Rural Thailand.* World Bank Staff Working Paper No. 790, Washington DC.

Chambers, R. (1988) *Managing Canal Irrigation: Practical Analysis from South Asia.* Press Syndicate of the University of Cambridge, Cambridge, UK.

Chandler, R.F. (1982) *An Adventure in Applied Science: A History of the International Rice Research Institute (IRRI).* International Rice Research Institute, Los Baños, Laguna, Philippines. 233 p. illus.

Chang, J.Y. (1991) Country papers: Republic of Korea. In: *Utilization of Farm Machinery in Asia.* Report of APO Multi-Country Study Mission, 19–29 June. Asian Productivity Organization, Tokyo, Japan.

Childs, N.W. (1990) *The World Rice Market: Government Intervention and Multi-lateral Policy Reforms.* Economic Research Service, US Department of Agriculture, Washington DC.

China Statistical Office (1993) *China Statistical Yearbook.* Beijing Agricultural Press.

Cho, K.H. (1983) Other country report on the status of farm mechanization – Republic of Korea. In: *Farm Mechanization in Asia.* Asian Productivity Organization, Tokyo, Japan.

Choudhury, N., Kabir, K.A., Biswas, S.K. and Islam, R. (1992) Influence of rice grain properties on market price in Bangladesh. In: Unnevehr, L.J., Duff, B. and Juliano, B.O. (eds) *Consumer Demand for Rice Grain Quality.* International Rice Research Institute and International Development Research Centre, Los Baños, Laguna, Philippines.

Coloma, A.G. (1984) *Management of Sedimentation Problem in Irrigation Systems in the Philippines.* East–West Environment and Policy Institute Paper, July (mimeographed).

Cordova, V.G., Maranan, C. and David, C.C. (1984) *Adoption of Wet-Seeding in Selected Rice Areas in the Philippines.* Paper presented at the 15th Annual Scientific Meeting of the Crop Science Society of the Philippines, 16–18 May, Batac, Ilocos Norte, Philippines.

Cruz, W., Francisco, H.A. and Conway, Z.T. (1988) *The On-Site and Downstream Costs of Soil Erosion.* Working Paper Series No. 88–11. Philippine Institute for Development Studies, Makati City, Philippines.

Dalrymple, D.G. (1986). *Development and Spread of High Yielding Rice Varieties in Developing Countries.* Metrotec, Inc., Washington DC.

Damardjati, D.S. and Oka, M. (1992) Evaluation of consumer preferences for rice quality characteristics in Indonesia. In: Unnevehr, L.J., Duff, B. and Juliano, B.O. (eds) *Consumer Demand for Rice Grain Quality.* International Rice Research Institute and International Development Research Centre, Los Baños, Laguna, Philippines.

Datta, S.K., Peterhaus, A., Datta, K., and Potrykus, I. (1990) Genetically engineered fertile Indica rice recovered from protoplasts. *Bio/Technology* 8, 736–740.

Datta, S.K., Datta, K., Soltanitor, N., Donn, G. and Potrykus, I. (1992) Herbicide-resistant Indica rice plants from IRRI breeding line IR-72 after PEG-mediated transformation of protoplasts. *Plant Molecular Biology* 20, 619–629.

David, C.C. and Barker, R. (1978) Modern rice varieties and fertilizer consumption. In: *Economic Consequences of the New Rice Technology.* International Rice Research Institute, Los Baños, Laguna, Philippines.

David, C.C. and Otsuka, K. (1994) *Modern Rice Technology and Income Distribution in Asia.* International Rice Research Institute, Los Baños, Laguna, Philippines and Lynne Rienner Publishers, Boulder and London.

De Datta, S.K. (1981) *Principles and Practices of Rice Production.* Wiley-Interscience Publication, John Wiley & Sons, New York.

De Datta, S.K. and Baltazar, A.M. (1996) Agronomic and ecological approaches to integrated weed management in rice in Asia. In: Naylor, R. (ed.) *Herbicide in Asian Rice: Transition in Weed Management.* Institute of International Studies, Stanford University, Palo Alto, California, and International Rice Research Institute, Los Baños, Laguna, Philippines.

De Datta, S.K. and Buresh, R.J. (1989) Integrated nitrogen management in irrigated rice. *Advances in Soil Science* 10, 143–169.

De Datta, S.K. and Flinn, J.C. (1986) Technology and economics of weed control in broadcast-seeded flooded tropical rice. In: Noda, K. and Mercado, B.L. (eds) *Weeds and the Environment in the Tropics.* Proceedings of the 10th Conference of Asian-Pacific Weed Science Society, Chiang Mai, Thailand, pp. 51–74.

De Datta, S.K., Garcia, F.V., Abilay, W.P., Jr, Alcantara, J.M., Mandac, A. and Marciano, V.P. (1979) *Constraints to High Yields, Nueva Ecija, Philippines.* International Rice Research Institute, Los Baños, Laguna, Philippines. pp. 191–234.

De Datta, S.K., Gomez, K.A. and Descalsota, J.P. (1988) Changes in yield response to major nutrients and in soil fertility under intensive rice cropping. *Soil Science* 146, 350–358.

De Vera, M.V.M. (1992) Impact of upper watershed destruction on the performance of national irrigation systems in the Philippines. An unpublished masteral thesis. College of Economics and Management, University of the Philippines at Los Baños, College, Laguna, Philippines. April.

DENR – Department of Environment and Natural Resources (1990) *Master Plan for Forestry Development.* Department of Environment and Natural Resources, Quezon City, Philippines. 445 p.

Department of Agriculture (1994) *GATT and Philippine Agriculture.* Quezon City, Philippines.

Desai, G.M. and Gandhi, V. (1989) Phosphorus for sustainable agriculture growth in Asia – an assessment of alternative sources and management. In: *Proceedings of the Symposium on Phosphorus Requirements for Sustainable Agriculture in Asia and Pacific Region.* International Rice Research Institute, Los Baños, Laguna, Philippines, 6–10 March.

Dey, M.M. and Hossain, M. (1994) Rice price intervention in rice importing countries of South and Southeast Asia: stabilization and protection. In: *Proceedings of the Third Workshop on Projections and Policy Implications of Medium and Long-Term Rice Supply and Demand,* Bangkok, Thailand. 12 p. illus. ref.

Dey, M.M. and Hossain, M. (1995) Yield potentials of modern rice varieties: an assessment of technological constraints to increase rice production. In: *Proceedings of the Final Workshop of the Projections and Policy*

Implications of Medium and Long-Term Rice Supply and Demand Project. Beijing, China, 23–26 April 1995.
Dogra, B. (1986) The Indian experience with large dams. In: Goldsmith, E. and Hildyard, N. (eds) *The Social and Environmental Effects of Large Dams.* Volume 2, Wadebridge Ecological Center, London, UK, pp. 201–208.
Doolette, J.B. and Macgrath, W.B. (eds) (1990) *Watershed Development in Asia – Strategies and Technologies.* World Bank Technical Paper Number 127. The World Bank, Washington DC.
Douthwaite, B., Quick, G.R. and Tado, C.J.M. (1993) Stripping: a new approach to small-area rice harvesting. *Philippine Journal of Crop Sciences* 18(2), 119–127.
Dove, M.R. (1987) The perception of peasant land rights in Indonesian development: causes and implications. In: Raintree, J.B. (ed.) *Land, Trees and Tenure.* Proceedings of an International Workshop on Tenure Issues in Agroforestry. Sponsored by Ford Foundation. Nairobi, Kenya, May 27–31, 1985. ICRAF and the Land Tenure Center, Nairobi and Madison.
Duff, B. (1986) *Changes in Small-Farm Rice Threshing Technology in Thailand and the Philippines.* Working Paper No. 120. International Rice Research Institute, Los Baños, Laguna, Philippines.
Durand, R.N., Pascoe, R. and Bingham, W. (1984) The handheld electrodyne sprayer: an operational tool for better crop management in developing countries. In: *Proceedings of the British Crop Protection Conference,* Brighton, England, 19–22 November.
Dyck, J.H., Huang, S.W. and Wailes, E. (1993) Structural change and competitiveness of the Asian rice economies in Taiwan, Korea and Japan. In: Yang-Boo Choe *et al.* (eds) *Agricultural Trade Reform and the Future of Asian Agriculture and Agribusiness.* Asian Society of Agricultural Economists, Seoul, Korea.
Eaqub, M., Buresh, R.J. and Xuan, V.T. (1988) *Integrated Nutrient Management in Rainfed Lowland Environment.* Paper presented at the International Rice Research Conference. (mimeo) International Rice Research Institute, Los Baños, Laguna, Philippines.
Ebron, L.Z. (1984) Changes in harvesting–threshing labor arrangements in Nueva Ecija. Unpublished masteral thesis, College of Economics and Management, University of the Philippines at Los Baños, College, Laguna, Philippines.
EMB – Environmental Management Bureau (1990) *The Philippine Environment in the Eighties.* Philippine Heart Center for Asia Building, East Avenue, Diliman, Quezon City, Philippines.
Engelhardt, T. (1984) Economics of traditional smallholder irrigation systems in the semi-arid tropics of South India. PhD Dissertation, University of Hohenheim, Hohenheim. July.
Erguiza, A., Duff, B. and Khan, C. (1990) Choice of rice crop establishment technique: transplanting vs wet seeding. *IRRI Research Paper Series* No. 139. February. International Rice Research Institute, Los Baños, Laguna, Philippines.
Escalada, M.M. and Heong, K.L. (1993) Communication and implementation of change in crop protection. In: Chadwick, D.J. and Marsh, J. (eds) *Crop*

Protection and Sustainable Agriculture. Ciba Foundation Symposium on World Food Production by Means of Sustainable Agriculture: The Role of Crop Protection, held in collaboration with the Centre for Research on Sustainable Agricultural and Rural Development, Fisherman's Cove Hotel, Madras, India, 30 November to 2 December 1992.

Evenson, R.E. (1974) *Comparative Evidence on Returns to Investment in National and International Research Institutions.* Staff Paper Series No. 133 (November). Department of Agricultural Economics, College of Agriculture, University of the Philippines at Los Baños, Laguna, Philippines.

FADINAP – Fertilizer Advisory, Development and Information Network for Asia and the Pacific (1994) *Agrochemical News in Brief,* Vol. 20. ESCAP/FAO/UNDP.

FAO – Food and Agriculture Organization. *FAO Trade Yearbook,* various issues. Rome.

FAO – Food and Agriculture Organization. *Food Balance Sheets,* various issues. Rome.

FAO – Food and Agriculture Organization (1984) *Institutional Aspects of Shifting Cultivation in Africa.* Human Resources, Institutions and Agrarian Reform Division, Rome.

FAO – Food and Agriculture Organization (1988) *Commodity Trade Statistics.* Statistical Papers Series D, Vol. 38, No. 1–19, United Nations.

FAO – Food and Agriculture Organization (1992) *Agrostat Database.* Rome.

FAO – Food and Agriculture Organization (1993) *Agriculture: Towards 2010.* Rome.

FAO – Food and Agriculture Organization (1994) *Uruguay Round Agreement: A Preliminary Assessment.* Commodities and Trade Division, Food and Agriculture Organization, Rome.

FAO – Food and Agriculture Organization (1995) *Agrostat Database.* Rome.

Feder, G. and Keck, A. (1996) Increasing competition for land and water resources: a global perspective. In: Pingali, P.L. and Paris, T.R. (eds) *Proceedings of the Workshop on Social Science Methods in Agricultural Systems: Coping with Increasing Resource Competition in Asia.* Chiang Mai, Thailand. 2–4 November. IRRI Discussion Paper Series No. 11. International Rice Research Institute, Los Baños, Laguna, Philippines.

Feder, G. and Onchan, T. (1987) Land ownership security and farm investment. *American Journal of Agricultural Economics* 69(1), 311–320.

Flinn, J.C. and De Datta, S.K. (1984) Trends in irrigated rice yields under intensive cropping at Philippine research stations. *Field Crops Research* 9, 1–15. Elsevier Science Publishers BV, Amsterdam, Netherlands.

Flinn, J.C., De Datta, S.K. and Labadan, E. (1982) An analysis of long-term rice yields in a wetland soil. *Field Crops Research* 5(3), 201–216.

Fredericksen, H.D., Berkoff, J. and Barber, W. (1993) *Water Resource Management in Asia.* Vol. I, Main Report. World Bank Technical Paper No. 212. The World Bank, Washington DC.

Fujimoto, H., Itoh, H., Yamamoto, M., Kyozuka, J. and Shimamoto, K. (1993) Insect resistant rice generated by introduction of a modified delta-endotoxin gene of *Bacillus thuringiensis. Bio/Technology* 11(10), 1151–1161 (October).

Fujisaka, J.S. (1990a) *Has Green Revolution Rice Research Paid Attention to Farmers' Technologies?* Social Sciences Division, International Rice Research Institute, Los Baños, Laguna, Philippines. 20 p.

Fujisaka, J.S. (1990b) *Targeting Research to Improve Sustainability and Productivity of Shifting Cultivation: Northern Laos.* Social Sciences Division, International Rice Research Institute, Los Baños, Laguna, Philippines.

Fujisaka, J.S. (1994) Learning from six reasons why farmers do not adopt innovations intended to improve sustainability of upland agriculture. *Agricultural Systems* 46(4), 409–425.

Fujisaka, J.S. and Capistrano, D.A. (1985) *Pioneer Shifting Cultivation in Calminoe, Philippines: Sustainability or Degradation from Changing Human–Ecosystem Interactions.* EAPI Working Paper, East–West Center, Honolulu.

Fujisaka, J.S. and Garrity, D.P. (1988) Developing sustainable food crop farming systems for the sloping acid uplands. In: *Proceedings of the Fourth Southeast Asian Universities Agroecosystems Network Research Symposium.* Khon Kaen, Thailand.

Fujisaka, J.S. and Wollenberg, E. (1991) From forest to agroforester: a case study. *Agroforestry Systems* 14(2), 113–130.

Garrity, D.P. and Flinn, J.C. (1987) Farm-level management systems for green manure crops in Asian rice environments. In: *Sustainable Agriculture: Green Manure in Rice Farming.* International Rice Research Institute, Los Baños, Laguna, Philippines.

Garrity, D.P., Garcia, A., Agustin, P.C. and Dacumos, R. (1988) Methods for the extrapolation of cropping systems technologies. In: *Proceedings of the 19th Asian Farming Systems Network Working Group Meeting.* 11–15 November, Baguio City, Philippines.

Garrity, D.P., Kummer, D.M. and Guiang, E.S. (1991) *The Upland Ecosystem in the Philippines: Alternatives for Sustainable Farming and Forestry.* A Study Commissioned by the National Research Council Project on Agricultural Sustainability and the Environment in the Humid Tropics, Washington DC.

GATT Secretariat (1992) *GATT, What It Is, What It Does.* Geneva

Ghani, M.A., Hakim, M.A. and Mondal, M.K. (1993) Water management for improving irrigation system performance in Bangladesh. In: Miranda, S.M. and Maglinao, A.R. (eds) *Irrigation Management for Rice-Based Farming Systems in Bangladesh, Indonesia and the Philippines: Proceedings of the Tri-Country Workshop* held in Colombo, Sri Lanka, 12–14 November. International Irrigation Management Institute and International Rice Research Institute, Colombo, Sri Lanka.

Gines, H.C., Moya, T.B., Pandey, R.K. and Carangal, V.R. (1989) Crop diversification: problems and prospects in partially irrigated rice-based farming systems. In: *Proceedings of a National Workshop on Crop Diversification in Irrigated Agriculture in the Philippines,* International Irrigation Management Institute, Sri Lanka.

Gleick, P.H. (ed.) (1993) *Water in Crisis: A Guide to the World's Freshwater Resources.* Oxford University Press, New York.

Gypmantasiri, P., Wiboonpongse, A., Rerkasem, B., Craig, I., Rerkasem, K., Ganjarapan, L., Titayawaan, M., Seetisarn, M., Thani, P., Jaisaard, R., Ongprasert, S. and Radanachaless, T. (1980) *An Interdisciplinary Perspective of Cropping Systems in the Chiang Mai Valley: Key Questions for Research.* Faculty of Agriculture, University of Chiang Mai, Thailand.

Hakim, M.A. and Parker, D.E. (1990) Role of farmer organizations in irrigation management. In: *Proceedings of the Workshop on Applied Research for Increasing Irrigation Effectiveness and Crop Production*, a collaborative project of Bangladesh Rice Research Institute, Bangladesh Water Development Board, International Irrigation Management Institute, and the International Rice Research Institute. Dhaka, Bangladesh.

Hakim, M.A., Parker, D.E. and Ghani, M.A. (1990) *Socioeconomic and Institutional Issues in Irrigation Management for Rice-Based Farming Systems in Bangladesh.* Paper presented at the IIMI–IRRI Intercountry Workshop on Irrigation Management for Rice-Based Farming Systems, Colombo, Sri Lanka, 12–14 November.

Hamilton, L.S. (1984) *A Perspective on Forestry in Asia and the Pacific.* EWEPI Reprint No. 67, East–West Environment and Policy Institute, Honolulu, Hawaii.

Han, K. (1992) Future development of feed industry in Asia and Pacific Region. In: Bunyavejchewin, P., Sangdid, S. and Hangsanet, K. (eds) *Proceedings of the Sixth AAAP Animal Science Congress.* Vol. I. Kasetsart University, Bangkok, Thailand. pp. 411–432.

Hayakaya, T., Zhu, Y., Itoh, K., Kimura, Y., Izawa, T., Shimamoto, K. and Toriyama, S. (1992) Genetically engineered rice resistant to rice stripe virus, an insect-transmitted virus. *Proceedings of the National Academy of Sciences* 89, 9865–9869.

Hayami, Y. and Herdt, R.W. (1977) Market price effects of technological change on income distribution in semi-subsistence agriculture. *American Journal of Agricultural Economics* 59(2), 245–256.

Hayami, Y. and Ruttan, V.W. (1985) *Agricultural Development: An International Perspective.* Revised and Extended Edition. Johns Hopkins University Press, Baltimore.

Hazell, P.B.R. and Rosegrant, M.W. (1994) Land and water management: property rights and institutions under agricultural deregulation. *Policy Brief.* International Food Policy Research Institute, Washington DC.

He Gui-Ting, Te. A., Xigang, Z., Travers, S.L., Xiufang, L. and Herdt, R.W. (1984) The economics of hybrid rice production in China. *International Rice Research Institute Research Paper Series* No. 101. International Rice Research Institute, Los Baños, Laguna, Philippines, 14 pp.

He Gui-Ting, Xigang, Z. and Flinn, J.C. (1987) Hybrid seed production in Jiangsu province, China. *Oryza* 24, 297–312.

Heinrichs, E.A. and Mochida, O. (1984) From secondary to major pest status: the case of insecticide-induced rice brown planthopper, *Nilaparvata lugens,* resurgence. *Prot. Ecol.* 7(2/3), 201–218.

Heong, K.L. (1990) Feeding rates of the rice leaffolder, *Cnaphalocrocis medinalis* (Lepidoptera: Pyralidae), on different plant stages. *Journal of Agricultural Entomology* 7, 81–90.

Heong, K.L. (1991) *Management of the Brown Planthopper in the Tropics.* Paper presented at the Symposium on Migration and Dispersal of Agricultural Insects, 25–28 September, Tsukuba, Japan.

Heong, K.L. (ed.) (1995) *Rice IPM Network Final Project Report, Phase II.* International Rice Research Institute, Los Baños, Laguna, Philippines.

Heong, K.L., Aquino, G.B. and Barrion, A.T. (1992) Population dynamics of plant and leafhoppers and their natural enemies in rice ecosystems in the Philippines. *Crop Protection* 11, 371–379.

Heong, K.L., Escalada, M.M. and Vo Mai (1994) An analysis of insecticide use in rice – case studies in the Philippines and Vietnam. In: *Rice IPM Network Workshop Report on Message Design for a Campaign to Encourage Farmers' Participation in Experimenting with Stopping Early Insecticide Spraying in Vietnam.* 25–28 May, Ho Chi Minh City, Vietnam. International Rice Research Institute, Los Baños, Laguna, Philippines.

Heong, K.L., Escalada, M.M. and Lazaro, A.A. (1995) Misuse of pesticides among rice farmers in Leyte, Philippines. In: Pingali, P.L. and Roger, P.A. (eds) *Impact of Pesticides on Farmer Health and the Rice Environment.* Kluwer Academic Publishers, Norwell, Massachusetts and International Rice Research Institute, Los Baños, Laguna, Philippines.

Herdt, R.W. (1983) Mechanization of rice production in developing Asian countries: perspective, evidence and issues. In: *Consequences of Small Farm Mechanization.* International Rice Research Institute, Los Baños, Laguna, Philippines.

Herdt, R.W. (1987) A retrospective view of technological and other changes in Philippine rice farming, 1965–1982. *Economic Development and Cultural Change* 35(2) (January).

Herdt, R.W. (1988) Increasing crop yields in developing countries. *Proceedings of the 1988 Meeting of the American Agricultural Economics Association*, 30 July–3 August, Knoxville, Tennessee.

Herdt, R.W. and Capule, C. (1983) *Adoption, Spread and Production Impact of Modern Rice Varieties in Asia.* International Rice Research Institute, Los Baños, Laguna, Philippines.

Herdt, R.W. and Riely, F.Z. (1987) International rice research priorities: implications for biotechnology initiatives. In: *Proceedings of the Rockefeller Workshop on Allocating Resources for Developing Country Agricultural Research*, 6–10 July, Bellagio, Italy.

Herdt, R.W., Castillo, L. and Jayasuriya, S. (1984) The economics of insect control in the Philippines. In: *Judicious and Efficient Use of Insecticides on Rice.* Proceedings of the FAO/IRRI Workshop. International Rice Research Institute, Los Baños, Laguna, Philippines.

Higgins, G.M. (1982) *Potential Population Supporting Capacities of Lands in the Developing World.* Technical Report of Project INT/75/P13. Food and Agriculture Organization, United Nations Fund for Population Activities, and International Institute for Applied Systems Analysis, Rome.

Ho, N.K. (1994) *Management Innovations and Technical Transfer in Wet-Seeded Rice – A Case Study of the Muda Irrigation Scheme, Malaysia.* Paper presented at the International Workshop on Constraints,

Opportunities and Innovations for Wet-Seeded Rice, 31 May – 3 June, Maruay Garden Hotel, Bangkok, Thailand.

Hobbs, P. and Morris, M. (1996) *Meeting South Asia's Future Food Requirements from Rice-Wheat Cropping Systems: Priority Issues Facing Researchers in the Post-Green Revolution Era.* Natural Resource Group Working Paper 96–01. CIMMYT, Mexico.

Hock, W.K. (1987) Pesticide use: the need for proper protection, application and disposal. In: Ragsdale, N.N. and Kuhr, R.J. (eds) *Pesticides: Minimizing the Risks.* Developed from a Symposium sponsored by the Division of Agrochemicals at the 191st Meeting of the American Chemical Society, 13–18 April 1986, New York. American Chemical Society, Washington DC.

Hopkins, A.G. (1973) *An Economic History of West Africa.* Columbia University Press, New York.

Hossain, M. (1993) Productivity, competitiveness and prospects of Asian rice economics: implications for rice research. In: Yang-Boo Choe *et al.* (eds) *Agricultural Trade Reform and the Future of Asian Agriculture and Agribusiness.* Asian Society of Agricultural Economists, Seoul, Korea.

Hossain, M. (1994) *Recent Development in Asian Rice Economy: Challenges for Rice Research.* Paper presented at the Workshop on Rice Research Prioritization in Asia, 21–22 February. International Rice Research Institute, Los Baños, Laguna, Philippines.

Hossain, M. (1995) *Sustaining Food Security for Fragile Environments in Asia: Achievements, Challenges and Implications for Rice Research.* Paper presented at the International Rice Research Conference, International Rice Research Institute, Los Baños, Laguna, Philippines, 13–17 February.

Hossain, M. (1996) *Asia Rice Economy: Recent Developments and Emerging Trends.* Paper presented for the International Conference on Food Neglect: Implications on Political and Social Stability of Nations, Kuala Lumpur, Malaysia, 27–28 May.

Hossain, M. and Laborte, A. (1993) *Asian Rice Economy: Recent Progress and Emerging Trends.* Extension Bulletin No. 378 (December), Food and Fertilizer Technology Center, Taipei, Taiwan.

Hossain, M. and Laborte, A. (1994) *Differential growth in rice production in Eastern India: agroecological and socioeconomic constraints.* Paper presented in the Workshop on the Physiology of Drought and Submergence, 28 February – 1 March, Lucknow, India.

Hossain, M. and Oo, M. (1995) *Myanmar Rice Economy: Policies, Performance and Prospects.* Paper presented at the Final Workshop on Projections and Policy Implications of Medium- and Long-Term Rice Supply and Demand, Beijing, 23–26 April. IRRI–IFPRI Collaborative Project. International Food Policy Research Institute and International Rice Research Institute, Los Baños, Laguna, Philippines.

Hossain, M. and Singh, V.P. (1995) Fertilizer use in Asia: recent trend and implications for food security. In: *Proceedings of the Final Workshop of the Projections and Policy Implications of Medium- and Long-Term Rice Supply and Demand Project,* Beijing, China, 23–26 April.

Huang, J. and David, C.C. (1992) *Demand for Cereal Grains in Asia: Effect of Urbanization.* Social Sciences Division, International Rice Research Institute, Los Baños, Laguna, Philippines.

Huang, J. and Rozelle, S. (1994) Environmental stress yields in China. *American Journal of Agricultural Economics* 77(4), 853–864.

Hung, Nguyen Tien (1977) *Economic Development of Socialist Vietnam, 1955–80.* Praeger Special Studies, New York.

IBRD – International Bank for Reconstruction and Development (1987) *Agricultural Mechanization: Issues and Options.* The World Bank, Washington DC.

INSURF – International Network for Sustainable Rice Fertility (1987) *Annual Report.* International Rice Research Institute, Los Baños, Laguna, Philippines.

IPCC – Intergovernmental Panel on Climate Change (1990) *Climate Change. The IPCC Scientific Assessment.* Cambridge University Press, Cambridge.

IRRI – International Rice Research Institute (1967) *Annual Report.* Los Baños, Laguna, Philippines.

IRRI – International Rice Research Institute (1972) *Annual Report.* Los Baños, Laguna, Philippines.

IRRI – International Rice Research Institute (1973) *Annual Report.* Los Baños, Laguna, Philippines.

IRRI – International Rice Research Institute (1986) *Annual Report.* Los Baños, Laguna, Philippines.

IRRI – International Rice Research Institute (1991) *World Rice Statistics 1990.* Los Baños, Laguna, Philippines.

IRRI – International Rice Research Institute (1992) *Annual Report.* Los Baños, Laguna, Philippines.

IRRI – International Rice Research Institute (1993a) *Program Report for 1992.* Los Baños, Laguna, Philippines.

IRRI – International Rice Research Institute (1993b) *IRRI Rice Almanac.* Los Baños, Laguna, Philippines.

IRRI – International Rice Research Institute (1993c) *Annual Report.* Los Baños, Laguna, Philippines.

IRRI – International Rice Research Institute (1995) *World Rice Statistics 1993–94.* Los Baños, Laguna, Philippines.

Islam, Z. (1992) Prospects for deepwater rice after boro in Bangladesh. *Deepwater Tidal Wetlands Rice* 19, 4, illus. February.

Ito, S., Peterson, W.W.F. and Grant, W.R. (1989) Rice in Asia: Is it becoming an inferior good? *American Journal of Agricultural Economics* 71, 32–42.

Jabbar, M.A., Bhuiyan, M.S.R., and Bari, A.K.M. (1983) Causes and consequences of power tiller utilization in two areas of Bangladesh. In: *Consequences of Small Farm Mechanization.* International Rice Research Institute, Los Baños, Laguna, Philippines.

Johnson, R.R. (1988) Putting soil movement into perspective. *Journal of Production Agriculture* 1(1).

Joos, H. and Morrill, W. (1989) *Control of insect pests on rice using Bacillus thuringiensis genes.* Paper presented at the Third Annual Meeting of the Rockefeller Foundation International Program on Rice Biotechnology, Colombia, Missouri, 8–10 March.

Joshi, R.C., Shepard, B.M., Kenmore, P.E. and Lydia, R. (1992) Insecticide-induced resurgence of brown planthopper (BPH) on IR-62. *International Rice Research Newsletter* 17(3), 9–10, illus. June.

Josling, T., Honma, M., MacLaren, D., Miner, B., Sumner, D., Tangermann, S. and Valdes, A. (1994) *The Uruguay Round Agreement on Agriculture: An Evaluation of the Outcome of the Negotiations.* IATRC Commissioned Paper No. 9. IATRC, Stanford.

Juarez, F. and Pathnopas, R. (1983) Comparative analysis of thresher adoption and use in Thailand and the Philippines. In: *Consequences of Small Farm Mechanization.* International Rice Research Institute, Los Baños, Laguna, Philippines.

Kang, B. T. and Juo, S.R. (1981) *Management of Low Activity Clay Soils in Tropical Africa for Food Crop Production.* Paper presented at the Fourth International Soil Classification Workshop, Kigali, Rwanda, 2–12 June.

Kenmore, P.E. (1991) *Indonesia's Integrated Pest Management: A Model for Asia.* FAO Intercountry IPC Rice Programme, Manila, Philippines.

Kenmore, P.E., Cariño, F.O., Perez, C.A., Dyck, V.A. and Gutierrez, A.P. (1984) Population regulation of the rice brown planthopper (*Nilaparvata lugens*) within ricefields in the Philippines. *Journal of Plant Protection in the Tropics* 1(1), 19–37.

Khan, A.U. (1986) The Asian axial-flow threshers. In: *Consequences of Small Farm Mechanization.* International Rice Research Institute, Los Baños, Laguna, Philippines.

Khan, M.H. (1975) *The Economics of the Green Revolution.* Praeger Publishers Inc., USA.

Khush, G.S. (1990) *Varietal Needs for Different Environments and Breeding Strategies.* International Rice Research Institute, Los Baños, Laguna, Philippines.

Khush, G.S. (1995) Modern varieties – their real contribution to food supply and equity. *Geojournal* 35(3), 275–284.

Khush, G. (1996) Rice varietal improvement for weed management. In: Naylor, R. (ed.) *Herbicide in Asian Rice: Transition in Weed Management.* Institute of International Studies, Stanford University, Palo Alto, California, and International Rice Research Institute, Los Baños, Laguna, Philippines.

Khush, G.S. and Toenniessen, G.H. (1991) *Rice Biotechnology.* CAB International, Wallingford, United Kingdom.

Kikuchi, M. and Hayami, Y. (1978) Agricultural growth against a land resource constraint: a comparative history of Japan, Taiwan, Korea, and the Philippines. *Journal of Economic History* 38 (December).

Kisu, M. (1983) Mechanization of rice farming. In: *Farm Mechanization in Asia.* Asian Productivity Organization, Tokyo, Japan.

Koppel, B. and Zurick, D. (1988) Rural transformation and the future of agricultural development policy in Asia. *Agricultural Administration and Extension* 28, 283–301.

Kropff, M.J., Peng, S., Setter, T.L., Matthews, R.B., Dionora, J. and Cassman, K.G. (1993) A quantitative understanding of rice yield potential. In: *Proceedings of the Workshop on Rice Yield Potential in Favorable*

Environments: Past, Present and Future, International Rice Research Institute, Los Baños, Laguna, Philippines.

Kshirsagar, K.G. and Pandey, S. (1995) Diversity of rice cultivars in a rainfed village in the Orissa State of India. *Proceedings of the Seminar on Using Diversity: Enhancing and Maintaining Genetic Resources On-Farm*, International Development Research Center, New Delhi, India, 19–21 June.

Kundu, D.K. and Ladha, J.K. (1995) Efficient management of soil and biologically fixed N2 in intensively cultivated ricefields. *Soil Biology and Biochemistry* 27(4/5), 431–439. illus. ref. April/May.

Lal, D. (1976) Agricultural growth, real wages, and the rural poor in India. *Economic and Political Weekly* 11(26), A47–61. June.

Lal, R. (1980) Soil erosion as a constraint to crop production. In: *Priorities for Alleviating Soil-Related Constraints to Food Production in the Tropics.* International Rice Research Institute, Los Baños, Laguna, Philippines, pp. 405–423.

Lal, R. (1983) *No-Till Farming: Soil and Water Conservation and Management in the Humid and Sub-Humid Tropics.* Monograph No. 2. International Institute of Tropical Agriculture, Ibadan, Nigeria.

Lim, G.S. and Heong, K.L. (1984) The role of insecticides in rice integrated pest management. In: *Proceedings of the FAO/IRRI Workshop on Judicious and Efficient Use of Insecticides on Rice.* International Rice Research Institute, Los Baños, Laguna, Philippines, pp. 19–39. Illus. Ref.

Lin, J.Y. (1990) *Hybrid Rice Innovation in China: A Study of Market Demand-Induced Technological Innovation in a Centrally Planned Economy.* Peking University, China.

Lin, J.Y. and Pingali, P.L. (1994) An economic assessment of the potential for hybrid rice in tropical Asia: lessons from the Chinese experience. In: Virmani, S.S. (ed.) *Hybrid Rice Technology: New Developments and Future Prospects.* Selected papers from the International Rice Research Conference. International Rice Research Institute, Los Baños, Laguna, Philippines.

Lin, W., Anuratha, C.S., Datta, K., Potykus, I., Muthukrishnan, S. and Datta, S.K. (1995) Genetic engineering of rice for resistance to sheath blight. *Bio/Technology* 13, 686–691.

Lindner, R. (1987) Towards a framework for evaluating agricultural economics research. *The Australian Journal of Agricultural Economics* 31(2): 95–111.

Litsinger, J.A. (1989) Second generation insect pest problems on high yielding rices. *Tropical Pest Management* 35(3), 235–242. Illus. Ref. July/Sept.

Litsinger, J.A. (1991) Crop loss assessment in rice. In: Heinrichs, E.A. *et al.* (eds) *Rice Insects: Management Strategies.* Springer-Verlag, New York, pp. 1–65.

Litsinger, J.A. Canapi, B.L., Bandong, J.P., Dela Cruz, C.G., Apostol, R.F., Pantua, P.C., Lumaban, M.D., Alviola III, A.L., Raymundo, F., Libetario, E.M., Loevinsohn, M.E. and Joshi, R.C. (1987) Rice crop loss from insect pests in wetland and dryland environments of Asia with emphasis on the Philippines. *Insect Science and Applications* 8(4/5/6), 677–692.

Lockwood, B., Munir, M., Hussain, K.A. and Gardezi, J. (1983) Farm mechanization in Pakistan: policy and practice. In: *Consequences of Small Farm Mechanization*. International Rice Research Institute, Los Baños, Laguna, Philippines.

Loevinsohn, M. (1985) *Agricultural Intensification and Rice Pest Ecology: Lessons and Implications*. Paper presented at IRRI Conference, 1–5 June, International Rice Research Institute, Los Baños, Laguna, Philippines.

Loevinsohn, M. (1987) Insecticide use and increased mortality in rural Central Luzon, Philippines. *Lancet* 1(8546), 1359–1362.

Lopes, A.S. (1980) Micronutrients in soils of the tropics as constraints to food production. In: *Priorities for Alleviating Soil-Related Constraints to Food Production in the Tropics*. Jointly sponsored and published by the International Rice Research Institute, Los Baños, Laguna, Philippines and the New York State College of Agriculture and Life Sciences, Cornell University.

Lynam, J.K. and Herdt, R.W. (1989) Sense and sustainability: sustainability as an objective in international agricultural research. *Agricultural Economics* 3, 381–398.

MacDicken, K.G. (1990) Agroforestry management in the humid tropics. In: MacDicken, K.G., and Vergara, N.T. (eds) *Agroforestry: Classification and Management*. John Wiley and Sons, New York, pp. 99–149.

MADECOR – Mandala Agricultural Development Corporation (1985) *Magat Watershed and Erosion Control Management Project*. Final project report submitted to the World Bank and the National Irrigation Administration. Umali Subdivision, Los Baños, Laguna, Philippines.

Maglinao, A.R. and Valdeavilla, A.D. (1992) Irrigation management for dry season production of corn in rice-based farming systems. In: Maglinao, A.R. (ed.) *Irrigation Management for Rice-Based Farming Systems in the Philippines*. PCARRD, IIMI, IRRI, Los Baños, Laguna, Philippines.

Maglinao, A.R., Cablayan, D.M., Undan, R.C., Moya, T.B. and Pascual, C.M. (1993) Main irrigation system management for rice-based farming systems in the Philippines. In: Miranda, S.M. and Maglinao, A.R. (eds) *Irrigation Management for Rice-Based Farming Systems in Bangladesh, Indonesia and the Philippines: Proceedings of the Tri-Country Workshop* held in Colombo, Sri Lanka, 12–14 November. International Irrigation Management Institute and International Rice Research Institute, Colombo, Sri Lanka.

Manasse, R.S. and Kareiva, P. (1990) Risk assessment of the escape of recombinant genes from *Brassica*. In: McCammon, S.L. and Dwyer, S.W. (eds) *Proceedings of the Workshop on Safeguards for Planted Introduction of Transgenic Oilseed Crucifers*. US Department of Agriculture, Washington DC.

Mandac, A.M., Magbanua, R.D. and Genesila, M.P. (1987) Multiple cropping system in Northern Mindanao, Philippines. *Philippine Journal of Crop Science* 12(2), 71–85.

Maranan, C.L., Duff, B. and Juliano, B.O. (1992) Quality preferences for modern and traditional rice at the retail level: the Philippine case. In: Unnevehr, L.J., Duff, B. and Juliano, B.O. (eds) *Consumer Demand for*

Rice Grain Quality. International Rice Research Institute and International Development Research Centre, Los Baños, Laguna, Philippines.

Masicat, P.B., De Vera, M.V. and Pingali, P.L. (1990) *Philippine Irrigation Infrastructure: Degradation Trends for Luzon 1966–1989*. IRRI Social Sciences Division Paper No. 90–03. International Rice Research Institute, Los Baños, Laguna, Philippines.

Masicat, P.B., Salandanan, S. and Pascual, C. (1993) Socioeconomic issues in irrigation management for rice-based farming systems in the Philippines. In: Miranda, S.M. and Maglinao, A.R. (eds) *Irrigation Management for Rice-Based Farming Systems in Bangladesh, Indonesia and the Philippines*. Proceedings of the Tri-Country Workshop on Irrigation Management for Rice-Based Farming Systems held in Colombo, Sri Lanka, 12–14 November 1990. International Irrigation Management Institute, Colombo, Sri Lanka. 357 pp.

McInerney, D.P. and Donaldson, G.F. (1975) *The Consequences of Farm Tractors in Pakistan*. World Bank Staff Working Paper 210. The World Bank, Washington DC.

McIntire, J., Bourzat, D. and Pingali, P.L. (1992) *Crop-Livestock Interaction in Sub-Saharan Africa*. The World Bank, Washington DC.

Meadows, M.P. (1993) *Bacillus thuringiensis* in the environment: ecology and risk assessment. In: Entwistle, P.F., Cory, J.S., Bailey, M.J. and Higgs, S. (eds) *Bacillus thuringiensis, An Environmental Biopesticide: Theory and Practice*. John Wiley and Sons, Chichester.

Mehra, S. (1976) Studies in Economics of Farm Management in Ferozepur District (Punjab) and in Mazaffarnagar District (Uttar Pradesh). Occasional Paper No. 88. Cornell University.

Metianu, A.A., Hobson, R.N. and Douthwaite, B. (1994) *Stripper Harvesting – The Application of Stripper Harvesting Technology Worldwide*. Silsoe Research Institute and The Overseas Development Administration, United Kingdom.

Miracle, M. (1967) *Agriculture in the Congo Basin*. University of Wisconsin Press, Madison.

Miranda, S.M. and Maglinao, A.R. (eds) (1993) *Irrigation Management for Rice-Based Farming Systems in Bangladesh, Indonesia and the Philippines*. Proceedings of the Tri-Country Workshop on Irrigation Management for Rice-Based Farming Systems held in Colombo, Sri Lanka, 12–14 November. International Irrigation Management Institute, Colombo, Sri Lanka. 357 pp.

Mitchell, D.O. (1987) *Factors Affecting Grain Prices*. The World Bank, Washington DC.

Mitchell, D.O. and Ingco, M.D. (1995) Global and regional food demand and supply prospects. In: Islam, N. (ed.) *Population and Food in the Early 21st Century: Meeting Future Demand of an Increasing Population*. International Ford Policy Research Institute, Washington DC. pp. 49–57.

Mizuno, T. (1991) Agricultural mechanization trends in Japan. In: *Utilization of Farm Machinery in Asia*. Report of APO Multi-Country Study Mission, 19–29 June. Tokyo, Japan.

Mohamed, B. (1991) Country Paper on Malaysia. In: *Utilization of Farm*

Machinery in Asia. Report of APO Multi-Country Study Mission, 19–29 June. Tokyo, Japan.

Mondal, M.K. et al. 1990. Water regimes and crop diversification. In: *Proceedings of the Workshop on Applied Research for Increasing Irrigation Effectiveness and Crop Production,* a collaborative project of BRRI, BWDB, IIMI and IRRI.

Monke, E. and Pearson, S. (1991a) Introduction. In: Pearson, S., Falcon, W., Heytens, P., Monke, E. and Naylor, R. (eds) *Rice in Indonesia.* Cornell University Press, Ithaca, pp. 1–7.

Monke, E. and Pearson, S. (1991b) The international rice market. In: Pearson, S., Falcon, W., Heytens, P., Monke, E. and Naylor, R. (eds) *Rice in Indonesia.* Cornell University Press, Ithaca. pp. 22–37.

Moody, K. (1991) Weed management in rice. In Pimentel, D. (ed.) *Handbook of Pest Management in Agriculture.* 2nd ed. CRC Press Inc., Boca Raton, Florida, pp. 301–328.

Moody, K. (1993) *Weed Seeds as Contaminants of Rice Seed.* Lecture presented to participants of the Fifth Rice Seed Health Training Course, 4 October–6 November, International Rice Research Institute, Los Baños, Laguna, Philippines, Mimeographed Papers No. 101.

Moody, K. (1994) Postplanting weed control in direct-seeded rice. In: Ali, A.H. (ed.) *Proceedings of the National Workshop on Direct Seeding Practices and Productivity.* Malaysian Agricultural Research and Development Institute (MARDI), Penang, Malaysia. pp. 249–265.

Moody, K. (1996) Economic and technological forces leading to increased herbicide use and some consequences. In: Naylor, R. (ed.) *Herbicide in Asian Rice: Transition in Weed Management.* Institute of International Studies, Stanford University, Palo Alto, California, and International Rice Research Institute, Los Baños, Laguna, Philippines.

Moormann, F.R. and van Breemen, N. (1978) *Rice: Soil, Water Land.* International Rice Research Institute, Los Baños, Laguna, Philippines.

Morgan, D.P. (1977) *Recognition and Management of Pesticide Poisonings.* 2nd ed. US Environmental Protection Agency, Office of Pesticide Programs, Washington DC.

Moya, P.F., Pingali, P.L., Pabale, D.L., Gerpacio, R.V. and Masicat, P.B. (1994) *Conversion of Agricultural Lands to Urban Uses: Who Gains and Who Loses?* Social Sciences Division Paper, International Rice Research Institute, Los Baños, Laguna, Philippines.

Moya, P.L., Pingali, P.L., Papag, A.M., Pabale, D.L. and Douthwaite, B. (1996) *The Stripper Gatherer System: Private Profitability versus the Social Costs of Labor Displacement.* Paper presented at the Federation of the Crop Sciences Societies of the Philippines Annual National Convention, held at Villa Victoria Resort, Davao City, 13–17 May.

Moya, T.B. (1992) Control of irrigation system water supplies for mixed cropping pattern. In: Maglinao, A.R. (ed.) *Irrigation Management for Rice-Based Farming Systems in the Philippines.* PCARRD, IIMI, IRRI, Los Baños, Laguna, Philippines.

Mustafa, U. (1991) Economic impact of land degradation (salt affected and waterlogged soils) on rice production in Pakistan's Punjab. PhD dissertation, College of Economics and Management, University of the Philippines at Los Baños (UPLB), College, Laguna, Philippines.

Nambiar, K.K.M. and Ghosh, A.B. (1984) *Highlights of Research of a Long Term Fertilizer Experiment in India (1971–1982)*. LTFE Research Bulletin No. 1, Indian Agricultural Research Institute, New Delhi, India.

NAS – National Academy of Sciences (1989) *Irrigation Induced Water Quality Problems – What Can Be Learned from the San Joaquin Valley Experience*. National Academy Press, Washington DC.

Naylor, R. (1994) The rural labor market in Indonesia. In: Pearson, S., Falcon, W., Heytens, P., Monke, P. and Naylor, R. (eds) *Rice Policy in Indonesia*. Cornell University Press, Ithaca, New York.

Naylor, R. (1996) (ed.) *Herbicide in Asian Rice: Transition in Weed Management*. Institute of International Studies, Stanford University, Palo Alto, California, and International Rice Research Institute, Los Baños, Laguna, Philippines.

Nemery, B. (1987) *The lungs as a target for the toxicity of some organophosphorus compounds*. NATO ASI Series, Vol. H13. Brussels, Belgium.

Neue, H.U. and Sass, R. (1996) Trace gas emissions from ricefields. In: Prinn, R.G. (ed.) *Global Atmospheric Biospheric Chemistry* 48, 119–147.

Neue, H.U., Lantin, R.S., Wassman, R., Alberto, M.C., Aduna, J.B. and Andales, M.S. (1994) Methane emission from rice soils of the Philippines. In: *CH_4 and N_2O*. National Institute of Agro-Environmental Sciences, Japan.

Neue, H.U., Ziska, L.H., Matthews, R.B. and Dai, Q. (1995) Reducing global warming – the role of rice. *Geojournal* 35(3), 351–362. March.

Norman, M.J.T. (1979) *Annual Cropping Systems in the Tropics: An Introduction*. University of Florida Press, Gainesville.

Noronha, R. (1985) *A Review of the Literature on Land Tenure Systems in Sub-Saharan Africa*. Discussion Paper, Report No. ARU 43, World Bank, Washington DC.

Norton, G.W. and Schuh, G.E. (1981) Evaluating returns to social science research, issues and possible methods. In: Fishel, W.L., Paulsen, A.A. and Sundquist, W.B. (eds) *Evaluation of Agricultural Research*. University of Minnesota Agricultural Experiment Station Miscellaneous Publication No. 8.

NRC – National Research Council (1982) *Ecological Aspects of Development in the Humid Tropics*. Committee on Selected Biological Problems in the Humid Tropics. National Academy Press, Washington DC.

O'Laughlin, D. (1985) *The Influence of Forest Roads on Erosion and Stream Sedimentation: Comparison Between Temperate and Tropical Forests*. Environmental and Policy Institute, Honolulu, Hawaii.

Okigbo, B.N. (1974) *Fitting Research to Farming Systems*. International Institute for Tropical Agriculture, Ibadan, Nigeria.

Pandey, S. and Medd, R.W. (1990) Integration of seed and plant kill tactics for control of wild oats: an economic evaluation. *Agricultural Systems* 34, 65–76.

Pandey, S. and Pingali, P.L. (1996) *Economic Aspects of Weed Management in*

Rice. Social Sciences Division, (mimeo) International Rice Research Institute, Los Baños, Laguna, Philippines.

Pandey, S., Lindner, R.K. and Medd, R.W. (1993) Towards an economic framework for evaluating potential benefits from research into weed control. *Journal of Agricultural Economics* 44(2), 322–333.

Pandey, S., Masicat, P.B., Velasco, L.E. and Gagalac, F. (1995) *An Economic Analysis of Rainfed Rice Farming in Tarlac, Central Luzon*. Mimeo. Social Sciences Division, (mimeo) International Rice Research Institute, Los Baños, Laguna, Philippines.

Paris, T.R., Diaz, C.P., Hossain, M. and Vasallo, A.B. (1995) Process of technology development and transfer to women: a case study of the micromill in Guimba, Nueva Ecija, Philippines. In: *Proceedings of an International Workshop on Enhancing Income of Women through Suitably Engineered Systems*. IRRI Discussion Paper Series No. 8. International Rice Research Institute, Los Baños, Laguna, Philippines, pp. 71–88.

Park, J.K. (1996) Economic modernization and impact of agricultural output and input markets in Korea. In: Pingali, P.L. and Paris, T.R. (eds) *Proceedings of the Workshop on Social Science Methods in Agricultural Systems: Coping with Increasing Resource Competition in Asia*. Chiang Mai, Thailand. 2–4 November. IRRI Discussion Paper Series No. 11. International Rice Research Institute, Los Baños, Laguna, Philippines.

Parks, R.W. (1967) Efficient estimation of a system of regression equations when disturbances are both serially and contemporaneously correlated. *Journal of the American Statistical Association* 62, 500–509.

Pasandaran, E. and Sayaka, B. (1996) Impact of economic development on resources allocation in Indonesia: sustaining agricultural development. In: Pingali, P.L. and Paris, T.R. (eds) *Proceedings of the Workshop on Social Science Methods in Agricultural Systems: Coping with Increasing Resource Competition in Asia*. Chiang Mai, Thailand. 2–4 November. IRRI Discussion Paper Series No. 11. International Rice Research Institute, Los Baños, Laguna, Philippines.

Pathak, M.D. and Khan, Z.R. (1984) *Insect Pests of Rice*. International Rice Research Institute and ICIPE, Los Baños, Laguna, 89 pp.

Payne, A.I. (1986) *The Ecology of Tropical Lakes and Rivers*. John Wiley and Sons, Great Britain.

Peng, S.B., Garcia, F.V., Laza, R.C., Sanico, A.L., Visperas, R.N. and Cassman, K.G. (1996) Increased nitrogen use efficiency using chlorophyll meter on high-yielding irrigated rice. *Field Crops Research* 47 (2–3), 243–252.

Peng, T.S. (1983) The development of agricultural mechanization and its strategies in Taiwan (tillage, transplanting, combine, grain dryer). *Farm Machinery Industrial Research Corporation, Tokyo, Japan* 14(3), 71–74.

Pernito, R. and Garrity, D.P. (1988) Mungbean response to surface drainage when grown as a pre-rice crop on waterlog-prone ricelands. In: *Proceedings of the Fourth Annual Scientific Meeting of the Federation of Crop Science Societies of the Philippines*. Apo View Hotel, Davao City, Philippines, 27–29 April.

Persley, G.J. (1990) *Beyond Mendel's Garden: Biotechnology in the Service of World Agriculture*. CAB International, Wallingford, UK.

Pingali, P.L. (1990) Institutional and environmental constraints to agricultural intensification. In: McNicoll, G. and Cain, M. (eds) *Rural Development and Population (Institutions and Policy)*. Population and Development Review – A Supplement to Volume 15, 1989. Oxford University Press, Oxford.

Pingali, P.L. (1992) Diversifying Asian rice farming systems: a deterministic paradigm. In: Barghouti, S., Garbux, L. and Umali, D. (eds) *Trends in Agricultural Diversification: Regional Perspectives*. Paper No. 180. The World Bank, Washington DC, pp. 107–126.

Pingali, P.L. (1994) Technological prospects for reversing the declining trend in Asia's rice productivity. In: Anderson, J. (ed.) *Agriculture Technology Policy Issues for the International Community*. Agricultural Policies Division, Agricultural and Rural Development Department, World Bank, Washington DC (Chapter 21) and CAB International, Wallingford, UK.

Pingali, P.L. (1995) *GATT and Rice: Do We Have Our Research Priorities Right?* Paper presented at the International Rice Research Conference, International Rice Research Institute, Los Baños, Laguna, Philippines, 13–17 February.

Pingali, P.L. and Binswanger, H.P. (1987) Population density and agricultural intensification: a study of the evolution of technologies in tropical agriculture. In: Johnson, D. Gale and Lee, R.D. (eds) *Population Growth and Economic Development – Issues and Evidence*. University of Wisconsin Press, London.

Pingali, P.L. and Binswanger, H.P. (1988) Population density and farming systems – the changing locus of innovations and technical change. In: Lee, R. (ed.) *Population, Food and Rural Development*. Clarendon Press, Oxford.

Pingali, P.L. and Carlson, G.A. (1985) Human capital, adjustments in subjective probabilities and the demand for pest controls. *American Journal of Agricultural Economics* (November).

Pingali, P.L. and Gerpacio, R.V. (1996) Living with reduced insecticide use in tropical rice. *Food Policy* (submitted).

Pingali, P.L. and Marquez, C.B. (1996) Herbicide and rice farmer health: a Philippine case study. In: Naylor, R. (ed.) *Herbicide in Asian Rice: Transition in Weed Management*. Institute of International Studies, Stanford University, Palo Alto, California, and International Rice Research Institute Los Baños, Laguna, Philippines.

Pingali, P.L. and Roger, P.A. (eds) (1995) *Impact of Pesticides on Farmer Health and The Rice Environment*. Kluwer Academic Publishers, Massachusetts, and the International Rice Research Institute, Los Baños, Laguna, Philippines.

Pingali, P.L. and Rola, A.C. (1994) *Public Regulatory Roles in Developing Markets: The Case of Pesticides*. International Rice Research Institute, Los Baños, Laguna, Philippines.

Pingali, P.L. and Rosegrant, M.W. (1993) Confronting the environmental consequences of the Green Revolution in Asia. In: *Proceedings of the 1993 American Agricultural Economist Association International Pre-Conference on 'Post Green Revolution Agricultural Development Strategies in the Third World: What Next?'* August.

Pingali, P.L. and Rosegrant, M.W. (1995) Agricultural commercialization and diversification: processes and policies. *Food Policy* 20(3), 171–185.

Pingali, P.L. and Siamwalla, A. (1993) Myanmar: rice policy reforms and the potential for exports. *TDRI Quarterly*, December. Thailand Development Research Institute, Bangkok, Thailand.

Pingali, P.L. and Xuan, V.T. (1992) Vietnam: decollectivization and rice productivity growth. *Economic Development and Cultural Change* 40(4), 697–718.

Pingali, P.L., Bigot, Y. and Binswanger, H. (1987) *Agricultural Mechanization and the Evolution of Farming Systems in Sub-Saharan Africa*. Johns Hopkins University Press, Baltimore.

Pingali, P.L., Masicat, P.B., Moya, P.F. and Papag, A.M. (1989) The micro-economics of crop diversification in a diversion irrigation system: a progress report from the UTRIS. In: Valera, A. (ed.) *Crop Diversification in Irrigated Agriculture in the Philippines*. International Irrigation Management Institute (IIMI), Digana Village via Kandy, Sri Lanka. August.

Pingali, P.L., Moya, P.F. and Velasco, L.E. (1990) *The Post-Green Revolution Blues in Asian Rice Production – The Diminished Gap Between Experiment Station and Farmer Yields*. Social Science Division Paper No. 90–01 (January), International Rice Research Institute, Los Baños, Laguna, Philippines.

Pingali, P.L., Marquez, C.B. and Palis, F.G. (1994) Pesticides and Philippine rice farmer health: a medical and economic analysis. *American Journal of Agricultural Economics* 76 (August), 587–592.

Pingali, P.L., Marquez, C.B., Palis, F.G. and Rola, A.C. (1995) Impact of pesticides on farmer health: a medical and economic analysis in the Philippines. In: Pingali, P.L. and Roger, P.A. (eds) *Impact of Pesticides on Farmer Health and the Rice Environment*. Kluwer Academic Publishers, Norwell, Massachusetts and International Rice Research Institute, Los Baños, Laguna, Philippines.

Pingali, P.L., Hossain, M., Pandey, S. and Price, L. (1996a) Economics of nutrient management in Asian rice systems: towards increasing knowledge intensity. *Proceedings of the Workshop on Nutrient Use Efficiency of the Cropping System*. International Rice Research Institute, Los Baños, Laguna, Philippines. 13–15 December.

Pingali, P.L., Xuan, V.T., Khiem, N.T. and Gerpacio, R.V. (1996b) Prospects for sustaining Vietnam's re-acquired rice exporter status. *Food Policy* (submitted).

Poapongsakom, N. (1996) Implications of modernization on the agricultural resource base in Thailand. In: Pingali, P.L. and Paris, T.R. (eds) *Proceedings of the Workshop on Social Science Methods in Agricultural Systems: Coping with Increasing Resource Competition in Asia*. Chiang Mai, Thailand. 2–4 November. IRRI Discussion Paper Series No. 11. International Rice Research Institute, Los Baños, Laguna, Philippines.

Ponnamperuma, F.N. (1974) Micronutrient limitations in acid tropical rice soils. In: Bornemisza, E. and Alvarado, A. (eds) *Soil Management in Tropical America*. Soil Science Department, North Carolina State University, Raleigh, pp. 330–347.

Postel, S. (1989) Water for agriculture: facing the limits. *Worldwatch Paper 93*, December.

Postel, S. (1993) Water and agriculture. In: Gleick, P.H. (ed.) *Water in Crisis: A Guide to the World's Freshwater Resources*. Oxford University Press, New York.

Poulsen, G. (1978) *Man and Tree in Tropical Africa: Three Essays on the Role of Trees in the African Environment*. International Development Research Corporation, Ottawa, Canada.

Pragtong, K. (1987) Land tenure and agroforestry in forest land in Thailand. In: Raintree, J.B. (ed.) *Land, Trees and Tenure*. Proceedings of an International Workshop on Tenure Issues in Agroforestry. Sponsored by the Ford Foundation, Nairobi, Kenya, 27–31 May. ICRAF, Nairobi and the Land Tenure Center, Madison.

Price, L.M.L. and Luis, J.S. (1995) *IPM–No Early Spray–Farmer Practice: Farmer Attitudes and Exploitation of Ricefield Flora and Fauna in the Philippines*. Paper presented at the International Plant Protection Congress, The Hague, The Netherlands, 2–7 July.

Prot, J.C., Soriano, I.R.S., Matias, D.M. and Savary, S. (1992) Use of green manure crops in control of *Hirschmanniella mucronata* and *H. oryzae* in irrigated rice. *Journal of Nematology* 24, 127–132.

Pusposutardjo, S., Murray-Rust, H. and Djunaedi, S. (1993) Main irrigation system management for rice-based farming systems in Indonesia. In: Miranda, S.M. and Maglinao, A.R. (eds) *Irrigation Management for Rice-Based Farming Systems in Bangladesh, Indonesia and the Philippines: Proceedings of the Tri-Country Workshop* held in Colombo, Sri Lanka, 12–14 November. International Irrigation Management Institute and International Rice Research Institute, Colombo, Sri Lanka.

Raintree, J.B. (ed.) (1987) *Land, Trees and Tenure*. Proceeding of an International Workshop on Tenure Issues in Agroforestry. Sponsored by the Ford Foundation, Nairobi, Kenya, 27–31 May. ICRAF, Nairobi and the Land Tenure Center, Madison.

Ranganathan, R., Neue, H.U. and Pingali, P.L. (1994) *Global Climate Change: Role of Rice in Methane Emissions and Prospects for Mitigation*. Paper presented at the International Symposium on Climate Change and Rice, 14–18 March, International Rice Research Institute, Los Baños, Laguna, Philippines.

Rao, A.N. and Moody, K. (1990) Weed seed contamination in rice seed. *Seed Science and Technology* 18, 139–146.

Rice IPM Network (1995) *Rice IPM in Review – Diagnostic Workshops Summary Reports*. 27 August–7 September. Project supported by the Swiss Development Cooperation. International Rice Research Institute, Los Baños, Laguna, Philippines.

Robinson, J.F. (1993) Rice production in Arkansas. In: *Proceedings of the International Seminar on Recent Trends and Future Prospects of Rice Farming in Asia*. NACF and FFTC, Seoul, Korea.

Rola, A.C. and Pingali, P.L. (1993) *Pesticide, Rice Productivity, and Farmer's Health: An Economic Assessment*. World Resources Institute, Washington DC, and the International Rice Research Institute, Los Baños, Laguna, Philippines.

Rola, A.C. and Pingali, P.L. (1992) Choice of crop protection technologies under risk: an expected utility maximization framework. *Philippine Journal of Crop Science* 17(1), 45–54.

Rosegrant, M.W. and Binswanger, H.P. (1994) Markets in tradable water rights: potential for efficiency gains in developing country water resource allocation. *World Development* 22(11), 1613–1625.

Rosegrant, M.W. and Mongkolsmai, D. (1990) *Trends and Determinants of Irrigation Investment in Thailand.* International Food Policy Research Institute, Washington DC (mimeo).

Rosegrant, M.W. and Pasandaran, E. (1990) *Irrigation in Indonesia: Trends and Determinants.* International Food Policy Research Institute, Washington DC (mimeo).

Rosegrant, M.W. and Pingali, P.L. (1994) Policy and technology for rice productivity growth in Asia. *Journal of International Development* 6(6), 665–688.

Rosegrant, M.W. and Roumasset, J.A. (1987) Economic feasibility of green manure in rice-based cropping systems. In: *Sustainable Agriculture: Green Manure in Rice Farming.* International Rice Research Institute, Los Baños, Laguna, Philippines

Rosegrant, M.W. and Svendsen, M. (1992) Irrigation investment and management in Asia: trends, priorities and policy directions. In: *Proceedings of the Planning Workshop on Projections and Policy Implications of Medium- and Long-Term Rice Supply and Demand.* Organized by International Rice Research Institute (IRRI) and the International Food Policy Research Institute (IFPRI). International Rice Research Institute, Los Baños, Laguna, Philippines.

Rosegrant, M.W. and Svendsen, M. (1993) Irrigation investment and management policy for Asian food production growth in the 1990s. *Food Policy* 18, 13–32

Rosegrant, M.W. and Yadav, S. (1993) Irrigation and agricultural diversification: technological vs market based solutions. *Proceedings of the Annual Meeting of the AAEA.* Orlando, Florida, USA, 4–7 August.

Rosegrant, M.W., Agcaoili, M. and Perez, N. (1995a) *Rice and the Global Food Economy: Projections and Policy Implications of Future Food Balances.* Paper presented at the Final Workshop of the IRRI/IFPRI Collaborative Project on Projections and Policy Implications of Medium- and Long-Term Rice Supply and Demand, Beijing, 23–26 April. International Food Policy Research Institute and International Rice Research Institute, Los Baños, Laguna, Philippines.

Rosegrant, M.W., Gazmuri Schleyer, R. and Yadav, S. (1995b) Water policy for efficient agricultural diversification: market-based approaches. *Food Policy* 20, 203–223.

Rosenzweig, C., Parry, M., Fischer, G. and Frohberg, F. (1992) *Climate Change and World Food Supply.* Environmental Change Unit, Oxford (mimeo).

Rozelle, S., Huang, J.K. and Rosegrant, M.W. (1995) Will China starve the world? *Far Eastern Agriculture.* July/August.

Ruddle, K. and Manshard, W. (1980) *Renewable Natural Resources and the Environment: Pressing Problems in the Developing World.* Published for the United Nations University. Tycooly International Publishing Limited, Dublin.

Ruthenberg, H. (1980) *Farming Systems in the Tropics.* 3rd Edition. Clarendon Press, Oxford.

Sajise, P.E. (1986) The changing upland landscape. In: Fujisaka, J.S., Sajise, P.E. and del Castillo, R. (eds) *Man, Agriculture and the Tropical Forest: Change and Development in the Philippine Uplands.* Winrock International Institute for Agricultural Development, Bangkok, Thailand, pp. 13–41.

Sajise, P.E. (1987) Stable upland farming in the Philippines: problems and prospects. In: Hadi, Y., Awang, K., Majid, N.M. and Mohamed, S. (eds) *Impact of Man's Activities on Tropical Upland Forest Ecosystems.* Faculty of Forestry, Universiti Pertanian Malaysia, pp. 633–644.

Sajise, P.E., Zafaralla, M.T. and Ganapin, D.J. Jr. (1988) *Resources and Environment: Saving the Present for the Future.* Paper presented at the University of the Philippines General Faculty Conference, Development Academy of the Philippines, Tagaytay City, Cavite, Philippines. 18–20 December.

Samad, M., Merrey, D., Vermillion, D., Fuchs-Casrsch, M., Mohtadullah, K. and Lenton, R. (1992) Irrigation management strategies for improving the performance of irrigated agriculture. *Outlook on Agriculture.* 21(4), 279–286.

Sanchez, P.A. (1976) *Properties and Management of Soils in the Tropics.* John Wiley and Sons, New York, pp. 198–216.

Sanchez, P.A. and Salinas, J.G. (1981) Low-input technology for managing oxisols and ultisols in Tropical America. *Advances in Agronomy* 34, 279–406.

Sanchez, P.A., Brandy, D.E., Villachica, J.H. and Nicholaides, J.J. (1982) Amazon Basin soils: management for continuous crop production. *Science* 216, 821–827.

Sanchez, P.A., Villachica, J.H. and Brandy, D.E. (1983) Soil fertility dynamics after clearing a tropical rainforest in Peru. *Soil Science Society of America Journal* 47(6), 1171–1178.

Schoenly, K., Cohen, J.E., Heong, K.L., Arida, G., Barrion, A.T. and Litsinger, J.A. (1996) Quantifying the impact of insecticides on food web structure of rice-arthropod populations in Philippine farmers' irrigated fields: a case study. In: Polis, G. and Winemiller, K. (eds) *Food Webs: Integration of Patterns and Dynamics.* Chapman and Hall, New York, pp. 343–351 (Chapter 32).

Schuh, G.E. and Barghouti, S. (1988) Agricultural diversification in Asia. *Finance and Development* 25(2), 41–44. June.

Sehgal, S.M. (1992) Opportunities in hybrid rice development. *Seed World* December, 20–25.

Shinawatra, B. and Pitackwong, P. (1996) Migration, gender roles and technological change. In: Pingali, P.L. and Paris, T.R. (eds) *Proceedings of the Workshop on Social Science Methods in Agricultural Systems: Coping with Increasing Resource Competition in Asia.* Chiang Mai, Thailand. 2–4 November. IRRI Discussion Paper Series No. 11. International Rice Research Institute, Los Baños, Laguna, Philippines.

Sidhu, D.S. and Byerlee, D. (1991) Technical change and wheat productivity in post-green revolution Punjab. *Economic and Political Weekly* 26(52).

Sillers, D.A. III (1980) Measuring risk preferences of rice farmers in Nueva Ecija, Philippines: an experimental approach. PhD dissertation, Yale University, New Haven, Connecticut.

Singh, U., Cassman, K.G., Ladha, J.K. and Bronson, K.F. (1995) Intensive nitrogen management strategies in lowland rice systems. In fragile lives in fragile ecosystems. *Proceedings of the International Rice Research Conference*, IRRI, Los Baños, pp. 229–254.

Soemarworto, O. and Soemarworto, I. (1984) The Javanese rural ecosystem. In: Rambo, T.A. and Sajise, P.E. (eds) *An Introduction to Human Ecology Research on Agricultural Systems in Southeast Asia*. University of the Philippines, Los Baños, Laguna, Philippines, pp. 254–287.

Sriarunrungreauang, S. (1989) The impacts of declining price of rice on production, allocation of farm resources and farm incomes: a case study of small rice farms in Amphoe Don Chedi and Amphoe U-Thong, Changwat Suphan Buri. MS Thesis. Department of Agricultural Economics, Kasetsart University, Bangkok, Thailand.

Stone, B. (1986) Chinese fertilizer application in the 1980s and 1990s: issues of growth, balances, allocation, efficiency and response. In: *China's Economy Looks Toward the Year 2000*. Volume 1. Congress of the United States, Washington DC, pp. 453–493.

Sukharomana, S. (1983) The impact of farm power strategies in Thailand. In: Farrington, J., Abeyratne, F. and Gill, G. (eds) *Farm Power and Employment in Asia*. Agricultural Development Council, Bangkok, Thailand. pp. 139–150.

Svendsen, M. and Ramirez, R. (1990) *Determinants of Irrigation Investment in the Philippines*. International Food Policy Research Institute, Washington DC (mimeo).

Tabbal, D.F., Lampayan, R.M. and Bhuiyan, S.I. (1990) Irrigation water control requirements and complementarities for rice and non-rice crops. In: Maglinao, A.R. (ed.) *Irrigation Management for Rice-Based Farming Systems in the Philippines*. PCARRD, IIMI, IRRI, Los Baños, Laguna, Philippines.

Tabbal, D.F., Undan, R.C., Alagcan, M.M., Lampayan, R., Bhuiyan, S.I. and Woodhead, T. (1993) Farm-level water management for rice-based farming systems in the Philippines. In: Miranda, S.M. and Maglinao, A.R. (eds) *Irrigation Management for Rice-Based Farming Systems in Bangladesh, Indonesia and the Philippines: Proceedings of the Tri-Country Workshop* held in Colombo, Sri Lanka, 12–14 November. International Irrigation Management Institute and International Rice Research Institute, Colombo, Sri Lanka.

Tandon, H.L.S. (1987) *Phosphorus Research and Agricultural Production in India*. Fertilizer Development and Consumption Organization, New Delhi, India.

Tangermann, S. (1994) *An Assessment of the Uruguay Round Agreement on Agriculture*. Paper prepared for the Directorate for Food, Agriculture and Fisheries and the Trade Directorate of OECD. Stanford.

Teng, P.S. (1990) *IPM in Rice: An Analysis of the Status Quo with Recommendations for Action.* Report to the International IPM Task Force (FAO/ACIAR/IDRC/USAID/NRI), IRRI, Los Baños, Laguna, Philippines.

Tichon, M. (1994) *Bacillus thuringiensis* – a viable alternative in cotton. In: Akhurst, R.J. (ed.) *Proceedings of the Second Canberra Meeting on Bacillus thuringiensis.* CSIRO, Canberra, Australia.

Timmer, C.P. (1988) The agricultural transformation. In: Chenery, H.B. and Srinivasan, T.N. (eds) *Handbook of Development Economics.* North Holland, Amsterdam.

Timmer, C.P. (1989a) *Agricultural and Structural Change: Policy implications of diversification in Asia and the Near East.* Harvard Institute for International Development Discussion Paper 291 AFP. 42 pp. illus. June.

Timmer, C.P. (1989b) Food price policy: the rationale for government intervention. *Food Policy* 14, 17–42.

Timmer, C.P. (1991) *Agriculture and the State – Growth, Employment and Poverty in Developing Countries.* Cornell University Press, Ithaca, NY.

Timmer, C.P. (1992) Agricultural diversification in Asia: lessons from the 1980s and issues for the 1990s. In: Barghouti, S., Garbux, L. and Umali, D.L. (eds) *Trends in Agricultural Diversification: Regional Perspectives.* World Bank Technical Paper Number 180, The World Bank, Washington DC.

Toquero, Z.F. and Duff, B. (1985) Physical losses and quality deterioration in rice post-production systems. *IRRI Research Paper Series* 107, International Rice Research Institute, Los Baños, Laguna, Philippines.

Unnevehr, L., Duff, B. and Juliano, B.O. (1992) *Consumer Demand for Rice Grain Quality.* Terminal Report of IDRC Projects. National Grain Quality (Asia) and International Grain Quality Economics (Asia). International Development Research Centre, Ottawa, Canada and International Rice Research Institute, Los Baños, Laguna, Philippines.

USDA – United States Department of Agriculture (1954) *Diagnosis and Improvement of Saline and Alkali Soils.* Agricultural Handbook No. 66, Washington DC.

USDA – United States Department of Agriculture (1994) *Effects of the Uruguay Round Agreement on US Agricultural Commodities.* Economic Research Service, United States.

Valayasevi, A. and Winichagoon, P. (1992) Contribution of animals to meet human nutritional needs in rural Asia. In: Bunyavejchewin, P., Sangdid, S. and Hangsanet, K. (eds) *Proceedings of the Sixth AAAP Animal Science Congress* Vol. I. pp. 41–56. Kasetsart University, Bangkok, Thailand.

Valerio, A.T. (1990) Determinants of fishery overexploitation in Laguna Lake. PhD dissertation. College of Economics and Management, University of the Philippines at Los Baños, College, Laguna, Philippines. April.

Valmonte, R.D. (1993) Impact of environmental externalities on fish productivity in Laguna de Bay, Philippines. An unpublished masteral thesis. College of Economics and Management, University of the Philippines at Los Baños, College, Laguna, Philippines.

Virmani, S.S. (1993) *Global Status of Hybrid Rice Research and Development.* Paper presented at the International Seminar on Hybrid Rice, MAHYCO Research Foundation, Bombay, India, 31 July – 1 August.

Virmani, S.S. (1996) Hybrid rice. *Advances in Agronomy* 57, 377–462.

Virmani, S.S. and Dedolph, C. (1994). Reaping the benefits of hybrid rice in the tropics. *World Agriculture* 2, 17–20.

Virmani, S.S. and Edwards, I.B. (1983) Current status and future prospects for breeding hybrid rice and wheat. *Advances in Agronomy* 36, 145–214.

Virmani, S.S., Khush, G.S. and Pingali, P.L. (1993) Hybrid rice for the tropics: potentials, research priorities and policy issues. In: Paroda, R.S. and Rai, M. (eds) *Hybrid Research and Development Needs in Major Cereals in the Asia-Pacific Region.* Food and Agriculture Organization of the United Nations, Regional Office for Asia and the Pacific, Bangkok, Thailand. pp. 61–86.

Waibel, H. (1986) *The Economics of Integrated Pest Control in Irrigated Rice: A Study from the Philippines.* Crop Protection Monographs. Springer-Verlag, Berlin.

Walker, T.S. and Kshirsagar, K.G. (1981) The village-level impact of machine threshing and implications for technology development in semi-arid tropical India. *ICRISAT Economics Program Progress Report* 27. ICRISAT, Hyderabad, India.

Walker, T.S. and Ryan, J.G. (1990) *Against the Odds: Village and Household Economics in India's Semi-Arid Tropics.* Johns Hopkins University Press, Baltimore.

Warburton, H., Palis, F.G. and Pingali, P.L. (1995) Farmer perceptions, knowledge and pesticide use practices. In: Pingali, P.L. and Roger, P.A. (eds) *Impact of Pesticides on Farmer Health and the Rice Environment.* Kluwer Academic Publishers, Norwell, Massachusetts, and International Rice Research Institute, Los Baños, Laguna, Philippines.

Wardana, P., Fagi, A.M., Lantican, M.A. and Bhuiyan, S.I. (1990) Water relations to dry season crop choice and profitability in the Cikeusik Irrigation System, West Java. *Proceedings of the Intercountry Workshop on Irrigation Management for Rice-Based Farming Systems.* IIMI, Colombo, 12–14 November.

Wardana, I.P., Syamsiah, I., Fagi, A.M., Juliardi, I., Sudamanto, Idjudin, A.A., Suwardjo, H., Sudjarat, W., Suganda, H., Lantican, M.A., Bhuiyan, S.I. and Woodhead, T. (1993) On-farm water management for rice-based farming systems in Indonesia. In: Miranda, S.M. and Maglinao, A.R. (eds) *Irrigation Management for Rice-Based Farming Systems in Bangladesh, Indonesia and the Philippines: Proceedings of the Tri-Country Workshop* held in Colombo, Sri Lanka, 12–14 November. International Irrigation Management Institute and International Rice Research Institute, Colombo, Sri Lanka.

Way, M.J. (1987) Integrated pest control strategies in food production and their bearing on disease vectors in agricultural lands. In: *Effects of Agricultural Development on Vector-Borne Diseases.* Food and Agriculture Organization, Rome. pp. 107–115.

Way, M.J. and Heong, K.L. (1994) The role of diversity in the dynamics and management of insect pests of tropical irrigated rice – a review. *Bulletin of Entomological Research* 84, 567–587.

WB – World Bank (1993) *World Development Report.* World Bank, Washington DC.

WB – World Bank (1995) *World Development Report.* Oxford University Press, New York.

Wickham, T.H. and Sen, C.N. (1978) Water management for lowland rice: water requirements and yield response. In: *Soils and Rice.* International Rice Research Institute, Los Baños, Laguna, Philippines.

Wijayaratna, C.M. (1991) Irrigation management in Asia. In: *Management of Irrigation Facilities in Asia and the Pacific.* Asian Productivity Organization, Tokyo, Japan. pp. 23–76.

Wong, L.C.Y., Husain, A.N., Ali, A. and Ithnin, B. (1992) Understanding grain quality in the Malaysian rice industry. In: Unnevehr, L.J., Duff, B. and Juliano, B.O. (eds) *Consumer Demand for Rice Grain Quality.* International Rice Research Institute and International Development Research Centre, Los Baños, Laguna, Philippines.

Wood Mackenzie Consultants Ltd (1993) *Global rice pesticide market.* A report submitted to the Plant Protection Department of Ciba-Geigy Limited, Basle, Switzerland.

WRI – World Resources Institute (1992) *World Resources: Toward Sustainable Development.* Oxford University Press, New York.

Yap, C.L. (1992) *A Comparison of Cost of Producing Rice in Selected Countries.* Economic and Social Development Paper No. 101, Food and Agriculture Organization, Rome.

Yuan, Long-Ping (1985) *A Concise Course in Hybrid Rice.* Hunan Technological Press, China. 168 pp.

Zandstra, H.G. (1992) Technological considerations in crop diversification. In: Barghouti, S., Garbux, L. and Umali, D.L. (eds) *Trends in Agricultural Diversification: Regional Perspectives.* World Bank Technical Paper Number 180, The World Bank, Washington DC.

Zeigler, R.S. and Puckridge, D.W. (1995) Improving sustainable productivity in rice-based rainfed lowland systems of South and Southeast Asia. *Geojournal* 35(3), 307–324.

Zhu Rong (ed.) (1988) *The Science of Crops in Modern China.* China Social Science Press, Beijing, China.

Index

Page numbers in **bold** indicate major references.
Page numbers in *italic* refer to figures and tables.